MATHEMATICS RESEARCH DEVELOPMENTS

BOUNDARY VALUE PROBLEMS FOR ANALYTIC AND HARMONIC FUNCTIONS IN NONSTANDARD BANACH FUNCTION SPACES

MATHEMATICS RESEARCH DEVELOPMENTS

Additional books in this series can be found on Nova's website under the Series tab.

Additional E-books in this series can be found on Nova's website under the E-book tab.

MATHEMATICS RESEARCH DEVELOPMENTS

BOUNDARY VALUE PROBLEMS FOR ANALYTIC AND HARMONIC FUNCTIONS IN NONSTANDARD BANACH FUNCTION SPACES

VAKHTANG KOKILASHVILI

AND

VAKHTANG PAATASHVILI

EDITOR

Nova Science Publishers, Inc.

New York

For permission to use material from this book please contact us:
Telephone 631-231-7269; Fax 631-231-8175
Web Site: http://www.novapublishers.com

LIBRARY OF CONGRESS CATALOGING-IN-PUBLICATION DATA

Kokilashvili, V. M. (Vakhtang Mikhailovich)
Boundary value problems for analytic and harmonic functions in nonstandard Banach function spaces / Vakhtang Kokilashvili, Vakhtang Paatashvili.
p. cm.
Includes bibliographical references and index.
ISBN 978-1-61942-301-5 (hardcover)
1. Boundary value problems. 2. Functional analysis. 3. Function spaces. 4. Harmonic functions. 5. Riemann-Hilbert problems. I. Paatashvili, Vakhtang. II. Title.
QA322.2.K65 2011
515'.732--dc23
2011045805

Published by Nova Science Publishers, Inc. † New York

Contents

Preface

The monograph is devoted to the boundary value problems (BVPs) for analytic and harmonic functions in domains with non-smooth boundaries within the framework of nonstandard Banach function spaces, namely, variable exponent Lebesgue spaces and grand Lebesgue spaces. This book might be considered as a continuation of the monograph G. Khuskivadze, V. Kokilashvili and V. Paatashvili, Boundary value problems for analytic and harmonic functions in domains with non-smooth boundaries. Application to conformal mappings. Mem. Diff. Equations Math. Phys., 14(1998), 1-195, where the Riemann problem of linear conjugation for analytic functions, the Dirichlet and Neumann problems for harmonic functions are studied in domains with non–smooth boundaries when the right–hand sides of boundary conditions are taken from classical Lebesgue spaces or its weighted versions.

In the present book we develope a new trend in the theory of BVPs for analytic and harmonic functions of two variables. The novelties of our book are the following: boundary value problems are studied when the right–hand sides of boundary condition belong to the nonstandard Banach function spaces larger than classical Lebesgue spaces and more characterizing a local behavior of function; for the first time we introduce and study variable exponent weighted Hardy and Smirnov classes of analytic and harmonic functions. It turns out that these classes are the right classes in which to give a complete picture of solutions is possible. The main feature of our book is to study BVP in domains with intricate geometric structure, to reveal the influence of boundary's geometry of solvability conditions, to discover non-Fredholmian cases depending on the boundary's geometry, to construct solutions in all cases of solvability. We managed to obtain a complete picture of solvability both for bounded simply and doubly-connected domains with non-smooth boundaries; to create the function theory bases for BVPs, namely, to prove boundedness criteria for weighted Cauchy singular integral operators in various new function spaces.

The monograph contains author's recent results published quite recently or accepted for publication in high rank international journals.

The problem of linear conjugation and one-sided boundary value problems for analytic functions were for the first time formulated by B. Riemann. Significant results in this direction have been obtained by D. Hilbert, Yu. Sokhotskiĭ, J. Plemelj, H. Poincaré, G. Bertrand, F. Noether, T. Carleman.

A further interest to these problems is caused by the wide applications in mechanics of continuum media. The problems of the plane theory of elasticity are reducible to the boundary value problems of the theory of analytic functions. Early in the 40-s of last century, N. Muskhelishvili constructed an elegant theory of plane elasticity basing on the development of the theory of BVPs for analytic functions, and related boundary singular integral equations.

During the last decade it became clear that classical function spaces are not able to solve different problems arising in various mathematical models of fluid mechanics, nonlinear theory of elasticity, in variational problems of mathematical physics, in differential equations with nonstandard growth conditions, in nonlinear partial differential equations etc. This reason stimulated an intensive study of some nonstandard Banach function spaces, namely variable exponent Lebesgue spaces and grand Lebesgue spaces.

In some problems of mathematical physics there arise variational problems with Lagrangians more complicated that is usually assumed in variational calculus when the character of nonlinearity varies from point to point. Investigation of variational problems with variable exponent started from the papers of V. Zhikov (1995), M. Růžička (2000) studied the problems in the so called rheological and electrorheological fluids which lead to the spaces with nonstandard growth condition.

Other spaces employed in this book are the grand Lebesgue spaces. These spaces were introduced by T. Iwaniec and C. Sbordone in 1992 to treat the integrability problem of Jacobian under the minimal hypothesis. In the theory of PDE's, it turns out that they are the right spaces in which some nonlinear partial differential equations have been considered in view of existence and uniqueness of solutions, of their regularity properties.

We intend to present the results concerning mapping properties of nonlinear harmonic analysis operators in nonstandard Banach function spaces, namely, boundedness criteria of the Cauchy singular operators in weighted variable exponent Lebesgue spaces and weighted grand Lebesgue spaces. The latter results are assumed as a base of the boundary value problems theory for analytic and harmonic functions. For the first time we introduce and study various type variable exponent Hardy and Smirnov classes of analytic functions. On the base of properties of these classes, as the sets of sought functions. Our aim is to give solutions of the Dirichlet, Neumann, Riemann, Riemann-Hilbert, Riemann-Hilbert-Poincaré, Haseman boundary value problems with data from nonstandard Banach function spaces. The main feature of our book is to study the above-mentioned boundary value problems in domains with piecewise smooth boundaries involving arbitrary cusps, to reveal the influence of boundary's geometry on solvability conditions and to construct solutions in explicit form in all solvability cases.

Nowadays the study of BVPs and related boundary integral equations, as well as the Sobolev's embedding and trace problems in domains with intricate geometrical structure of the boundary ("bad" domains) is one of the challenging problems (V.

Maz'ya, A. A. Soloviev, S. V. Poborchy, V. A. Kozlov, O. V. Besov, P. Hajlash, P. Koskela, T. Kilpeläinen, J. Malý). In the past decade, the authors of this book published more than 20 papers related to solutions of various boundary value problems in domains with non–smooth boundaries within the frame of classical Lebesgue and Zygmund classes and weighted variable exponent Lebesgue spaces. The book intends to present the complete solution of the above-mentioned BVPs focusing on construction of solutions in explicit form.

In the book the non-linear harmonic analysis method is developed, in particular, the Cauchy type integrals method for BVPs of analytic and harmonic functions. We present the boundedness criteria for Cauchy singular integral operator in some nonstandard Banach function spaces. The other important tool of our investigation in the domain with non–smooth boundaries are the results on behavior of conformal mapping derivatives in the neighborhood of boundary's angular points. Extension of classical results in this direction plays a crucial role in the study of BVP's in "bad" domains.

The book may be considered as a systematic and detailed analysis of various boundary value problems of analytic and harmonic functions in the domains with complicate geometrical structure ("bad" domains).

We explore new aspects of the following problems:

(i) The Riemann problem: find a function Φ from a given class of functions, analytic in the plane, cut along Γ– a rectifiable curve, which boundary values satisfy the conjugacy condition

$$\Phi^+(t) = G(t)\Phi^-(t) + g(t),$$

where G and g are functions prescribed on Γ, and Φ^+ and Φ^- are boundary values of Φ on Γ;

(ii) The Riemann-Hilbert problem: find an analytic function $\Phi(z)$ such that its boundary value Φ^+ satisfies the one-sided boundary condition

$$\mathrm{Re}[G(t)\Phi^+(t)] = f(t), \quad t \in \Gamma,$$

where G is a function given on Γ, and also, more general, the Riemann-Hilbert-Poincaré boundary value problem, when one-sided boundary condition contains derivatives of unknown functions.

(iii) The Dirichlet and Neumann problems. Though the Dirichlet problem for the real parts of Smirnov classes analytic functions (or Cauchy type integrals) is a particular case of the Riemann-Hilbert problem, nevertheless the results related to this problem take up a considerable deal space of this book as its solution is obtained not only in simply-connected but also in doubly-connected domains with piecewise smooth boundary. Along with this the results in this direction are more informative and transparent.

Depending on the assumptions on the unknown functions, BVPs for analytic functions are conventionally divided into two groups: continuous and piecewise continuous problems when the continuity is violated only at a finite number of boundary points and all other problems of discontinuous type. As it was mentioned above in the continuous statement the solution of the problem of linear conjugation has been given by F. D. Gakhov. N. Muskhelishvili has constructed the theory of solutions of the above-cited problems and related boundary singular integral equations in the piecewise continuous statement.

Discontinuous BVPs when the right side of boundary condition belongs to the Lebesgue space L^p $(1 < p < \infty)$ and the sought functions are from the set of Cauchy type integrals in the domains with sufficiently smooth boundaries has been studied by B. Khvedelidze, I. B. Simonenko, I. Danilyuk, I. Gohberg and N. Krupnik, A. Böttcher and Yu. I. Karlovich, G. Manjavidze, etc.

This book intends to make a breakthrough in the following directions: even in the case of classical Lebesgue spaces to establish uniquely solvability criteria of the Dirichlet problem in the simply connected domain with an arbitrary rectifiable boundary when the range of unknown functions is the set of real parts of analytic functions from Smirnov classes; to enlarge the spaces of admissible functions for the right sides; to study the cases when output data of boundary value problems are taken from nonstandard Banach function spaces; for the first time we introduce and study various type variable exponent Hardy and Smirnov classes of analytic functions both in simply connected and in doubly-connected domains; to give solutions of BVPs in domains with piecewise smooth boundaries. We are focused on a complete description of solvability picture, revealing the influence of boundary's geometry on solvability conditions and constructing of the solutions in explicit form provided that they exist. For example, in the case of the Dirichlet problem for harmonic functions being real parts of functions from variable exponent Smirnov classes in domains with non-smooth boundaries it is shown that depending on the boundary's geometry and values of a space exponent at angular points, this problem may turn out unsolvable, solvable uniquely or non-uniquely. In unsolvable case, the necessary and sufficient conditions are found, for boundary functions which govern the solvability.

Solving the Riemann-Hilbert problem in simply-connected domains we heavily rely on the Muskhelishvili's well-known method of reducing the above-mentioned problem to the Riemann boundary value problem of linear conjugation. For double-connected domains this method does not work. Therefore for this case we have to elaborate a new approach.

We have succeeded in obtaining the solution of the Riemann-Hilbert-Poincaré problem via generalization of I. Vekua's well-known representation of analytic functions possessing higher order Hölder-continuous derivatives up to the Lyapunov boundary. In the case of such a representation for analytic functions having higher order derivatives represented by Cauchy type integrals with densities from weighted variable exponent Lebesgue spaces in the domains with nonsmooth boundaries we found that the violation of their smoothness has a considerable influence on the possibility to represent analytic functions by I. Vekua's integrals.

It should be emphasized that the authors results in fact are pioneering, for example, in singular integral operator theory in nonstandard Banach function spaces, the Hardy classes theory in non-classical settings, boundary value problems with data from the larger function spaces than studied previous, in the study of boundary value problems in domains with non-smooth boundaries.

The book is aimed at a wide audience, ranging from researchers in BVPs of function theory and mathematical physics to experts in theory of elasticity and post-graduate students. The text of the book is self-contained. It can be read without looking frequently into other sources, however a knowledge of elements of function spaces, operator theory and complex analysis is needed; therefore the book can be used for teaching purposes.

The authors gratefully acknowledge a lot of helpful suggestions of Prof. Alexander Meskhi during the preparation of this book.

We wish to express our sincere gratitude to Dr. Tsira Tsanava and Dr. Maia Kvinikadze from A. Razmadze Mathematical Institute for the energy and patience they devoted to the production of the TEX master of the book.

VAKHTANG KOKILASHVILI
I. Javakhishvili Tbilisi State University,
A. Razmadze Mathematical Institute,
2 University Str.,
0186 Tbilisi, Georgia
International Black Sea University,
2 Agmashenebeli Alley,
0131 Tbilisi, Georgia
e-mail: kokil@rmi.ge

VAKHTANG PAATASHVILI
I. Javakhishvili Tbilisi State University,
A. Razmadze Mathematical Institute,
2 University Str., 0186 Tbilisi, Georgia
Georgian Technical University,
77 M. Kostava Str.,
0175 Tbilisi, Georgia
e-mail: paatashvilitam@gmail.com
Tbilisi, 2011

Basic Ingredients

In this section we present definitions, notation and propositions which will be used throughout the paper. Definitions within separate chapters or section will be given in appropriate places.

0.1. $\mathbb{C}$ is the extended complex plane. $\mathbb{R} = (-\infty, +\infty)$.

0.2. $U = \{z : |z| < 1\}$, $\gamma = \{z : |z| = 1\}$.

0.3. $H_\alpha(E)$, $0 < \alpha \leq 1$ is the class of functions satisfying Hölder condition on the set E with the exponent α; $H(E) = \bigcup\limits_{\alpha \in (0,1]} H_\alpha(E)$ is the Hölder class; $H_1(E)$ is the Lipschitz class.

0.4. $C(E)$ is the Banach space of all continuous on E functions with usual norm.

0.5. Under a curve is meant an oriented, rectifiable, Jordan curve on which as a parameter the arc length is chosen starting from any fixed point. The equation of the curve in this case is $t = t(s)$, $0 \leq s \leq l$, where l is its length. The parameter s is called the arc abscissa. The function $t = t(s)$ is assumed to be periodically extended on $\mathbb{R}$.

Every curve on the plane $\mathbb{C}$ generates the set of Γ-images of the segment $[0, l]$ for the mapping $t = t(s)$. The curve will often be identified with this set. In particular, by saying that a closed Jordan curve divides the plane into two connected sets we mean that the set Γ possesses this property. An analogous remark refers to the expressions such as "a curve bounds a domain D", "a function is given on a curve", etc.

0.6. We will say that the function $f = f(t)$ defined on Γ is almost everywhere finite, integrable, and so on, if the function $f_*(s) = f(t(s))t'(s)$ possesses the corresponding property on $[0, l]$. If f is integrable, we assume

$$\int\limits_\Gamma f(t)dt = \int\limits_0^l f(t(s))t'(s)ds.$$

0.7. $L^p(\Gamma)$, $p \geq 1$ is the space of summable in the p-th degree functions f with the norm

$$\|f\|_p = \left(\int\limits_\Gamma |f(t(s))|^p ds\right)^{1/p}.$$

If w is a measurable, measurable almost everywhere finite and different from zero function, then $L^p(\Gamma, w)$ is the space of functions with the norm

$$\|f\|_{p,w} = \|fw\|_p, \quad L^\infty(\Gamma) = \{f : \|f\|_\infty = \operatorname*{ess\,sup}_{t\in\Gamma}|f(t)| < \infty\}.$$

Suppose $L(\Gamma) = L^1(\Gamma)$, $L(\Gamma; w) = L^1(\Gamma; w)$.

The spaces $L^p(\Gamma; w)$ and $L^{p'}(\Gamma; w^{-1})$, $p > 1$, $p' = \frac{p}{p-1}$ are conjugate ones.

0.8. Let $f \in L(\Gamma)$ and $t_0 = t(s_0) \in \Gamma$. Denote by $\Gamma_\varepsilon(t_0)$, $\varepsilon < \frac{l}{4}$ the portion of the curve Γ left after removing a small arc with the ends $t(s_0 - \varepsilon)$ and $t(s_0 + \varepsilon)$. If

$$\lim_{\varepsilon\to 0} \frac{1}{\pi i} \int\limits_{\Gamma_\varepsilon(t_0)} \frac{f(t)dt}{t - t_0} = \lim_{\varepsilon\to 0} \frac{1}{\pi i} \int\limits_{s_0+\varepsilon}^{l-s_0-\varepsilon} \frac{f(t(s))t'(s)ds}{t(s) - t(s_0)} = \frac{1}{\pi i} \int\limits_\Gamma \frac{f(t)dt}{t - t_0},$$

then we say that there exists the singular Cauchy integral (or the singular integral) of the function f at the point t_0.

In the sequel we put

$$(S_\Gamma f)(t_0) = \frac{1}{\pi i} \int\limits_\Gamma \frac{f(t)dt}{t - t_0}, \quad t_0 \in \Gamma.$$

0.9. $D(S_\Gamma)$ is the set of those functions $f \in L(\Gamma)$ for which $(S_\Gamma f)(t_0)$ exists for almost all $t_0 \in \Gamma$. For rectifiable Γ we have

$$D(S_\Gamma) = L(\Gamma).$$

[20].

0.10. The operator

$$S_\Gamma : f \to S_\Gamma f$$

is called the Cauchy operator. A norm of the operator S_Γ, when it acts boundedly (continuously) from $L^p(\Gamma)$ to $L^s(\Gamma)$, $p \geq s > 0$, will be denoted by $\|S_\Gamma\|_{p,s}$. Assume $\|S_\Gamma\|_p = \|S_\Gamma\|_{p,p}$.

0.11. **Classes of Curves.** Let $t = t(s)$, $0 \leq s \leq l$, be the equation of the curve Γ. It will be called: (a) a smooth curve if t' is continuous (and in the case of its closedness, $t'(0) = t'(l)$); (b) Lyapunov curve, if $t' \in H([0, \ell])$; (c) a piecewise smooth curve, if t' is piecewise continuous, and a piecewise Lyapunov curve, if t' is a piecewise Hölder function. If Γ is a piecewise smooth (piecewise Lyapunov) closed, oriented Jordan curve with angular points A_k bounding the finite domain D and $\pi\nu_k$, $0 \leq \nu_k \leq 2$, $k = \overline{1, n}$ is sizes of interior with respect to D angles at these points, we say that $\Gamma \in C^1_D(A_1, A_2, \ldots, A_n; \nu_1, \nu_2, \ldots \nu_n)$ ($\Gamma \in C^{1,L}_D(A_1, A_2, \ldots, A_n; \nu_1, \nu_2, \ldots \nu_n)$) $\Gamma \in C^{1,L}_D(A_1, A_2, \ldots, A_n; \nu_1, \nu_2, \ldots \nu_n)$; often we write $C^1(A_1, A_2, \ldots, A_n; \nu_1, \nu_2, \ldots \nu_n)$ ($C^{1,L}(A_1, A_2, \ldots, A_n; \nu_1, \nu_2, \ldots \nu_n)$) instead of $C^1_D(A_1, A_2, \ldots, A_n; \nu_1, \nu_2, \ldots \nu_n)$ ($C^{1,L}_D(A_1, A_2, \ldots, A_n; \nu_1, \nu_2, \ldots \nu_n)$); (d) A closed Jordan curve Γ is called Smirnov curve, if the finite domain D bounded by this curve is a Smirnov domain, i.e., the function $w = \ln|z'(w)|$ is representable by a Poisson integral, where $z = z(w)$ is a conformal mapping of U onto D.

0.12. $R_{p,s}$, $p \geq 1$, $s \leq p$, is the set of all Jordan curves Γ for which the operator S_Γ is continuous from $L^p(\Gamma)$ to $L^s(\Gamma)$. $R_p = R_{p,p}$, $R = \bigcap\limits_{p>1} R_p$.

A closed rectifiable curve Γ is said to be regular (Carleson) if for the measure defined by the arc–length of the set $\Gamma \cap B(z;r)$ we have

$$\sup_{r>0,\, z\in\Gamma} \frac{|\Gamma \cap B(z;r)|}{r} < \infty,$$

where $B(z;r)$ is the disk of radius r centered at z.

In order for the curve Γ to belong to the class R it is necessary and sufficient that Γ be regular ([14]).

0.13. The operator S is of a strong type (p,s), if it is continuous from $L^p(\Gamma)$ to $L^s(\Gamma)$, and of a weak type (p,p), if for arbitrary $f \in L^p(\Gamma)$ and $\lambda > 0$,

$$m(t : |(Sf)(t)| > \lambda) < \frac{c}{\lambda^p} \int\limits_\Gamma |f|^p ds,$$

where c does not depend on λ and f, and m is the Lebesgue measure.

If Γ is regular, then the operator S_Γ is of weak type $(1,1)$ (see [14]).

If for any converging in $L^p(\Gamma;\omega)$, $p \geq 1$, sequence f_n the sequence $S_\Gamma f_n$ converges in measure, then we will say that S_Γ is continuous in measure in $L^p(\Gamma;\omega)$..

If Γ is a rectifiable Jordan curve, $p \geq 1$ and $\frac{1}{\omega} \in L^{p'}$, then the operator S_Γ is continuous in measure in $L^p(\Gamma;\omega)$ (see [101], [40, p. 21]).

0.14. $W^p(\Gamma)$, $p > 1$, is the set of weighted (i.e., measurable, almost everywhere finite and different from zero) functions w such that for arbitrary $f \in L^p(\Gamma;w)$,

$$\int\limits_\Gamma |(S_\Gamma f)w|^p ds \leq M_p \int\limits_\Gamma |fw|^p ds,$$

where M_p does not depend on f.

A_p is the Muckenhoupt class, the set of given on γ measurable, almost everywhere positive functions w such that

$$\sup_I \Big(\frac{1}{|I|}\int\limits_I w ds\Big)\Big(\frac{1}{|I|}\int\limits_I w^{-\frac{1}{p-1}} ds\Big)^{p-1} < \infty,$$

where $I \subset \gamma$ is an arbitrary arc of length less than 2π, $|I|$ is its length.

0.15. Let Γ be a closed Jordan curve bounding a finite domain D^+, and let D^- be a complement to $\mathbb{C}$ of the set $D^+ \cup \Gamma$. A curve lying in D^+ (D^-) and ending at the point t is said to be a non-tangential path, if in the neighbourhood of t it lies in some angle of size less than π, with the vertex at the point t and the bisectrix coinciding with the normal to Γ. If the function $\phi = \phi(z)$ defined in $D^+(D^-)$ tends along any non-tangential path to the limit $\phi^+(t)$ $(\phi^-(t))$ as $z \in D^+(D^-)$, $z \to t$, then this limit is called the angular boundary value from the left (from the right) of the function ϕ at the point t. If $\phi^+(t)$ $(\phi^-(t))$ exists almost everywhere on Γ, then the function $\phi^+ = \phi^+(t)$ will be called an angular boundary value of ϕ on Γ from the left (from the right).

0.16. Let Γ be a closed, rectifiable Jordan curve, and let $f \in L(\Gamma)$. The function

$$(K_\Gamma f)(z) = \frac{1}{2\pi i} \int\limits_\Gamma \frac{f(t)dt}{t-z}, \quad z \overline{\in} \Gamma,$$

is termed a Cauchy-type integral with density f.

If $\phi(z) = (K_\Gamma f)(z)$ and $\phi^+(t) = f(t)$ almost everywhere on Γ, that is,

$$\phi(z) = (K_\Gamma \phi^+)(z), \quad z \in D^+,$$

then we will say that ϕ is represented in the domain D^+ by the Cauchy integral. The Cauchy integral in D^- is defined analogously.

The Cauchy type integral has angular boundary values ϕ^+ and ϕ^- which are given by the Plemelj–Sokhotskiĭ formulas

$$\phi^\pm(t) = \pm\frac{1}{2} f(t) + \frac{1}{2\pi i} \int\limits_\Gamma \frac{f(\tau)d\tau}{\tau - t}, \quad t \in \Gamma.$$

In order for a Cauchy type integral ϕ to be representable in D^+ (D^-) by a Cauchy integral, it is necessary and sufficient that ϕ^+ belong to $L(\Gamma)$ and the equality $S_\Gamma \phi^+ = \phi^+$ $(S_\Gamma \phi^- = -\phi^-)$ take place ([42, p.100]).

0.17.

$$K^p(\Gamma, w) = \left\{ \phi : \phi(z) = \frac{1}{2\pi i} \int\limits_\Gamma \frac{\varphi(t)dt}{t-z}, \ z \overline{\in} \Gamma, \ \varphi \in L^p(\Gamma, w) \right\},$$

$$K^p(\Gamma) = K^p(\Gamma, 1).$$

$$\widetilde{K}^p(\Gamma, w, n) = \left\{ \phi : \phi = \phi_0 + q_n, \quad \phi_0 \in K^p(\Gamma, w), \ q_n = \sum_{k=0}^n a_k z^k \right\},$$

$$n \in \{0, 1, 2, \ldots\}.$$

$$\widetilde{K}^p(\Gamma, \omega) = \bigcup_{n=0}^\infty \widetilde{K}^p(\Gamma, w, n), \ \widetilde{K}^p_n(\Gamma, \omega) = \widetilde{K}^p(\Gamma, \omega, n), \ \widetilde{K}^p_0(\Gamma) = \widetilde{K}^p_0(\Gamma, 1),$$

$$\widetilde{K}^p(\Gamma) = \widetilde{K}^p(\Gamma; 1).$$

0.18. H^p, $p > 0$, is the Hardy class, the set of analytic in U functions ϕ for which

$$\sup_{r\in(0,1)} \int_0^{2\pi} |\phi(re^{i\theta})|^p d\theta < \infty.$$

0.19. Let D be a simply connected domain bounded by a rectifiable curve Γ. Then $E^p(D)$ or E^p, $p > 0$, is the Smirnov class of analytic in D functions ϕ for which there exists a sequence of closed curves $\Gamma_n \subset D$ converging to Γ such that

$$\sup_n \int_{\Gamma_n} |\phi(z)|^p |dz| < \infty.$$

If D is an infinite domain bounded by Γ, then $\widetilde{E}^p(D)$ is a set of analytic in D functions ϕ for which $\Psi(w) = \phi\left(\frac{1}{w} + z_0\right) \in E^p(D_1)$, where D_1 is the finite domain into which the function $w = \frac{1}{z-z_0}$, $z_0 \overline{\in} D$, maps D (this class obviously does not depend on the choice of z_0).

0.20. Basic properties of the functions from E^p:

(1) $\phi \in E^p(D)$ if and only if

$$\sup_r \int_{\Gamma_r} |\phi(z)|^p |dz| < \infty,$$

where Γ_r is the image of the circumference $|z| = r$ under a conformal mapping of U onto D (see, e.g., [108, p.203]).

(2) The function $\phi \in E^p(D)$ possesses angular boundary values on Γ, and the boundary function belongs to $L^p(\Gamma)$ ([108, p. 205]).

(3) The class $E^1(D)$ coincides with the class of functions representable in D by the Cauchy integral ([108], p. 205–206).

(4) Smirnov's theorem. If D is a Smirnov domain, $\phi \in E^p(D)$ and $\phi^+ \in L^q(D)$, $q > p$, then $\phi \in E^q(D)$.

(5) Let D be a simply connected domain with the boundary Γ, $e \subset \Gamma$, $\operatorname{mes} e > 0$ and let ϕ_n be a sequence of analytic in D functions. If $\phi_n \in E^p(D)$, $\sup \|\phi_n^+\|_p < \infty$ and ϕ_n^+ converges in measure on e to the function f, then ϕ_n converges in D to the function $\phi \in E^p(D)$, and on e we have $\phi^+ = f$ (see, e.g., [108], pp. 268–269).

0.21. If the curve Γ, bounding the finite domain D, belongs to $R_{s,p}$, $s \geq 1$, then (i) for any $\varphi \in L^s(\Gamma)$ the Cauchy type integral $K_\Gamma\varphi$ belongs to $E^p(D)$ (see [40], p. 29); (ii) Γ is the Smirnov curve [31].

If $\Gamma \in R$, $\varphi \in L^p(\Gamma)$ then $K_\Gamma\varphi \in E^p(D)$. In particular, if $\varphi \in L^\infty(\Gamma)$, then $K_\Gamma\varphi \in \cap_{p>1} E^p(D^+)$.

If $\Gamma \in R_{1,s}$, $s < 1$, (in particular, if $\Gamma \in R$ (see 0.13), then for any $\varphi \in L(\Gamma)$ the Cauchy type integral $K_\Gamma\varphi \in E^s(D)$.

If Γ is regular curve bounded domains D^+ and D^- and $\Phi(z) = (K_\Gamma\varphi)(z)$ is Cauchy type integral whose restrictions on D^+ and D^- belong $E^1(D^+)$ and $E^1(D^-)$ respectively then

$$\Phi(z) = K_\Gamma(\Phi^+ - \Phi^-)(z).$$

0.22. Let a linear operator A map the Banach space X into the Banach space Y. We will say that A is a Noetherian operator if: (i) the equation $Ax = y$ is solvable for any right-hand side of y which satisfies the condition $f(y) = 0$, where f is an arbitrary solution of the conjugate homogeneous equation $A^*f = 0$; (ii) zero subspaces $N(A)$ and $N(A^*)$ of the operators A and A^* are finite-dimensional. If A is a Noetherian operator, and λ and μ are the dimensions of the subspaces $N(A)$ and $N(A^*)$, then the difference $\varkappa = \lambda - \mu$ is called the index of the operator A ($\operatorname{ind} A$). If $\varkappa = 0$ then operator A is called Fredholmian operator.

If A is a Noetherian operator, and V is a compact operator acting from X to Y, then the operator $A + V$ is likewise Noetherian, and $\operatorname{ind}(A+V)=\operatorname{ind} A$ ([3]).

If a and b are bounded functions on the regular curve Γ and operator

$$A\varphi = a\phi + bS_\Gamma\varphi$$

is considered in $L^p(\Gamma)$, $p > 1$, then

$$A^*\psi = a\psi - S_\Gamma b\psi, \quad \psi \in L^{p'}(\Gamma).$$

0.23. **The Class $A(p)$ of functions. The index of the function $G \in A(p)$.** We will say that a measurable on Γ function G belongs to the class $A(p)$, $1 < p < \infty$, if the following conditions are fulfilled:

(1) $0 < \operatorname{ess\,inf} |G|$; $\operatorname{ess\,sup} |G| < \infty$

(2) for all $t \in \Gamma$, there exists on Γ a neighbourhood in which the values of G lie in some sector with the vertex at the origin and the angle less than $a(p) = \frac{2\pi}{\max(p,p')}$.

Suppose that $\Gamma_k = [c_k, c_{k+1})$, $c_{n+1} = c_1$, $k = \overline{1,n}$, are half-open arcs of the curve Γ connecting the points c_k and c_{k+1}, where $c_{n+1} = c_1$. Given $\varepsilon > 0$, there exist arcs $[t(s_k), t(s_k+\eta_k))$ and $(t(s_{k+1}-\eta_{k+1}), t(s_{k+1}))$, η_k, $\eta_{k+1} \in (0, \frac{1}{2}(s_{k+1}-s_k))$ such that the values of G lie in a sector with the vertex at the origin and the angle less than ε. Assume that the numbers η_k correspond to the choice of $\varepsilon = a(p)$, and let τ_k and τ_{k+1} be some points from these intervals. Every point of the closed arc $[\tau_k, \tau_{k+1}]$ possesses a neighbourhood in which the values of the function G are located in a sector with the angle less than $a(p)$. These neighbourhoods cover $[\tau_k, \tau_{k+1}]$, and therefore we can choose a finite covering. Adding to the set of intervals of this covering the intervals $[t(s_k), t(s_k + \eta_k)]$ and $(t(s_{k+1} - \eta_{k+1}), t(s_{k+1}))$, we obtain the arc covering $[c_k, c_{k+1})$. Since $\Gamma = \cup\Gamma_k$, we finally conclude that there exists a finite covering B of the curve Γ consisting either of open arcs or of those of the kind $[t(s_k), t(s_k + \eta_k))$; on every arc, the values of the functions are located in a sector with the vertex at the origin and the angle less than $a(p)$.

Let now c be an arbitrary point on Γ. We form a new covering B' of the curve Γ replacing the arc $(a, b) \ni c$ in the covering B by the arcs $(a, c]$ and $[c, b)$. For a fixed covering B', we select arbitrarily an argument of the number $G(c)$ and denote it by $\arg_p G(c)^+$. Going along Γ in the positive direction, we define the argument $\arg_p G(t)$ on all arcs from B' so that if t_1 and t_2 belong to the same arc, then $|\arg_p G(t_1) - \arg_p G(t_2)| < a(p)$. Continuing the process of defining the argument, after going around the curve we come at the point c to a new value of $\arg_p G(c)^-$.

The integer

$$\varkappa = \varkappa_p = \varkappa_p(G) = \frac{1}{2\pi}[\arg_p G(c)^- - \arg_p G(c)^+]$$

does not depend on the choice of the covering B' and on the point c; we will call it the index of the function G in the class $K^p(\Gamma)$ and denote by $\operatorname{ind}_p G$.

Let φ bounded measurable function on Γ and

$$\Omega(\varphi, t) = \inf_{l(t)\subset\Gamma} \{ \sup_{\tau\in l(t)} \varphi(\tau) - \inf_{\tau\in l(t)} \varphi(\tau)\}.$$

If Γ is simple closed Lyapunov curve and $G \in A(p)$, $p > 1$ then exist the functions G_1, G_2 such that $G_1 \in H(\Gamma)$, $\operatorname{ind}G_1 = \operatorname{ind}G$; $G_2 = \exp(i\varphi_2)$, $\sup_{t\in\Gamma} \Omega(\varphi_2, t) < a(p)$ and$G = |G|G_1G_2$, (see [119]).

The classes $A(p)$ where introduced and studied by I. Simonenko (see [119]).

0.24. If for a function G, $\inf |G| > 0$, one can construct a function X which is analytic on the plane cut along Γ, and for which we have: (i) $X \in \widetilde{\mathcal{K}}^p(\Gamma)$; (ii) $\frac{1}{X} \in \widetilde{\mathcal{K}}_{p'}$, (Γ); (iii) $X^+ = GX^-$, (iv) $X^+ \in W^p(\Gamma)$ then we say that G is factorable and X is called a factor-function for G in the class $K^p(\Gamma)$. Number $\varkappa$ defined by the relation $\lim_{z\to\infty} X(z)z^{\varkappa} \neq 0$ is called the index of G in the class $K^p(\Gamma)$.

If $G \in H(\Gamma)$, $G \neq 0$ and $X(z)$ is nonzero piecewise analytic function satisfying the condition iii), then $X(z)$ is called a canonical function of G.

0.25. Suppose $\Gamma \in C^{1,L}(A_1, \dots, A_n; \nu_1, \dots \nu_n)$, $0 < \nu_k \le 2$ and $z = z(w)$ conformally map U on D, and $w = w(z)$ is it inverse function then

$$z'(w) \sim \prod_{k=1}^{n} (w - a_k)^{\nu_k - 1} z_0(w), \quad a(k) = w(A_k),$$

$$(z(w) - z(a_k)) \sim (w - a_k)^{\nu_k} z_{0,k}(w),$$

$$(w(z) - w(A_k)) \sim (z - A_k)^{\frac{1}{\nu_k}} w_k(z),$$

where z_0, $z_{0,k}$, w_k are nonzero continuous functions on $\overline{U}$ and $\overline{D}$ respectively, and $f \sim g$ means that $0 < m \le \left|\frac{f}{g}\right| \le M < \infty$ (see [133]).

In [38] was proved that $z_0(w)$ is a Hölder class function on $\overline{U}$ (see also [1]).

0.26. When $\Gamma \in C^1(A_1, \ldots, A_n; \nu_1, \ldots \nu_n)$, $0 \le \nu_k \le 2$, then

$$z'(w) \sim \prod_{k=1}^{n} (w - a_k)^{\nu_k - 1} \exp\Big(\frac{1}{\pi} \int_{\gamma} \frac{\delta(\tau)d\tau}{\tau - w}\Big), \quad a_k = w(A_k),$$

where $\delta = \delta(t)$ is a real continuous on γ function ([38], see also [40, p.146]).

Throughout of this book constants (often different constants in the same series of inequalities) will generally be denoted by c or C.

Chapter 1

The Smirnov Classes of Harmonic Functions and the Dirichlet Problem

This chapter deals with criterion for the domain in which the Dirichlet problem for harmonic functions from the Smirnov class is uniquely solvable. We introduce the Smirnov type two classes of harmonic functions in simply connected domains and prove of their non coincidence. A difference between solvability pictures of the Dirichlet problem in the above mentioned classes is determined.

Let us recall the definition of the Hardy class of harmonic functions in a unit disk U (see 0.2). A harmonic function u is said to be of h^p $(p > 0)$ class in U if

$$\sup_{0<r<1} \int_0^{2\pi} |u(re^{i\theta})|^p \, d\theta < \infty, \tag{1}$$

(see, for e.g., [108], Chapter III). If $u \in h^p$, $p > 1$, then for almost all ξ lying on the unit circumference γ there exist angular boundary values which form the function $u(\xi)$, summable in the p-th degree on γ, $(u \in L^p(\gamma))$, and the function $u(z) = u(re^{i\theta})$ is representable by the Poisson integral of that boundary function

$$\begin{aligned} u(re^{i\theta}) &= \frac{1}{2\pi} \int_0^{2\pi} u(e^{it}) \frac{1 - r^2}{1 + r^2 - 2r\cos(t - \theta)} \, dt \\ &= \frac{1}{2\pi} \int_0^{2\pi} u(e^{it}) P(r, t - \theta) \, dt, \quad P(r, x) = \frac{1 - r^2}{1 + r^2 - 2r\cos x}. \end{aligned} \tag{2}$$

If $f \in L^p(\gamma)$, $p \geq 1$, then the harmonic function

$$u(re^{i\theta}) = \frac{1}{2\pi} \int_0^{2\pi} f(e^{it}) P(r, t-\theta)\, dt \tag{3}$$

belongs to h^p and its angular boundary values coincide almost everywhere with f (see, for e.g., [27], Chapter IX).

Consequently, the function given by equality (3) is a solution of the following Dirichlet problem:

$$\begin{cases} \Delta u = 0, & u \in h^p, \;\; p > 1, \\ u|_\gamma = f, & f \in L^p(\gamma). \end{cases} \tag{4}$$

Since for the function $u \in h^p$ for $p > 1$ the equality (2) is valid, problem (4) has a unique solution. M. Riesz has proved the theorem: the function v, harmonically conjugate to the function $u \in h^p$, $p>1$, belongs to h^p (see [113]). Hence the analytic function $\Phi(z) = u(z) + iv(z)$ satisfies the condition

$$\sup_{0<r<1} \int_0^{2\pi} |\Phi(re^{i\theta})|^p\, d\theta < \infty. \tag{5}$$

That is Φ belongs to the Hardy class H^p. By virtue of the mentioned result we have

$$h^p = \operatorname{Re} H^p, \quad p > 1, \tag{6}$$

where $\operatorname{Re} H^p = \{u : u = \operatorname{Re}\Phi,\ \Phi \in H^p\}$.

Thus the class h^p is, on one side, characterized by condition (1) and on the other side by equality (6).

When studying harmonic functions in domains different from a circle, we have to consider different classes of harmonic functions analogous to h^p. Before formulating our definitions, we recall generalization of the Hardy classes due to Smirnov (see 0.20).

Let D be a simply connected finite domain bounded by the Jordan rectifiable curve Γ; moreover, let $z = z(w)$ be a function conformal mapping of the circle U onto D and Γ_r be the images of circumferences $\gamma_r = \{z : |z| = r\}$ for that mapping. The function Ψ, analytic in D, belongs to the Smirnov class $E^p(D)$ if

$$\sup_{0<r<1} \int_{\Gamma_r} |\Psi(z)|^p |dz| < \infty. \tag{7}$$

By analogy with the classes h^p and $\operatorname{Re} H^p$, in the domain D we can introduce the following Smirnov classes of harmonic functions:

$$\widetilde{e}^p(D) = \left\{ u : \Delta u = 0,\ \sup_{0<r<1} \int_{\Gamma_r} |u(z)|^p |dz| < \infty \right\}, \tag{8}$$

and

$$e^p(D) = \Big\{u : u = \operatorname{Re}\Phi,\ \Phi \in E^p(D)\Big\} = \operatorname{Re} E^p(D). \tag{9}$$

Obviously,

$$e^p(D) \subseteq \widetilde{e}^p(D). \tag{10}$$

When $D = U$, we have $\widetilde{e}^p(U) = h^p$, $e^p(U) = \operatorname{Re} H^p$ and by virtue of (6), in this case $\widetilde{e}^p(U) = e^p(U)$.

In the Section 1.1 we show that in a general case for the classes $\widetilde{e}^p(D)$ and $e^p(D)$ the equality $\widetilde{e}^p = e^p(D)$ is invalid. The domains D for which $\widetilde{e}^p(D)$ is wider than the class $e^p(D)$ are indicated and one condition for the function z' guaranteeing the equality $\widetilde{e}^p(D) = e^p(D)$ is found.

1.1 Classes $\widetilde{e}^p(D)$ and $e^p(D)$

One of the basic aims of the given Section is to prove following

Theorem 1. *Let D be a finite domain bounded by the Jordan rectifiable curve Γ and $z = z(w)$ be a conformal mapping of U onto D. If the function $\rho(t) = \sqrt[p]{z'(t)}$, $t \in \gamma$ belongs to $W^p(\gamma)$, then the classes $\widetilde{e}^p(D)$ and $e^p(D)$ coincide.*

(Definition of the class $W^p(\gamma)$ see 0.14).

To prove the theorem we need some lemmas.

Lemma 2. *If the function $f(w) = u(w) + iv(w)$ belongs to the Hardy class H^1, then the equality*

$$f(w) = \frac{1}{2\pi i}\int\limits_{\gamma} u(\xi)\frac{\xi + w}{\xi - w}\frac{d\xi}{\xi} + iv(0) \tag{11}$$

is valid.

Proof. Consider the function

$$\Phi(w) = \frac{1}{2\pi i}\int\limits_{\gamma} u(\xi)\frac{\xi + w}{\xi - w}d\xi. \tag{12}$$

Assuming $w = re^{i\theta}$ we have

$$\Phi(w) = \frac{1}{2\pi i}\int\limits_{0}^{2\pi} u(e^{i\alpha})P(r, \theta - \alpha)d\alpha + ih(w), \tag{13}$$

where $h(w)$ is a real harmonic function. Since $f \in H^1$, by the Fihtengoltz theorem (see e.g., [27, p. 392] or [108, p. 97]), it is representable by the Poisson integral with density $f(e^{i\alpha}) = u(e^{i\alpha}) + iv(e^{i\alpha})$ and therefore

$$u(w) = \frac{1}{2\pi i}\int\limits_{0}^{2\pi} u(e^{i\alpha})P(r, \theta - \alpha)\, d\alpha. \tag{14}$$

It follows from equalities (13)–(14) that

$$f(w) - \Phi(w) = ih_1(w), \tag{15}$$

where $h_1(w)$ is the real harmonic function. Since $f - \Phi$ is analytic, this implies that $h_1(w) = \text{const}$. Moreover, from the equalities $f(0) = u(0) + iv(0)$ and $\Phi(0) = u(0)$ we have that $h_1(0) = v(0)$ and hence $h_1(w) \equiv v(0)$. Now we can conclude from (15) and (13) that equality (11) is valid. □

Remark 3. Lemma 2 is evidently well-known. However, we could not find it in the references and therefore we cite its proof.

Lemma 4. *Let $z = z(w)$ be a conformal mapping of the unit circle U onto a finite domain D which is bounded by a rectifiable curve and $\frac{1}{z'(\xi)} \in L^\sigma(\gamma)$, for some $\sigma > 0$. Then: (a) D is Smirnov's domain; (b) $\frac{1}{z'(w)}$ belongs to the Hardy class H^σ.*

Proof. (a) Since $z' \in H^1$ (see e.g., [27], p. 405),

$$z'(w) = \exp\left\{\frac{1}{2\pi}\int_0^{2\pi} \ln|z'(e^{it})|\frac{e^{it}+w}{e^{it}-w}\,dt\right\}\exp\left\{\frac{1}{2\pi}\int_0^{2\pi}\frac{e^{it}+w}{e^{it}-w}\,d\mu(t)\right\}, \tag{16}$$

where μ is a non-increasing function with bounded variation and $\mu' = 0$ almost everywhere on $[0, 2\pi]$ (see e.g., [108], p. 226). Consider an analytic in U function $\Phi(w) = \ln z'(w)$ and show that $\Phi \in H^{1+\delta}$ for some δ from $(0,1)$. We have

$$\begin{aligned} |\Phi(w)| = |\ln z'(w)| &\le \left|\frac{1}{2\pi}\int_0^{2\pi} \ln|z'(e^{it})|\frac{e^{it}+w}{e^{it}-w}\,dt\right| \\ &+ \left|\frac{1}{2\pi}\int_0^{2\pi}\frac{e^{it}+w}{e^{it}-w}\,d\mu(t)\right| = |\Phi_1(w)| + |\Phi_2(w)|. \end{aligned} \tag{17}$$

Since for positive ε we have

$$\big|\ln|z'(e^{it})|\big| \le \begin{cases} \dfrac{1}{\varepsilon}|z'(e^{it})|^\varepsilon, & \text{when } |z'| > 1, \\ \dfrac{1}{\varepsilon}\dfrac{1}{|z'(e^{it})|^\varepsilon}, & \text{when } |z'| < 1, \end{cases}$$

and $\frac{1}{z'(\xi)} \in L^\sigma(\gamma)$, by choosing δ such that $\varepsilon(1+\delta) < \min(1, \sigma)$ we will get $\ln|z'(e^{i\theta})| \in L^{1+\delta}(\gamma)$.

Estimate the function Φ_1 from (17). We have

$$\Phi_1(w) = \frac{1}{2\pi}\int_0^{2\pi} \ln|z'(e^{it})|\frac{e^{it}+w}{e^{it}-w}\,dt = \frac{1}{2\pi}\int_0^{2\pi} \ln|z'(e^{it})|\,dt$$

$$+\frac{w}{\pi}\int_0^{2\pi}\frac{\ln|z'(e^{it})|dt}{e^{it}-w} = \text{const} + \frac{w}{2\pi i}\int_\gamma \frac{\ln|z'(\xi)|}{\xi}\frac{d\xi}{\xi - w}$$

Since $\ln|z'| \in L^{1+\delta}(\gamma)$, by the well-known property of the Cauchy type integral in a circle (see 0.21), it follows from the above inequality that $\Phi_1 \in H^{1+\delta}$. For Φ_2 we have

$$\operatorname{Re}\Phi_2(re^{it}) = \frac{1}{2\pi}\int_0^{2\pi} P(r, t-\theta)\,d\mu(t),$$

and since μ is a non-increasing function, $\operatorname{Re}\Phi_2 < 0$ if $\mu \neq \text{const}$ (if $\mu = \text{const}$, then $\Phi_2 = 0$). As is known, in this case $\Phi_2 \in \bigcap_{q<1} H^q$ (see [108], p. 93). This and the inclusion $\Phi_1 \in H^{1+\delta}$ imply that $\Phi \in \bigcap_{q<1} H^q$. Moreover, $\Phi^+ = \Phi_1^+$, therefore $\Phi^+ \in L^{1+\delta}(\gamma)$ and hence $\operatorname{Re}\Phi \in h^{1+\delta}$. But $\operatorname{Re}\Phi = \ln|z'(w)|$, therefore it is representable by the Poisson integral. Consequently, D is Smirnov's domain.

(b) Let $w = w(z)$ be an inverse to $z = z(w)$ function. Since $z' \in H^1$ we have $w' \in E^1(D)$. Moreover,

$$\int_\Gamma |w'(z)|^{1+\sigma}|dz| = \int_\gamma \left|w'(z(\xi))\right|^{1+\sigma}|z'(\xi)|\,|d\xi| = \int_\gamma \frac{|d\xi|}{|z'(\xi)|^\sigma}.$$

Thus

$$w' \in E^1(D), \quad w'(t) \in L^{1+\sigma}(\Gamma). \tag{18}$$

By virtue of the above-proven D is Smirnov's domain and we can apply Smirnov's theorem according to which in conditions (18) we have $w' \in E^{1+\sigma}(D)$. This is equivalent to the fact that $\frac{1}{z'} \in H^\sigma$. □

Corollary 5. *If $\sqrt[p]{z'} \in W^p(\gamma)$, $p > 1$, then $\frac{1}{\sqrt[p]{z'}} \in H^{p'+\delta} > 0$, for some $\delta > 0$.*

Indeed, if $\sqrt[p]{z'} \in W^p(\gamma)$, then there exists $\delta > 0$ such that $\frac{1}{\sqrt[p]{z'}} \in L^{p'+\delta}(\gamma)$ (see, for e.g., [21]) and by the lemma, $\frac{1}{\sqrt[p]{z'}} \in H^{p'+\delta}$.

Remark 6. As it follows from the proof of statement (a), our lemma remains valid if we replace the condition $\frac{1}{\sqrt[p]{z'}} \in L^\sigma(\gamma)$, $\sigma > 0$ by $\ln|z'(\xi)| \in L^{1+\delta}(\gamma)$, $\delta > 0$.

Proof of Theorem 1. By virtue of (10), it suffices to establish that $\widetilde{e}^p(D) \subset e^p(D)$.

Let $u \in \widetilde{e}^p(D)$. Consider the harmonic in U function $u^*(w) = u(z(w))$ and show that $u^* \in h^{1+\varepsilon}$, for some $\varepsilon > 0$. Since $\sqrt[p]{z'(\xi)} \in W^p(\gamma)$, according to the Corollary 5, we have $\frac{1}{\sqrt[p]{z'(w)}} \in H^{p'+\delta}$. Further,

$$\int\limits_{\gamma_r} |u^*(w)|^{1+\varepsilon}|dw| \leq \Big(\int\limits_{\gamma_r} |u(z(w))\sqrt[p]{z'(w)}|^p|dw|\Big)^{\frac{1+\varepsilon}{p}} \Big(\int\limits_{\gamma_r} \frac{|dw|}{|\sqrt[p]{z'(w)}|^{\frac{(1+\varepsilon)p}{p-\varepsilon-1}}}\Big)^{\frac{p-1-\varepsilon}{p}}$$

$$= \Big(\int\limits_{\Gamma_r} |u(z)|^p|dz|\Big)^{\frac{1+\varepsilon}{p}} \Big(\int\limits_{\gamma_r} \frac{|dw|}{|\sqrt[p]{z'(w)}|^{\frac{(1+\varepsilon)p}{p-\varepsilon-1}}}\Big)^{\frac{p-1-\varepsilon}{p}}.$$

The first multiplier here is bounded with respect to $r \in (0,1)$ because $u \in \widetilde{e}^p(D)$. The second multiplier will also be bounded if we choose ε such that $p(1+\varepsilon)(p-1-\varepsilon)^{-1} < p' + \delta$. Thus $u^* \in h^{1+\varepsilon}$. The function $v^*(w)$, $v^*(0) = 0$, conjugate harmonically to u^*, by virtue of the M. Riesz theorem is also contained in $h^{1+\varepsilon}$ and hence the function $\Phi^*(w) = u^*(w) + iv^*(w)$ belongs to $H^{1+\varepsilon}$. Using Lemma 1 we can write

$$\Phi^*(w) = \frac{1}{2\pi i}\int\limits_{\gamma} u^*(\xi)\frac{\xi + w}{\xi - w}\frac{d\xi}{\xi} = \frac{1}{2\pi i}\int\limits_{\gamma} \frac{u^*(\xi)d\xi}{\xi - w} + \frac{w}{2\pi i}\int\limits_{\gamma} \frac{u^*(\xi)}{\xi}\frac{d\xi}{\xi - w}.$$

Let us show that $\Psi(w) = \Phi^*(w)\sqrt[p]{z'(w)}$ is the function of the class H^p.

Indeed,

$$\Psi(w) = \frac{\sqrt[p]{z'(w)}}{2\pi i}\int\limits_{\gamma} \frac{u^*(\xi)d\xi}{\xi - w} + \frac{\sqrt[p]{z'(w)}w}{2\pi i}\int\limits_{\gamma} \frac{u^*(\xi)}{\xi}\frac{d\xi}{\xi - w}.$$

Here $\sqrt[p]{z'(w)} \in H^p$ and its multipliers being the Cauchy type integrals with density from $L^{1+\varepsilon}(\gamma)$ belong to $H^{1+\varepsilon}$. Therefore $\Psi \in H^{p_1}$, $p_1 = p(1+\varepsilon)(p+1+\varepsilon)^{-1}$. Next,

$$\Psi^+(\tau) = \frac{1}{2}\sqrt[p]{z'(\tau)}\Big[u^*(\tau) + \frac{1}{\pi i}\int\limits_{\gamma} \frac{u^*(\xi)\sqrt[p]{z'(\xi)}}{\sqrt[p]{z'(\xi)}}\frac{d\xi}{\xi - \tau} + u^*(\tau)$$

$$+ \frac{\tau}{\pi i}\int\limits_{\gamma} \frac{u^*(\xi)\xi\sqrt[p]{z'(\xi)}}{\xi\sqrt[p]{z'(\xi)}}\frac{d\xi}{\xi - w}\Big]. \tag{19}$$

Since $u^*(\xi)\xi\sqrt[p]{z'(\xi)} \in L^p(\gamma)$ and $\sqrt[p]{z'(\tau)} \in W^p(\gamma)$ (and hence $\tau\sqrt[p]{z'(\tau)} \in W^p(\gamma)$), from equality (19) we conclude that $\Psi^+ \in L^p(\gamma)$. By Smirnov's theorem $\Psi \in H^p$. But then $\Phi(z) = \Phi^*(w(z))$ belongs to $E^p(D)$, and since $u = \operatorname{Re}\Phi$, we have $u \in e^p(D)$.

□

If $\sqrt[p]{z'(\xi)}\overline{\in} W^p(\gamma)$, then, generally speaking, $\widetilde{e}^p(D) \neq e^p(D)$. In particular, this takes place in the case of domains bounded by piecewise smooth curves. Below we present the appropriate examples.

Examples of the domains for which $\widetilde{e}^p(D) \neq e^p(D)$

Example A. Let $z(w) = \frac{1}{3-\ln(1-w)}$, where $\ln(1-w)$ is that branch of the function, analytic in the plane cut along the beam $[1,+\infty)$ which at the point $w = 0$ is equal to zero. This function is schlicht in U. It maps U onto such a domain D_0 for which $z = 0$ is a cusp of the boundary. Consider in D_0 the function $F(z) = -i\ln(1-w(z))$, where $w = w(z)$ the inverse to $z(w)$ function. We have $F(z) = \arg(1-w(z))+i\ln|1-w(z)| = u(z)+iv(z)$. The function $u(z)$ here is bounded and therefore $u \in \bigcap\limits_{p>0} \widetilde{e}^p(D)$.

Let us show that $u\overline{\in} \cup_{q>1} e^q(D)$. Let $q > 1$. Assuming $C_r = \sqrt{\frac{1-r}{r}}$ sufficiently close to zero for $\theta \in [0, C_r]$ we will have

$$|\ln\sqrt{2}(1-r)| \leq \left|\ln\sqrt{(1-r)^2 + 4r\frac{1}{4}C_r^2}\right| < \left|\ln\sqrt{(1-r)^2 + 4r\sin^2\frac{\theta}{2}}\right| < |\ln(1-r)|,$$

$$\left|\ln\sqrt{(1-r)^2 + 4r\sin^2\frac{\theta}{2}}\right| > \max\left|\arg(1-re^{i\theta})\right|.$$

Therefore it is easy to verify the validity of the following inequalities:

$$I_r = \int\limits_{\Gamma_r} |\operatorname{Im} F(z)|^q|dz| = \int\limits_{\Gamma_r} \left|\ln|1-w(z)|\right|^q|dz|$$

$$= \int\limits_{\gamma_r} |\ln(1-w)|^q|z'(w)|\,|dw| \geq m\int\limits_0^{C_r} \frac{|\ln|1-re^{i\theta}||^q d\theta}{|1-re^{i\theta}|\,|\ln(1-re^{i\theta})|^2}$$

$$= m\int\limits_0^{C_r} \frac{|\ln\sqrt{(1-r)^2+4r\sin^2\frac{\theta}{2}}|^q d\theta}{|1-re^{i\theta}|\left(\ln^2\sqrt{(1-r)^2+4r\sin^2\frac{\theta}{2}} + (\arg(1-re^{i\theta})^2\right)}$$

$$> \frac{m}{2}\int\limits_0^{C_r} \frac{|\ln\sqrt{(1-r)^2+4r\sin^2\frac{\theta}{2}}|^{q-2}}{\sqrt{(1-r)^2+4r\sin^2\frac{\theta}{2}}}\,d\theta > m_1\left|\ln\sqrt{2}(1-r)\right|^{q-2}$$

$$\times \int\limits_0^{C_r} \frac{d\theta}{\sqrt{(1-r)^2+\theta^2}} \geq m_1\left|\ln\sqrt{2}(1-r)\right|^{q-2}\ln\frac{3C_r}{1-r}$$

$$\geq m_2|\ln(1-r)|^{q-1}, \quad m, m_1, m_2 > 0.$$

which imply that $I_r \to \infty$ as $r \to 1$. Thus $F(z)\overline{\in} H^q$. Consequently, $u\overline{\in} e^q(D)$. Taking into account that $q \in (1,\infty)$ is arbitrary, we conclude $u\overline{\in} \cup_{q>1} e^q(D)$.

Example B. Let Γ be an arbitrary piecewise Lyapunov closed curve bounding the finite domain D and having only one angular point A at which the size of the angle, interior with respect to D, is equal to $\pi\nu$, $1<\nu\le 2$. Further, let $w=w(z)$ be the conformal mapping of D onto U and $w(A)=1$.

Consider the function $f(z)=\frac{1}{1-w(z)}=u(z)+iv(z)$ and prove that $u\in\widetilde{e}^\nu(D)$ and $f\,\overline{\in}\,E^\nu(D)$. This will prove that $u\,\overline{\in}\,e^\nu(D)$.

Under our assumption for the domain D we have $z'(w)\sim(1-w)^{\nu-1}z_0(w)$, where $z_0(w)$ is a continuous, different from zero function in $\overline{U}$ (this is Warschawski's theorem, see 0.25.). Assuming again $C_r=\sqrt{\frac{1-r}{r}}$, we have

$$I_r=\int\limits_{\Gamma_r}|f(z)|^\nu|dz|=\int\limits_{\gamma_r}|f(z(w))|^\nu|z'(w)|\,|dw|=\int\limits_{\gamma_r}\frac{|1-w|^{\nu-1}|z_0(w)|}{|1-w|^\nu}|dw|$$

$$\ge m\int\limits_{\gamma_r}\frac{|dw|}{|1-w|}>m\int\limits_0^{C_r}\frac{rd\theta}{\sqrt{(1-r)^2+4r\sin^2\frac{\theta}{2}}}$$

$$>m\ln|\theta+\sqrt{(1-r)^2+\theta^2}|\Big|_0^{C_r}=m\ln\Big|\frac{C_r\sqrt{(1-r)^2+C_r^2}}{1-r}\Big|,\quad m>0,$$

from which we can see that $I_r\to\infty$ as $r\to1$. Hence $u\,\overline{\in}\,e^\nu(D)$.

Let us show that $u\in\widetilde{e}^\nu(D)$. We have

$$\int\limits_{\Gamma_r}|u(z)|^\nu|dz|=\int\limits_{\gamma_r}\Big|\operatorname{Re}\frac{1}{1-w}\Big|^\nu|z'(w)|\,|dw|\le M\int\limits_{\gamma_r}\Big|\operatorname{Re}\frac{1}{1-w}\Big|^\nu|1-w|^{\nu-1}|dw|$$

$$=M\int\limits_0^{2\pi}\Big|\frac{1-r\cos\theta}{(1-r)^2+4r\sin^2\frac{\theta}{2}}\Big|^\nu\Big((1-r)^2+4r\sin^2\frac{\theta}{2}\Big)^{\frac{\nu-1}{2}}rd\theta$$

$$=Mr\int\limits_0^{2\pi}\frac{(1-r+2r\sin^2\frac{\theta}{2})^\nu}{\big((1-r)^2+4r\sin^2\frac{\theta}{2}\big)^{\frac{\nu+1}{2}}}\,d\theta$$

$$=Mr\Big(\int\limits_0^{\delta}+\int\limits_{\delta}^{2\pi-\delta}+\int\limits_{2\pi-\delta}^{2\pi}\Big)=Mr(I_1+I_2+I_3).\tag{20}$$

Estimate the values I_k.

$$I_1=\int\limits_0^{\delta}\frac{(1-r+2r\sin^2\frac{\theta}{2})^\nu}{\big((1-r)^2+4r\sin^2\frac{\theta}{2}\big)^{\frac{\nu+1}{2}}}d\theta\le2^\nu\int\limits_0^{\delta}\frac{(1-r)^\nu+(2r\sin^2\frac{\theta}{2})^\nu}{\big((1-r)^2+4r\sin^2\frac{\theta}{2}\big)^{\frac{\nu+1}{2}}}d\theta$$

$$\le2^\nu\Big(\frac{(1-r)^\nu}{(1-r)^{\nu+1}}\int\limits_0^{\delta}\frac{d\theta}{\big(1+\frac{4r\sin^2\frac{\theta}{2}}{(1-r)^2}\big)^{\frac{\nu+1}{2}}}+\int\limits_0^{\delta}\frac{(2r\sin^2\frac{\theta}{2})^\nu}{(4r\sin^2\frac{\theta}{2})^{\frac{\nu+1}{2}}}\Big).\tag{21}$$

Since for sufficiently small θ, $k_1\theta < \sin\theta < k_2\theta$, $k_1, k_2 > 0$, from (21) follows

$$I_1 \le 2^\nu\Big(\frac{1}{1-r}\int_0^\delta \frac{d\theta}{\big(1+\frac{rk_1^2\theta^2}{(1-r)^2}\big)^{\frac{\nu+1}{2}}} + \int_0^\delta \frac{(2rk_2\frac{\theta^2}{4})^\nu}{\big(4rk_1^2\frac{\theta^2}{4}\big)^{\frac{\nu+1}{2}}}d\theta\Big)$$

$$\le 2^\nu\Big(\frac{1}{1-r}\int_0^{\frac{\delta}{1-r}} \frac{(1-r)dt}{(1+rk_1^2t^2)^{\frac{\nu+1}{2}}} + \frac{r^{\nu-1}k_2^{2\nu}}{2^\nu k_1^{\nu+1}}\int_0^\delta \Big(\frac{\theta}{2}\Big)^{\frac{\nu-1}{2}} d\theta\Big)$$

$$\le 2^\nu\Big(\int_0^\infty \frac{dt}{(1+rk_1^2t^2)^{\frac{\nu+1}{2}}} + M\Big) < C_1$$

Next, if $r > \frac{1}{2}$, then

$$I_2 = \int_\delta^{2\pi-\delta} \frac{(1-r+2r\sin^2\frac{\theta}{2})^\nu}{((1-r)^2+4r\sin^2\frac{\theta}{2})^{\frac{\nu+1}{2}}}d\theta \le \int_\delta^{2\pi-\delta} \frac{(1-r+2r\sin^2\frac{\theta}{2})^\nu}{(4r\sin^2\frac{\delta}{2})^{\frac{\nu+1}{2}}}d\theta$$

$$\le \frac{3^\nu(2\pi-\delta)}{(4r\sin^2\frac{\delta}{2})^{\frac{\nu+1}{2}}}$$

For $r < \frac{1}{2}$ it is obvious that $I_2 \le 3^\nu 2^{\frac{\nu+1}{2}}$. In both cases there exists C_2 such that $I_2 < C_2$.

In the integral I_3 assuming $\theta = 2\pi - t$, we obtain

$$I_3 = \int_0^\delta \frac{(1-r+2r\sin^2\frac{2\pi-t}{2})^\nu dt}{((1-r)^2+4r\sin^2\frac{2\pi-t}{2})^{\frac{\nu+1}{2}}} = \int_0^\delta \frac{(1-r+2r\sin^2\frac{t}{2})^\nu dt}{((1-r)^2+4\sin^2\frac{t}{2})^{\frac{\nu+1}{2}}} = I_1.$$

Hence $I_3 < C_1$.

From the above-obtained estimates for the integrals I_k, $k = 1, 2, 3$ it follows that $\int\limits_{\Gamma_r} |u(z)|^\nu |dz| < 2C_1 + C_2$. Consequently, $u \in \tilde{e}^\nu(D)$.

1.2 The uniquely solvability criterion of the Dirichlet problem in $e^p(D)$

Our aim is to prove the following

Theorem 7. *Let D be internal domain bounded by the rectifiable Jourdan curve Γ and let $z = z(w)$ be a conformal mapping of the unit circle U onto D. In order for* *the Dirichlet problem*

$$\begin{cases} \Delta u = 0, & u \in e^p(D), \quad p > 1, \\ u\big|_\Gamma = f, & f \in L^p(\Gamma), \end{cases} \tag{22}$$

to be uniquely solvable for any real function $f \in L^p(\Gamma)$, it is necessary and sufficient that the function $\rho = \sqrt[p]{z'(\xi)}$ to belong to the class $W^p(\gamma)$.

Theorem below is also valid.

Theorem 8. *In order for problem (22) to be solvable for any function $f \in L^p(\Gamma)$ and the corresponding homogeneous problem*

$$\begin{cases} \Delta u = 0, & u \in e^p(D), \quad p > 1, \\ u\big|_\Gamma = 0, \end{cases} \tag{22_0}$$

to have n $(n \geq 1)$ linearly independent solutions forming its general solution, it is necessary and sufficient that the polynomial $P(w)$ of order n with zeros on γ such that $\rho(\xi) = \frac{\sqrt[p]{z'(\xi)}}{P(\xi)} \in W^p(\gamma)$ to exist.

If in this theorem we suppose that $n \geq 0$, then it will also cover the statement of Theorem 7. In proving the theorems 7 and 8 we have to apply different methods, so we present here their separate statements.

Since, according to Theorem 1, the classes $e^p(D)$ and $\widetilde{e}^p(D)$ coincide for $\sqrt[p]{z'(\xi)} \in W^p(\gamma)$, we have that under this assumption the Dirichlet problem is uniquely solvable in the class $\widetilde{e}^p(D)$ as well. However, if $\rho \,\overline{\in}\, W^p(\gamma)$, then there occur the domains in which the picture of the solvability in classes $e^p(D)$ and $\widetilde{e}^p(D)$ is different, (see below Section 1.4).

Reduction of the problem (22) to the Riemann problem

If Ω is the analytic function in the plane, cut along the circumference γ, then by Ω_* we denote the function defined by the equality (see [98], p. 142)

$$\Omega_*(w) = \overline{\Omega\Big(\frac{1}{\overline{w}}\Big)}, \quad |w| \neq 1. \tag{23}$$

Lemma 9. *If $u = \operatorname{Re}\Phi$ is a solution of problem (22), $\Psi(w) = \sqrt[p]{z'(w)}\Phi(w)$ and*

$$\Omega(w) = \begin{cases} \Psi(w), & |w| < 1, \\ \overline{\Psi\Big(\frac{1}{\overline{w}}\Big)}, & |w| > 1, \end{cases} \tag{24}$$

then $\Omega \in \widetilde{K}_0^p(\gamma)$ and satisfies the conditions

$$\begin{cases} \Omega^+(\tau) = -\dfrac{\sqrt[p]{z'(\tau)}}{\overline{\sqrt[p]{z'(\tau)}}}\Omega^-(\tau) + 2f(z(\tau))\sqrt[p]{z'(\tau)}, \\ \Omega(w) = \Omega_*(w), \quad |w| \neq 1, \end{cases} \tag{25}$$

where $\Omega^+(\tau)$ $(\Omega^-(\tau))$ are boundary angular values of the function Ω as w tends to τ inside U (outside U). Conversely, if Ω is a solution of problem (25) of the class $\widetilde{K}_0^p(\gamma)$, then the function $u(z) = \operatorname{Re}\Omega(w(z))\sqrt[p]{w'(z)}$ is a solution of problem (22).

This lemma is the generalization for the case of classes $\widetilde{K}^p(\gamma)$ of the analogous result obtained in [98], (see §41) for piecewise analytic functions. Its proof can be found in [54].

Lemma 10. *If problem (22) is solvable for any real function $f \in L^p(\Gamma)$, $p > 1$, then the boundary value problem*

$$\Omega^+(\tau) = -\frac{\overline{\sqrt[p]{z'(\tau)}}}{\sqrt[p]{z'(\tau)}}\Omega^-(\tau) + \mu(\tau), \quad \mu \in L^p(\gamma), \quad \Omega \in \widetilde{K}_0^p(\gamma) \tag{26}$$

is solvable for any complex function $\mu \in L^p(\Gamma)$.

Proof. By Lemma 9 problem (26) is solvable for functions $\mu(\tau) = 2f(z(\tau))\sqrt[p]{z'(\tau)}$, where f is any real function from $L^p(\Gamma)$. It is obvious that (26) will also be solvable for any functions μ of the type $\mu = if(z(\tau))\sqrt[p]{z'(\tau)}$. Let now μ be an arbitrary complex function from $L^p(\gamma)$. Then the conclusion of the lemma follows from the representation

$$\begin{aligned}\mu(\tau) = \mu_1(\tau) + i\mu_2(\tau) &= \frac{\mu_1(\tau) + i\mu_2(\tau)}{\sqrt[p]{z'(\tau)}}\sqrt[p]{z'(\tau)} \\ &= [f_1(z(\tau)) + if_2(z(\tau))]\sqrt[p]{z'(\tau)}, \quad f_k(w(\xi)) \in L^p(\Gamma).\end{aligned}$$

□

One condition for which the equality $\Omega(w) = \Omega_*(w)$ holds

Lemma 11. *Let*

$$\Omega(w) = \frac{1}{2\pi i}\int_\gamma \frac{f(\xi)d\xi}{\xi - w} + a, \quad |w| \neq 1, \quad f \in L^p(\gamma), \quad a = \text{const}.$$

Then in order for the equality $\Omega(w) = \Omega_*(w)$, $|w| \neq 1$, to take place it is necessary and sufficient that the conditions

$$\operatorname{Re} f(\xi) = 0, \quad \int_\gamma f(e^{i\theta})d\theta = -i4\pi \operatorname{Im} a. \tag{27}$$

be fulfilled.

Proof. *Necessity.* We have

$$\Omega_*(w) = \overline{\Omega\Big(\frac{1}{\overline{w}}\Big)} = \overline{\frac{1}{2\pi i}\int\limits_{\gamma}\frac{f(\xi)d\xi}{\xi - \frac{1}{\overline{w}}}} + \overline{a} = \frac{1}{2\pi i}\int\limits_{\gamma}\frac{\overline{f(\xi)}}{\frac{1}{\xi}-\frac{1}{w}}\frac{d\xi}{\xi^2} + \overline{a}$$

$$= \frac{w}{2\pi i}\int\limits_{\gamma}\frac{\overline{f(\xi)}}{w-\xi}\frac{d\xi}{\xi} + \overline{a} = -\frac{1}{2\pi i}\int\limits_{\gamma}\frac{\overline{f(\xi)}d\xi}{\xi - w} + \frac{1}{2\pi i}\int\limits_{\gamma}\frac{\overline{f(\xi)}d\xi}{\xi} + \overline{a}.$$

Thus

$$\Omega_*(w) = -\frac{1}{2\pi i}\int\limits_{\gamma}\frac{\overline{f(\xi)}d\xi}{\xi - w} + \frac{1}{2\pi i}\int\limits_{\gamma}\frac{\overline{f(\xi)}d\xi}{\xi} + \overline{a} \tag{28}$$

and the condition $\Omega(w) = \Omega_*(w)$ results in

$$0 = \frac{1}{2\pi i}\int\limits_{\gamma}\frac{f(\xi)+\overline{f(\xi)}}{\xi - w}d\xi - \frac{1}{2\pi i}\int\limits_{\gamma}\frac{\overline{f(\xi)}}{\xi}d\xi + a - \overline{a}. \tag{29}$$

Passing in the above equality to the limit as $w \to \infty$, we obtain

$$-\frac{1}{2\pi i}\int\limits_{\gamma}\frac{\overline{f(\xi)}}{\xi}d\xi + a - \overline{a} = 0. \tag{30}$$

Putting here $f = f_1 + if_2$, $a = a_1 + ia_2$, where f_1, f_2, a_1, a_2 are real, we find that

$$-\frac{1}{2\pi i}\int\limits_0^{2\pi} f_1(e^{i\theta})d\theta = 0, \quad \frac{1}{2\pi i}\int\limits_0^{2\pi} f_2(e^{i\theta})d\theta = -2a_2. \tag{31}$$

Further, by virtue of (30), from (29) we conclude that

$$F(w) = \frac{1}{2\pi i}\int\limits_{\gamma}\frac{f(\xi)+\overline{f(\xi)}}{\xi - w}d\xi = 0, \quad |w| \neq 1.$$

By Plemelj–Sokhotskiĭ formula we get $f(\xi) + \overline{f(\xi)} = F^+(\xi) - F^-(\xi)$. This implies that $f(\xi) + \overline{f(\xi)} = 0$ a.e. on γ, i.e. $\operatorname{Re} f(\xi) = f_1(\xi) = 0$. Moreover, since $f(\xi) \equiv if_2(\xi)$, the second of equalities (31) is the second of the proven equalities (27).

Sufficiency. According to (27) we have

$$f(\xi) = if_2(\xi), \quad \int\limits_0^{2\pi} if_2(e^{i\theta})d\theta = -4\pi a_2 i,$$

where $a_2 \in \mathbb{R}$ and f_2 is the real function. Then

$$\Omega(w) = \frac{1}{2\pi i}\int\limits_{\gamma}\frac{if_2(\xi)d\xi}{\xi - w} + a_1 + ia_2,$$

and

$$\frac{1}{2\pi i}\int\limits_{\gamma}\frac{\overline{f(\xi)}}{\xi}d\xi = -\frac{1}{2\pi}\int\limits_{0}^{2\pi} if_2(e^{i\theta})d\theta = -4a_2 i.$$

Therefore equality (28) takes the form

$$\Omega_*(w) = \frac{1}{2\pi i}\int\limits_{\gamma}\frac{if_2(\xi)d\xi}{\xi - w} + 2\pi a_2 i + a_1 - ia_2 = \frac{1}{2\pi i}\int\limits_{\gamma}\frac{if_2(\xi)d\xi}{\xi - w} + a_1 + ia_2.$$

Consequently, $\Omega_*(w) = \Omega(w)$. □

Investigation of the Homogeneous Problem

Lemma 12. *If problem* (22_0) *has linearly independent solutions* $u_1, u_2, \dots u_n$*, which form the basis of a set of all its solutions, then a general solution of the problem*

$$\mathrm{Re}[\Phi^+(z)] = 0, \quad t \in \Gamma, \quad \Phi \in E^p(D), \quad p > 1, \tag{32}$$

is given by the equality

$$\Phi(z) = \sum_{k=1}^{n} C_k\Phi_k(z) + iC = \sum_{k=1}^{n} C_k\Phi_k(z) + C\Phi_0(z), \tag{33}$$

where $\Phi_0(z) = i$ *and* C_k, C *are arbitrary real constants,* $\Phi_k(z) = u_k(z) + iv_k(z)$ *where* $v_k(z)$ *are harmonically conjugate to* $u_k(z)$ *functions,* $v_k(0) = 0$*. The functions* Φ_k*,* $k = 1, n$ *and* $\Phi_0 = i$ *are linearly independent over the field of real numbers. In particular, if* (22_0) *is uniquelyuniquely solvable, then* $\Phi(z) = iC$*.*

Proof. If $\Phi(z) = u(z) + iv(z)$ is a solution of problem (32), then $u \in e^p(D)$ and is a solution of problem (22_0). Therefore there exist constants C_k such that $u(z) = \sum_{k=1}^{n} C_k u_k(z)$. Let $\Phi_k(z) = u_k(z) + iv_k(z)$, $v_k(0) = 0$. Since $u_k \in e^p(D)$, $\Phi_k(z) \in E^p(D)$. The functions Φ_0, Φ_k, $k = \overline{1, n}$ are linearly independent because if

$$\sum_{k=1}^{n} d_k\Phi_k(z) + d_0 i \equiv 0, \quad d_k \in \mathbb{R},$$

then $\sum_{k=1}^{n} d_k u_k(z) = 0$. This contradicts the linear independence of the functions u_k.

The analytic function $\Psi(z) = \Phi(z) - \sum_{k=1}^{n} C_k\Phi_k(z)$ belongs to $E^p(D)$ and $\mathrm{Re}\,\Psi(z) = 0$, $z \in D$. Therefore $\Psi(z) = iC$, $C \in \mathbb{R}$. This proves equality (33). However, if $u \equiv 0$, then all $C_k = 0$ and hence $\Phi(z) = iC$. □

Now we start to study of homogeneous problem.

Lemma 13. *Every solution* Φ *of problem* (32) *generates a solution of the problem*

$$\operatorname{Re}\frac{\Psi^+(\tau)}{\sqrt[p]{z'(\tau)}}=0,\quad \Psi\in H^p,\quad p>1,\quad \tau\in\gamma, \tag{34}$$

given by the equality $\Psi(w)=\Phi(z(w))$. *Conversely, if* Ψ *is a solution of problem* (34), *then* $\Phi(z)=\Psi(w(z))\sqrt[p]{w'(z)}\in E^p(D)$ *and is a solution of problem* (32).

Proof. If $\Phi\in E^p(D)$, then $\Psi(w)=\Phi(z(w))\sqrt[p]{z'(w)}$ belongs to H^p. It is evident that Ψ satisfies condition (34). If Ψ is an arbitrary solution of problem (34), then $\Psi(w(z))\sqrt[p]{w'(z)}=\Phi(z)\in E^p(D)$ (as $\Psi\in H^p$) and satisfies condition (32). □

Corollary 14. *If problem* (22_0) *has linearly independent solutions* $u_1,\dots,u_n$ *forming the basis of a set of all its solutions, then a general solution of problem* (34) *is given by the equality*

$$\begin{gathered}\Psi(w)=\sum_{k=1}^{n}(C_k\Phi_k(z(w))+iC)\sqrt[p]{z'(w)},\\ \Phi_k(z)=u_k(z)+iv_k(z),\quad v_k(0)=0;\end{gathered} \tag{35}$$

in particular, if problem (22_0) *is uniquely solvable, then all the solutions of problem* (34) *are given by the equality* $\Psi(w)=iC\sqrt[p]{z'(w)}$.

The validity of our corollary follows immediately from Lemma 12.

Lemma 15. *Under the assumptions of Lemma* 12, *all solutions of the problem*

$$\begin{cases}\Omega^+(\tau)=-\dfrac{\overline{\sqrt[p]{z'(\tau)}}}{\sqrt[p]{z'(\tau)}}\Omega^-(\tau),\ \ \Omega\in\widetilde{K}_0^p(\gamma), & p>1,\\ \Omega(w)=\Omega_*(w), & |w|\neq 1,\end{cases} \tag{36}$$

are given by the equality

$$\Omega(w)=\begin{cases}\Big[\displaystyle\sum_{k=1}^{n}C_k\Phi_k(w)+iC\Big]\sqrt[p]{z'(w)}, & |w|<1,\\ \overline{\Big[\displaystyle\sum_{k=1}^{n}C_k\Phi_k\Big(\frac{1}{\overline{w}}\Big)-iC\Big]}\ \overline{\sqrt[p]{z'\Big(\frac{1}{\overline{w}}\Big)}}, & |w|>1,\end{cases} \tag{37}$$

where $\Phi_k(w)\equiv\Phi_k(z(w))$, $k=\overline{1,n}$ *and* $\Phi_k(z)$ *are linearly independent functions mentioned in Lemma* 12, $C\in\mathbb{R}$.

If problem (22_0) *is uniquely solvable, then a general solution of problem* (36) *is given by the equality*

$$\Omega(w)=C\Omega_0(w),\quad C\in\mathbb{R}, \tag{37_0}$$

where

$$\Omega_0(w) = \begin{cases} i\sqrt[p]{z'(w)}, & |w| < 1, \\ -i\sqrt[p]{z'\Big(\dfrac{1}{\overline{w}}\Big)}, & |w| > 1. \end{cases} \tag{38_0}$$

Proof. Let $\Psi(w)$, $|w| < 1$, be a narrowing of the function Ω on U. Since $\Omega \in \widetilde{K}_0^p(\gamma)$, $p > 1$, then $\Psi \in H^p$. As $\Omega(w) = \Omega_*(w)$, we have $\Omega^-(\xi) = \overline{\Omega^+(\xi)}$ (see e.g., [98], p. 143 and also [54]), and the boundary condition from (36) can be written in the form

$$\frac{\Psi^+(\xi)}{\sqrt[p]{z'(\xi)}} = -\frac{\overline{\Psi^+(\xi)}}{\overline{\sqrt{z'(\xi)}}}, \quad \xi \in \gamma.$$

This expression results in the equality $\operatorname{Re} \frac{\Psi^+(\xi)}{\sqrt[p]{z'(\xi)}} = 0$, $\xi \in \gamma$. Consequently, Ψ is a solution of problem (34). By the corollary 14, Ψ is representable by formula (35). Since $\Omega_*(w) = \Omega(w)$, for $|w| > 1$ we must have $\Omega(w) = \overline{\Psi(\frac{1}{\overline{w}})}$.

This yields the second of equalities (37).

If problem (22_0) is uniquely solvable, then $C_k = 0$, $k = \overline{1,n}$ and from (37) we get (37_0). □

Remark 16. If we put $F_0(w) = \Omega_0(w)$ and for $k = \overline{1,n}$

$$F_k(w) = \begin{cases} \Phi_k(w)\sqrt[p]{z'(w)}, & |w| < 1, \\ \overline{\Phi_k\Big(\dfrac{1}{\overline{w}}\Big)\sqrt[p]{z'\Big(\dfrac{1}{\overline{w}}\Big)}}, & |w| > 1, \end{cases}$$

then formula (37) can be written as follows:

$$\Omega(w) = \sum_{k=0}^{n} C_k F_k(w), \tag{38}$$

where C_k, $k = \overline{0,n}$ are arbitrary real constants.

Since the functions Φ_k, $k = \overline{0,n}$ are linearly independent in the U, it is obvious that the functions F_k in representation (38) are linearly independent, and hence problem (36) has $n+1$ linearly independent solutions which form its general solution.

Theorem 17. *If problem (22_0) has linearly independent solutions $u_1, u_2, \ldots, u_n$ forming the basis of a set of all its solutions, then a general solution of the problem*

$$\Omega^+(\tau) = -\frac{\overline{\sqrt[p]{z'(\tau)}}}{\sqrt[p]{z'(\tau)}}\Omega^-(\tau), \quad \Omega \in \widetilde{K}_0^p(\gamma), \ p > 1, \tag{39}$$

is given by the equality

$$\Omega(w)=\sum_{k=0}^{n}\frac{C_k-iC_k'}{2}F_k(w), \tag{40}$$

where C_k and C_k' are arbitrary real constants. When problem (22_0) is uniquely solvable, then a general solution of problem (39) can be obtained by the equality

$$\Omega(w)=\frac{1}{2}(B_1-iB_2)\Omega_0(w). \tag{40_0}$$

where Ω_0 is the function defined by (38_0) and $B_i\in\mathbb{R}$, $i=1,2$.

Proof. First we suppose that problem (22_0) is uniquely solvable. By Lemma 15 the functions given by equality (40_0) are the solutions of problem (39). Let us show that any solution Ω of problem (39) is of the form (40_0).

Let Ω be a solution of problem (39). Then its solutions will be Ω_* and hence the function $\Omega+\Omega_*$ which is already a solution of problem (36) (see [98], p. 146). By Lemma 15, there exists a real constant B_1 such that

$$\Omega(w)+\Omega_*(w)=B_1\Omega_0(w),\quad |w|\neq 1. \tag{41}$$

If $\Omega(w)=(K_\gamma f)(w)+a$, then, taking into account equalities (28) and (41), we obtain

$$\frac{1}{2\pi i}\int_\gamma\frac{f(\xi)-\overline{f(\xi)}}{\xi-w}d\xi+\frac{1}{2\pi i}\int_\gamma\frac{\overline{f(\xi)d\xi}}{\xi}+a+\overline{a}=B_1\Omega_0(w). \tag{41_0}$$

Whence letting w approach ∞ we conclude that

$$\frac{1}{2\pi i}\int_\gamma\frac{\overline{f(\xi)}}{\xi}d\xi++a+\overline{a}=B_1\Omega_0(\infty), \tag{42}$$

and therefore the equality (41_0) can be written as

$$\frac{1}{2\pi i}\int_\gamma\frac{f(\xi)-\overline{f(\xi)}}{\xi-w}d\xi=B_1\big(\Omega_0(w)-\Omega_0(\infty)\big). \tag{43}$$

It is obvious that together with Ω the solution of problem (39) is $i\Omega$ and hence $(i\Omega)_*$. But $(i\Omega)_*=\overline{i\Omega(\frac{1}{\overline{w}})}=-i\overline{\Omega(\frac{1}{\overline{w}})}=-i\Omega_*(w)$. The function $i\Omega(w)+[i\Omega(w)]_*=i\Omega(w)-i\Omega_*(w)$ is the solution of problem (36) and therefore by Lemma 15 there exists $B_2\in\mathbb{R}$ such that

$$i\big(\Omega(w)-\Omega_*(w)\big)=B_2\Omega_0(w), \tag{44}$$

i.e.,

$$i\left(\frac{1}{2\pi i}\int_\gamma\frac{f(\xi)+\overline{f(\xi)}}{\xi-w}d\xi-\frac{1}{2\pi i}\int_\gamma\frac{\overline{f(\xi)d\xi}}{\xi}+a-\overline{a}\right)=B_2\Omega_0(w). \tag{45}$$

As $w \to \infty$ the latter equality yields

$$\frac{1}{2\pi i}\int_\gamma \frac{\overline{f(\xi)d\xi}}{\xi} + a - \overline{a} = -iB_2\Omega_0(\infty). \tag{46}$$

This equality together with equality (45) allows us to conclude that

$$\frac{1}{2\pi i}\int_\gamma \frac{f(\xi)+\overline{f(\xi)}}{\xi - w}d\xi = -iB_2\big(\Omega_0(w) - \Omega_0(\infty)\big). \tag{47}$$

Combining the above equality with (43) we obtain

$$\frac{1}{2\pi i}\int_\gamma \frac{f(\xi)d\xi}{\xi - w} = \frac{B_1 - iB_2}{2}\big(\Omega_0(w) - \Omega_0(\infty)\big).$$

From equalities (42) and (46) it follows that $a = \frac{B_1 - iB_2}{2}\Omega_0(\infty)$. We now have

$$\Omega(w) = \frac{1}{2\pi i}\int_\gamma \frac{f(\xi)d\xi}{\xi - w} + a = \frac{B_1 - iB_2}{2}\big(\Omega_0(w) - \Omega_0(\infty)\big)$$
$$+\frac{B_1 - iB_2}{2}\Omega_0(\infty) = \frac{B_1 - iB_2}{2}\Omega_0(w).$$

This proves equality (40_0).

In the general case, relying on equality (38) as when obtaining equality (43), we establish the existence of constants C_k, $k = \overline{0,n}$, such that

$$\frac{1}{2\pi i}\int_\gamma \frac{f(\xi)-\overline{f(\xi)}}{\xi - w}d\xi = \sum_{k=0}^{n} C_k\big(F_k(w) - F_k(\infty)\big). \tag{43_0}$$

Taking into account the equalities (43) and (43_0) from (47) follows

$$\frac{1}{2\pi i}\int_\gamma \frac{f(\xi)+\overline{f(\xi)}}{\xi - w}d\xi = -i\Big[\sum_{k=0}^{n} C'_k\big(F_k(w) - F_k(\infty)\big)\Big],\ \ C'_k = \frac{B_2}{B_1}C_k.$$

This and the previous equality result in

$$\frac{1}{2\pi i}\int_\gamma \frac{f(\xi)d\xi}{\xi - w} = \sum_{k=0}^{n}\frac{C_k - iC'_k}{2}\big(F_k(w) - F_k(\infty)\big).$$

For the number a this time we have

$$a = \frac{1}{2}\sum_{k=0}^{n}\frac{C_k - iC'_k}{2}F_k(\infty),$$

and hence

$$\Omega(w)=\frac{1}{2\pi i}\int\limits_{\gamma}\frac{f(\xi)d\xi}{\xi-w}+a=\sum_{k=0}^{n}\frac{C_k-iC'_k}{2}F_k(w). \qquad \square$$

Corollary 18. *If problem* (22_0) *is uniquely solvable, then the problem*

$$\Omega^+(\xi)=\frac{-\sqrt[p]{z'(\xi)}}{\overline{\sqrt[p]{z'(\xi)}}}\Omega^-(\xi), \quad \Omega\in K^p(\gamma),$$

has only zero solution. However, if problem (22_0) *has a different from zero solution, then this problem has the same solution.*

The first of the above-said statements follows from (40_0), since $\Omega_0(\infty)=\overline{\sqrt{z'(0)}}\neq 0$, and in order for $\Omega\in K^p(\gamma)$ it is necessary to take $B_1=B_2=0$. The second statement follows from (40) because we can choose C_k, C'_k, among which there may exist different from zero constants, such that $\sum\limits_{k=0}^{n}\frac{C_k-iC'_k}{2}F_k(\infty)=0$.

On singular integral equations related with problems (25) **and** (26)

In the sequel, the emphasis will be given to problems (25) and (26) which are tightly connected with singular integral equations.

Let S_γ be the Cauchy singular operator. It is not difficult to state that if a and b are bounded measurable functions, then the operator

$$T^*:\psi\to T^*\psi, \;\; T^*\psi=a\psi-S_\gamma(b\psi), \;\; \psi\in L^{p'}(\gamma), \;\; p'=\frac{p}{p-1},$$

is conjugate to the operator

$$T:\varphi\to T\varphi, \;\; T\varphi=a\varphi+bS_\gamma\varphi, \;\; \varphi\in L^p(\gamma).$$

(see e.g., [43], p. 137).

Lemma 19. *Let* $a(t)$ *and* $b(t)$ *be bounded measurable functions on* γ*. Moreover, let the function* $b(t)$ *be almost everywhere different from zero and* $\inf_{t\in\gamma}|a(t)-b(t)|>0$*. If* $p>1$ *and* $\psi\in L^{p'}(\gamma)$*,* $p'=\frac{p}{p-1}$*, is a solution of equation*

$$a\psi-S_\gamma(b\psi)=\mu, \quad \mu\in L^{p'}(\gamma), \tag{48}$$

then the function $\Psi(w)=K_\gamma(b\psi)(w)$ *is a solution of the following problem of linear conjugation:*

$$\Psi^+=\frac{a+b}{a-b}\Psi^-+\frac{\mu b}{a-b}, \quad \Psi\in K^{p'}(\gamma;\frac{1}{b}). \tag{49}$$

Conversely, if Ψ *is a solution of problem* (49)*, then the function* $\psi(t)=\frac{\Psi^+(t)-\Psi^-(t)}{b(t)}$ *is the solution of equation* (48).

Proof. Let $\psi \in L^{p'}(\gamma)$ be a solution of equation (48). Consider the function Ψ defined in the statement of the lemma. Obviously, $\Psi \in K^{p'}(\gamma; \frac{1}{b})$. Moreover, by the formulas due to Plemelj–Sokhotskiĭ, almost everywhere on γ we have

$$\Psi^+(t) - \Psi^-(t) = b(t)\psi(t), \quad \Psi^+ + \Psi^-(t) = (S_\gamma(b\psi))(t).$$

Inserting the values ψ and $S_\gamma(b\psi)$ into (48), we obtain

$$ab^{-1}(\Psi^+ - \Psi^-) - (\Psi^+ + \Psi^-) = \mu.$$

Taking now into account the fact that $b(t)$ almost everywhere on γ is different from zero and $\inf |a - b| > 0$, we can see that the last equality is equivalent to equality (49).

Let us prove the converse statement. Let

$$\Psi(w) = \frac{1}{2\pi i} \int_\gamma \frac{\lambda(\tau)d\tau}{\tau - w}, \quad |w| \neq 1, \quad \lambda \in L^{p'}(\gamma; \frac{1}{b}),$$

be a solution of problem (49). Then $\Psi^+ - \Psi^- = \lambda$, $S_\gamma \lambda = \Psi^+ + \Psi^-$ and the function $\psi = b^{-1}(\Psi^+ - \Psi^-) = b^{-1}\lambda$ belongs to $L^{p'}(\gamma; \frac{1}{b})$. According to (49) we have

$$\frac{\lambda + S_\gamma \lambda}{2} = \frac{a + b}{a - b} \frac{-\lambda + S_\gamma \lambda}{2} + \frac{\mu b}{a - b},$$

which by virtue of the condition $\inf |a - b| > 0$ is equivalent to the equation

$$a\lambda - bS_\gamma \lambda = \mu b.$$

Since $\lambda = b\psi$, we obtain

$$b(a\psi - S_\gamma(b\psi)) = \mu b.$$

But $b(t)$ is almost everywhere different from zero, hence the above equation is equivalent to (48). □

Remark 20. In what follows, for the sake of conciseness, instead of the statement of Lemma 19 we will say that equation (48) in the class $L^{p'}(\gamma)$ is equivalent to problem (49) in the class $K^{p'}(\gamma; \frac{1}{b})$.

Lemma 21. *If problem* (22) *is uniquely solvable for any real* $f \in L^p(\Gamma)$, *then the problem*

$$\Omega^+(\tau) = -\frac{\sqrt[p]{z'(\tau)}}{\sqrt[p]{z'(\xi)}} \Omega^-(\tau) + g(\tau), \quad \tau \in \gamma, \quad g \in L^p(\gamma), \quad \Omega \in K^p(\gamma), \tag{50}$$

is uniquely solvable.

Proof. Under the assumptions of the lemma we can use Lemma 10 and conclude that problem (26) is solvable for any $\mu \in L^p(\gamma)$. By virtue of Theorem 17, all its solutions are given by the formula

$$\Omega = \Omega_\mu + \frac{1}{2}(B_1 - iB_2)\Omega_0,$$

where Ω_0 is defined by equality (38_0), and Ω_μ is a particular solution of problem (26). Since $\Omega_0(\infty) = \overline{\sqrt[p]{z'(0)}} \neq 0$, by choosing the constants B_1, B_2 we can achieve the fulfilment of the condition $\Omega(\infty) = 0$. Thus problem (50) is solvable for any $\mu \in L^p(\gamma)$. The uniqueness of the solution follows from the Corollary 18. □

Now we will investigate a connection between uniquely solvability of the Dirichlet problem and corresponding Riemann problem.

Remark 22. It is not difficult to verify that the boundary value problems

$$\Psi^+ = G\Psi^- + g, \quad \Psi \in K^p(\Gamma),$$

and

$$F^= \frac{1}{2} GF^- + g, \quad F \in K^p(\Gamma),$$

are solvable simultaneously.

Indeed, equalities

$$F(w) = \begin{cases} \Phi(w), & |w| < 1, \\ \dfrac{1}{2}\Phi(w), & |w| > 1, \end{cases} \qquad \Phi(w) = \begin{cases} F(w), & |w| < 1, \\ 2F(w), & |w| > 1, \end{cases}$$

allow one to find solutions for one of these problems, provided the solutions of the other problem are known. In particular, if one of the problems is uniquely solvable, then the other problem will also be uniquely solvable.

Therefore from Lemma 21 we have

Corollary 23. *If problem* (22) *is uniquely solvable, then the problem*

$$\Omega^+ = -\frac{1}{2}\frac{\sqrt[p]{z'}}{\overline{\sqrt[p]{z'}}}\Omega^- + g, \quad g \in L^p(\gamma), \ \Omega \in K^p(\gamma), \tag{51}$$

is uniquely solvable as well.

Proof of Theorem 7

Sufficiency. Let $\rho(\xi) = \sqrt[p]{z'(\xi)} \in W^p(\gamma)$, and we prove that problem (22) is uniquely solvable for any $f \in L^p(\Gamma)$.

The function $\varphi(\tau) = f(z(\tau))\rho(\tau)$ belongs to $L^p(\gamma)$. From the condition $\rho \in W^p(\gamma)$ it follows that $\frac{1}{\rho} \in W^{p'}(\gamma)$, $\frac{1}{\rho} \in L^{p'}(\gamma)$. Therefore $f(z(\tau)) \in L(\gamma)$. Consider the function

$$\Phi(w) = \frac{1}{2\pi i}\int_\gamma \frac{f(z(\tau))}{\tau}\frac{\tau+w}{\tau-w}d\tau = \frac{1}{2\pi i}\int_\gamma \frac{\varphi(\tau)}{\tau\rho(\tau)}\frac{\tau+w}{\tau-w}d\tau, \quad |w|<1.$$

This function, being the Cauchy type integral, belongs to the set $\bigcap\limits_{q<1} H^q$ (see e.g., [108], p. 93). Since $\sqrt[p]{z'(w)} \in H^p$, we have $\Psi(w) = \Phi(w)\sqrt[p]{z'(w)} \in H^\eta$ for some $\eta > 0$. As far as

$$\Psi^+(\xi) = \rho(\xi)\left[\frac{1}{2}\frac{\varphi(\xi)}{\rho(\xi)} + \frac{1}{2\pi i}\int_\gamma \frac{\varphi(\tau)d\tau}{\rho(\tau)(\tau-\xi)} + \frac{1}{2}\frac{\varphi(\xi)}{\rho(\xi)} + \frac{\xi}{2\pi i}\int_\gamma \frac{\varphi(\tau)}{\tau\rho(\tau)}\frac{d\tau}{\tau-\xi}\right],$$

and $\rho \in W^p(\gamma)$, hence $\Psi^+ \in L^p(\gamma)$. By Smirnov's theorem we conclude that $\Psi \in H^p$. But then $\Psi(w(z)) \in E^p(D)$. Moreover, we have

$$\left[\operatorname{Re}\Phi(w(t))\right] = \frac{\varphi(w(t))}{\rho(w(t))} = f(t), \quad t \in \Gamma.$$

and hence the function $u(z) = \operatorname{Re}\Phi(w(z))$ is the solution of problem (22). Let us show that the solution is unique. By Lemma 9 all the solutions of problem (22_0) are obtained by the equality

$$u(z) = \operatorname{Re}\Omega(w(z))\sqrt[p]{w'(z)}, \tag{52}$$

where $\Omega(w)$ is the restriction on U of the solution of the problem

$$\Omega^+ = -\frac{\sqrt[p]{z'}}{\overline{\sqrt[p]{z'}}}\Omega^-, \quad \Omega \in \widetilde{K_0^p}, \quad \Omega = \Omega_*. \tag{53}$$

Since $\frac{1}{\sqrt[p]{z'(\xi)}} \in W^{p'}(\gamma)$, by virtue of Lemma 4 we have $\frac{1}{\sqrt[p]{z'(w)}} \in H^{p'}$ and hence the function $\frac{\Omega(w)}{\sqrt[p]{z'(w)}}$, $|w|<1$, belongs to H^1. According to Lemma 9 we conclude that the function

$$F(w) = \begin{cases} \dfrac{\Omega(w)}{\sqrt[p]{z'(w)}}, & |w|<1, \\ -\dfrac{\Omega(w)}{\overline{\sqrt[p]{z'(\frac{1}{\overline{w}})}}}, & |w|>1. \end{cases}$$

belongs to $\widetilde{K_0^1}(\gamma)$. By (53) we have $F^+ = F^-$ and therefore $F = \text{const}$. Consequently,

$$\Omega(w) = \begin{cases} C\sqrt[p]{z'(w)}, & |w| < 1, \\ -C\sqrt[p]{\overline{z'\dfrac{1}{(\overline{w})}}}, & |w| > 1. \end{cases}$$

where C is an arbitrary complex constant. The condition $\Omega = \Omega_*$ provides $C = -\overline{C}$. This implies that $C = i\beta$, $\beta \in \mathbb{R}$. From (52) we now obtain

$$u(z) = \operatorname{Re}\left[i\beta\sqrt[p]{z'(w)}\sqrt[p]{w'(z)}\right] = \operatorname{Re} i\beta = 0.$$

Thus the solution of problem (22_0) is only the function $u(z) \equiv 0$. This completes the proof of a part of the sufficiency.

Necessity. By virtue of the assumption of our theorem, from Lemma 21 it follows that problem (50) is unique solvable. By the corollary 23, problem (51) is also uniquely solvable. Denote $G = -\frac{1}{2}\frac{\sqrt[p]{z'}}{\sqrt[p]{\overline{z'}}}$ and write (51) in the form

$$\Omega^+ = G\Omega^- + g, \quad \Omega \in K^p(\gamma). \tag{54}$$

From the unique solvability of that problem it follows that the integral equation

$$\frac{1+G}{2}\varphi + \frac{1-G}{2}S_\gamma\varphi = g$$

has for any $g \in L^p(\gamma)$ the unique solution φ of the class $L^p(\gamma)$. Suppose $a = \frac{1+G}{2}$ and $b = \frac{1-G}{2}$. Then this equation takes the form

$$T\varphi = a\varphi + bS_\gamma\varphi = g. \tag{55}$$

It is evident that the functions a and b are bounded, and $a - b = G = -\frac{\sqrt[p]{z'}}{2\sqrt[p]{\overline{z'}}}$. Consequently, $|a - b| = \frac{1}{2} > 0$. Moreover, since $|G| = \frac{1}{2}$, for any function $b = \frac{1-G}{2}$ we have $|b| \geq \frac{1}{4}$. This allows us to apply Lemma 19 and we can conclude that the equation

$$T^* \equiv a\psi - S_\gamma(b\psi) = \mu, \quad \mu \in L^{p'}(\gamma),$$

is equivalent to the problem of linear conjugation

$$\Psi^+ = \frac{1}{G}\Psi^- + \frac{\mu(1-G)}{2G}, \quad \Psi \in K^{p'}(\gamma; \frac{2}{1-G}). \tag{56}$$

Owing to the fact that $|G| = \frac{1}{2}$, we have $K^{p'}(\gamma; \frac{2}{1-G}) = K^{p'}(\gamma)$. Show that the problem

$$\Psi^+ = \frac{1}{G}\Psi^-, \quad \Psi \in K^{p'}(\gamma),$$

i.e., the problem

$$\Psi^{+} = -2\frac{\overline{\sqrt[p]{z'}}}{\sqrt[p]{z'}}\Psi^{-}, \quad \Psi \in K^{p'}(\gamma), \tag{56_0}$$

has only zero solution. Suppose

$$\Omega(w) = \begin{cases} \Psi(w)\sqrt[p]{z'(w)}, & |w| < 1, \\ \overline{-2\Psi\Big(\frac{1}{\overline{w}}\Big)} \; \overline{\sqrt[p]{z'\Big(\frac{1}{\overline{w}}\Big)}}, & |w| > 1. \end{cases}$$

Then $\Omega \in K^1(\gamma)$ (see [40], p. 131). From (56_0) we now obtain $\Omega^+ = \Omega^-$, $\Omega \in K^1(\gamma)$, whence $\Omega \equiv 0$. Thus $\Psi \equiv 0$ and therefore equation (48) has in $L^{p'}(\gamma)$ the solution $\psi \equiv 0$ only.

Consequently, for the operator T we have that $T(L^p(\gamma)) = L^p(\gamma)$ equations $T\varphi = 0$ and $T^*\psi = 0$ are uniquely solvable in $L^p(\gamma)$ and $L^{p'}(\gamma)$, respectively. This implies that T is the Noether operator in the space $L^p(\gamma)$. Therefore, according to Simonenko's theorem [120] we conclude that the function $G = -\frac{1}{2}\frac{\overline{\sqrt[p]{z'}}}{\sqrt[p]{z'}}$ is factorable. This means the existence of the function $X \in \widetilde{K^p}(\gamma)$ with the properties $\frac{1}{X} \in \widetilde{K^{p'}}(\gamma)$, $X^+ \in W^p(\gamma)$ and $G = X^+/X^-$. It can be easily shown that all the solutions of problem $\Omega^+ = G\Omega^-$ of the class $\widetilde{K^{p'}}(\gamma)$ are given by the equality $\Omega(w) = CX(w)$, where C is an arbitrary constant. Obviously, the function

$$\Omega_0(w) = \begin{cases} \sqrt[p]{z'(w)}, & |w| < 1, \\ -\frac{1}{2}\overline{\sqrt[p]{z'\Big(\frac{1}{\overline{w}}\Big)}}, & |w| > 1. \end{cases}$$

is one of the solutions of mentioned problem. Therefore there exists the constant $C_0 \neq 0$ such that $\Omega_0(w) = C_0X(w)$, i.e., for $|w| < 1$ we have $\sqrt[p]{z'(w)} = C_0X(w)$ and since $X^+ \in W^p(\gamma)$, we have that $\sqrt[p]{z'} \in W^p(\gamma)$. □

Incidentally, when proving the sufficiency we have established that the following statement is valid.

Statement 24. *If $\sqrt[p]{z'(\xi)} \in W^p(\gamma)$, then problem* (22) *has the unique solution given by the equality*

$$u(z) = \operatorname{Re}\Phi(w(z)),$$

where

$$\Phi(w) = \frac{1}{2\pi i}\int_{\gamma} \frac{f(z(\tau))}{\tau}\frac{\tau + w}{\tau - w}d\tau. \tag{57}$$

Proof of Theorem 8

Sufficiency. Let $P(w)$ be the indicated in the conditions of the theorem polynomial with zeros on γ such that $\frac{\sqrt[p]{z'(\xi)}}{P(\xi)} \in W^p(\gamma)$.

Assume

$$R(w) = \begin{cases} \dfrac{\sqrt[p]{z'(w)}}{P(w)}, & |w| < 1, \\ -\dfrac{\overline{\sqrt[p]{z'\left(\frac{1}{\overline{w}}\right)}}}{P(w)}, & |w| > 1. \end{cases}$$

then $\frac{R^+}{R^-} = -\frac{\sqrt[p]{z'}}{\overline{\sqrt[p]{z'}}}$, and problem (26) can be rewritten in the form

$$\frac{\Omega^+}{R^+} = -\frac{\Omega^-}{R^-} + \frac{\mu}{R^+}, \quad \mu = 2f(z(\xi))\sqrt[p]{z'(\xi)}.$$

Evidently, $\frac{\Omega}{R} \in \widetilde{K}^p(\gamma)$. On the basis of the above-said and taking into consideration the inclusion $\frac{\sqrt[p]{z'}}{P} \in W^p(\gamma)$, we can conclude that all the solutions of problem (26) are given by the equality

$$\begin{aligned} \Omega(w) &= \frac{R(w)}{2\pi i} \int\limits_{\gamma} \frac{2f(z(\xi))\sqrt[p]{z'(\xi)}}{R^+(\xi)} \frac{d\xi}{\xi - w} + R(w)q_n(w) \\ &= \Omega_f(w) + R(w)q_n(w), \end{aligned} \tag{58}$$

where $q_n(w)$ is an arbitrary polynomial of order n (see, e.g., [43], p. 114). The function

$$u_f(z) = \frac{1}{2}\,\mathrm{Re}\left[\Omega_f(w(z)) + \left(\Omega_f(w(z))_*\right)\right] \tag{58_f}$$

is the particular solution of problem (22). All the solutions of problem (22_0) are given by the equality

$$u_0(z) = \mathrm{Re}\left[\frac{R(w(z))q_n(w(z))}{\sqrt[p]{z'(w)}}\right] = \mathrm{Re}\left[\frac{q_n(w(z))}{P(w(z))}\right], \tag{59}$$

where $q_n(w)$ are polynomials such that

$$[R(w)q_n(w)]_* = R(w)q_n(w). \tag{60}$$

We have to prove that the set of functions (59) with condition (60) is formed by means of n linearly independent harmonic functions.

Equality (60) implies

$$\overline{q_n\left(\frac{1}{\overline{w}}\right)}P(w) = \overline{P\left(\frac{1}{\overline{w}}\right)}q_n(w). \tag{61}$$

As far as all the roots of the polynomial P are located on γ, we can consider that

$$P(w)=\prod_{m=1}^{n}(w-w_m),\quad w_m=e^{i\alpha_k}.$$

holds.

Suppose $q_n(w)=\sum\limits_{k=0}^{n}b_kw^k$. Then it follows from (61) that

$$\begin{aligned}\sum_{k=0}^{n}\frac{\overline{b_k}}{w^k}\prod_{m=1}^{n}(w-w_m)&=\sum_{k=0}^{n}b_kw^k\prod_{m=1}^{n}\left(\frac{1}{w}-\overline{w}_m\right)\\ &=\sum_{k=0}^{n}b_kw^k\prod_{m=1}^{n}\left(\frac{1}{w}-\frac{1}{w_m}\right)=\sum_{k=0}^{n}b_kw^k\prod_{m=1}^{n}\frac{w_m-w}{ww_m}\\ &=\frac{1}{w^n}\frac{(-1)^n}{w_1\cdot w_2\cdots w_n}\sum_{k=0}^{n}b_kw^k\prod_{m=1}^{n}(w-w_m).\end{aligned}$$

Since

$$\sum_{k=0}^{n}\frac{\overline{b_k}}{w^k}=\frac{1}{w^n}\sum_{k=0}^{n}\overline{b_k}w^{n-k},$$

we have

$$(-1)^nw_1\cdot w_2\cdots w_n\sum_{k=0}^{n}\overline{b_k}w^{n-k}=\sum_{k=0}^{n}b_kw^k.$$

This implies

$$A\overline{b_k}=b_{n-k},\quad k=\overline{0,n},\tag{62}$$

where $A=(-1)^nw_1\cdot w_2\cdots w_n$, $w_k=e^{i\alpha_k}$.

Suppose $A=a_1+ia_2$, $b_k=d_k+ie_k$. Then from (62) follows $d_{n-k}=a_1d_k+a_2e_k$, $e_{n-k}=a_2d_k-a_1e_k$, where d_k, e_k are arbitrary constants. We write now the polynomial q_n in the form

$$q_n(w)=\begin{cases}\sum\limits_{k=0}^{j-1}(b_kw^k+\overline{b}_{n-k}w^{n-k})+b_jw^j & \text{for}\quad n=2j,\\ \sum\limits_{k=0}^{j-1}(b_kw^k+\overline{b}_{n-k}w^{n-k}) & \text{for}\quad n=2j-1,\end{cases}$$

and assume $w^k=S_k+ir_k$. Let $n=2j-1$. Then

$$q_n(w)=\sum_{k=0}^{j-1}(d_k+ie_k)(S_k+ir_k)$$

$$
\begin{aligned}
&+\sum_{k=0}^{j-1}\big[(a_1d_k+a_2e_k)+i(a_2d_k-a_1e_k)\big](S_{n-k}+ir_{n-k})\\
=&\sum_{k=0}^{j-1}\big[d_k(S_k+a_1S_{n-k}-a_2r_{n-k})+e_k(-r_k+a_2S_{n-k}+a_1r_{n-k})\big]\\
+i&\sum_{k=0}^{j-1}\big[d_k(r_k+a_1r_{n-k}+a_2S_{n-k})+e_k(S_k+a_2r_{n-k}-a_1S_{n-k})\big].
\end{aligned}
$$

Moreover, if

$$
\begin{aligned}
D_k &= S_k+a_1S_{n-k}-a_2r_{n-k},\\
E_k &= -r_k+a_2S_{n-k}+a_1r_{n-k},\\
g_k &= r_k+a_1r_{n-k}+a_2S_{n-k},\\
\varphi_k &= S_k+a_2r_{n-k}-a_1S_{n-k},
\end{aligned}
\qquad k=\overline{0,j-1} \tag{63}
$$

then

$$
q_n(w)=\sum_{k=0}^{j-1}(d_kD_k+e_kE_k)+i\sum_{k=0}^{j-1}(d_kg_k+e_k\varphi_k).
$$

All the solution of problem (36) for $|w|<1$ can now be written as

$$
\begin{aligned}
\Omega(w)&=\frac{q_n(w)}{P(w)}\sqrt[p]{z'(w)}=\frac{q_n(w)}{P_1+iP_2}\sqrt[p]{z'(w)}\\
&=\frac{(P_1-iP_2)q_n(w)}{P_1^2+P_2^2}\sqrt[p]{z'(w)}=F(w)\sqrt[p]{z'(w)},\\
F(w)&=\sum_{k=0}^{j-1}\Big[(d_kU_k+e_kU_{j+k})+i(d_kV_k+e_kV_{j+k})\Big]\\
&=\sum_{k=0}^{j-1}\Big[d_k(U_k+iV_k)+e_k(U_{j+k}+iV_{j+k})\Big]\\
&=\sum_{k=0}^{j-1}(d_k\Phi_k+e_k\Phi_{j+k}),
\end{aligned} \tag{64}
$$

where

$$
U_k=\frac{D_kP_1-g_kP_2}{P_1^2+P_2^2},\quad U_{j+k}=\frac{E_kP_1-\varphi_kP_2}{P_1^2+P_2^2},\quad k=\overline{0,j-1} \tag{65}
$$

and $V_k(w)$, $V_k(0)=0$ are harmonically conjugate to U_k functions.

If we assume that the functions U_k, $k=\overline{0,2j-1}$ are linearly dependent, then it follows from (64) that Φ_k are linearly dependent as well. This means that a number of linearly independent solutions of problem (36) is less than $2j(=n+1)$. This contradicts out statement given in the conclusion of Lemma 15. Thus the functions U_k, $k = \overline{0, 2j-1}$ are linearly independent.

It follows from (64) that

$$u_0(z) = \sum_{k=0}^{j-1} (d_k U_k + e_k U_{j+k}), \tag{66}$$

where d_k, e_k are arbitrary real constants, provides us with a general solution of problem (22_0).

The case $n = 2j$ is considered analogously.

Necessity. The solvability of problem (22) implies that of problems (25) and (26). The homogeneous to (26) is problem (39). Its general solution is given by equality (40). Then the general solution of problem (26) can be given in the form

$$\Omega = \Omega_\mu + \Omega_0,$$

where Ω_μ is its particular solution and Ω_0 is given by equality (40). Since $n \geq 1$, for any μ we can choose the constants C_k, C'_k in such a way that $\Omega(\infty) = 0$, i.e., $\Omega \in K^p(\gamma)$. Therefore problem (50) is also solvable for any right-hand side. Moreover, the corresponding homogeneous problem

$$\Omega^+ = -\frac{\overline{\sqrt[p]{z'}}}{\sqrt[p]{z'}} \Omega^-, \quad \Omega \in K^p(\gamma) \tag{50_0}$$

has n linearly independent solutions. Indeed, according to the Remark 16, the problem (50_0) has in the class $\widetilde{K_0^p}$ $n+1$ linearly independent solutions Ω_k, $k = \overline{0, n}$.

It can be easily verified that the functions

$$\Omega_{k,0}(w) = \Omega_k(w) - \frac{\Omega_k(\infty)}{\Omega_0(\infty)} \Omega_0(w), \quad k = 1, n$$

are linearly independent, they belong to $K^p(\gamma)$ and, hence, are the solutions of problem (50_0).

Let us now show that the conjugate to problem (39)

$$F^+ = -\frac{\overline{\sqrt[p]{z'}}}{\sqrt[p]{z'}} F^-, \quad F \in K^{p'}(\gamma) \tag{50_0^*}$$

has only zero solution. Indeed, let $\Omega \neq 0$ be a solution of problem (50_0) (such a solution exists, since $n \geq 1$) and F be an arbitrary solution of problem (50_0^*). Then $\Omega F \in K^1(\gamma)$ and $(\Omega F)^+ = (\Omega F)^-$. Therefore $\Omega F \equiv 0$, and hence $F \equiv 0$.

Thus: problem (50) is solvable in $K^p(\Gamma)$ for any right-hand side; homogeneous problem (50_0) has n linearly independent solutions; the homogeneous conjugate problem has only zero solution. This means that the singular operator

$$Tf = \frac{1}{2} f + \frac{1}{2} S_\gamma f + \frac{\overline{\sqrt[p]{z'}}}{\sqrt[p]{z'}} \Big(-\frac{1}{2} f + S_\gamma f \Big)$$

generated by condition (50_0) is the Noether operator in the space $L^p(\gamma)$, and its index is equal to n. By Simonenko's theorem (see [120]), this implies that the function $G = -\frac{\sqrt[p]{z'}}{\sqrt[p]{\bar{z}'}}$ is factorable in $K^p(\gamma)$, i.e., there exists the function $X(w) \in \widetilde{K}^p_n(\gamma)$ such that $G = \frac{X^+}{X^-}$, $X^+ \in W^p(\gamma)$ and the general solution of problem (50_0) is given by the equality $\Omega = XP_n$, where P_n is an arbitrary polynomial of order n (see, for e.g., [43], p. 114). Since the function

$$\Omega(w) = \begin{cases} \sqrt[p]{z'}, & |w| < 1, \\ -\sqrt[p]{\overline{z'\Big(\dfrac{1}{w}\Big)}}, & |w| > 1. \end{cases}$$

is one of the solutions of problem (50_0), there exists a polynomial P_n such that

$$\Omega(w) = X(w)P_n(w). \tag{67}$$

Hence for $|w| < 1$ we will have $\sqrt[p]{z'} = X(w)P_n(w)$, and taking into account the fact that $X^+ \in W^p(\gamma)$, we obtain $\frac{\sqrt[p]{z'(\xi)}}{P_n(\xi)} \in W^p(\gamma)$. Since $z'(w) \neq 0$ (and, consequently, $X(w) \neq 0$, $|w| \neq 1$), all zeros of the polynomial P_n mast lie on γ. Thus the necessity considered in the theorem is proved. □

Corollary 25. *If $\sqrt[p]{z'}$ is such that there exists a polynomial P_n with zeros on γ such that $\frac{\sqrt[p]{z'(\xi)}}{P_n(\xi)} \in W^p(\gamma)$, then all the solutions of problem (22) are given by the equality*

$$u(z) = u_f + u_0(z), \tag{68}$$

where u_f is prescribed by formula (58_f) and u_0 by equalities (66), (65) and (63).

1.3 Some sets of Domains D for which the Dirichlet problem in $e^p(D)$ is Uniquely Solvable

A class $T(\mu)$ of curves. Let $t = t(s)$, $0 \leq s \leq l$ be the equation of a simple, rectifiable curve Γ, and let $\alpha(s)$ be the angle formed by a oriented tangent at the point $t(s)$ and the x-axis.

Definition 26. We say that Γ belongs to the class $T(\mu)$, $\mu \in (0, \pi]$ if for every point $t \in \Gamma$ there exists its arc neighbourhood on which the values of the function $e^{i\alpha(s)}$ lie in an angle of size μ and with the vertex at the origin.

Let $\Gamma \in T(\pi)$, and define specifically the tangential function $\alpha(\theta) \equiv \alpha(s(\theta))$. Here the use will be made of the rule of selection of an argument's branch of the function from the class $A(p)$ (see 0.23).

Since $\Gamma \in T(\pi)$, the curve Γ can be covered by a finite number of arcs $\Gamma_k = (t_{k,1}, t_{k,2}) = (t(s_{k,1}); t(s_{k,2}))$, $k = \overline{1, n}$, $t_{k+1,1} \in \Gamma_k$ such that for $s \in (s_{k,1}, s_{k,2})$ the values of $e^{i\alpha(s)}$ lie in an angle less than π with the vertex at the origin. Add to the points $t(s_k)$ the point $t_0 = z(e^{i\theta_0})$ at which the equality

$$\lim_{w \overset{\wedge}{\to} \exp i\theta} z'(w) = -ie^{-i\theta} \frac{dz\left(e^{i\theta}\right)}{d\theta}$$

holds (see [27], p. 405) and let

$$\alpha_0 = \lim_{w \overset{\wedge}{\to} \exp i\theta_0} \arg z'(w) + \theta_0 + \frac{\pi}{2}.$$

Without restriction of generality, we may assume that the curve Γ at the points $t_{k,1}$, $t_{k,2}$ has tangents. For the sake of definiteness, assume that t_0 lies on the arc Γ_1, and replace it by the arcs $(t_{1,1}, t_0)$ and $(t_0, t_{1,2})$. Define on $(t_0, t_{1,2})$ a function $\alpha(t) = \alpha(t(s)) = \alpha(s)$ such that $\alpha(s_0) = \alpha_0$ and $|\alpha(s) - \alpha_0| < \pi$ for all s for which $t(s) \in (t_0, t_{1,2})$. Thus we have defined the value $\alpha(s_{2,1})$. Next, on the arc Γ_2 we define the function $\alpha(s)$, keeping to the condition $|\alpha(s) - \alpha(s_{2,1})| < \pi$. Continuing this process, we define the function $\alpha(s)$ on $[0, l]$ (hence the function $\alpha(s(\theta))$ on $[0, 2\pi]$). Consequently, we have defined the tangential function $\alpha(s)$ for the curves from $T(\pi)$.

Theorem 27. *Let a simply connected domain D is bounded by simple close rectifiable curve Γ. Then the Dirichlet problem is uniquely solvable in $e^p(D)$*

I. *If $\Gamma \in T(\pi)$ and $p \geq 2$*

or

II. *If $\Gamma \in T(\pi(p-1))$ and $1 < p < 2$.*

Proof. Let

$$\Omega(w) = \begin{cases} \sqrt[p]{z'(w)}, & |w| < 1, \\ \overline{\sqrt[p]{z'(\frac{1}{\overline{w}})}}, & |w| > 1. \end{cases}$$

It can be easily verified that $\Omega \in \widetilde{\mathcal{K}}^p(\gamma)$ and Ω is a solution of the boundary problem

$$\phi^+(e^{i\theta}) = \exp\left\{\frac{i2\beta(\theta)}{p}\right\}\phi^-(e^{i\theta}), \tag{69}$$

where $\beta(\theta) = \alpha(\theta) - \theta - \frac{\pi}{2}$, and $\alpha(\theta) = \alpha(s(\theta))$ is the tangential function defined as above. Suppose

$$G(\tau) = G(e^{i\theta}) = \exp\left\{\frac{i2\beta(\theta)}{p}\right\}. \tag{70}$$

I. It follows from the condition $\Gamma \in T(\pi)$ and the fact that the function $G_1(\theta) = \exp[-i(\frac{\pi}{2} + \theta)]$ is continuous that every point $\tau \in \gamma$ possesses a neighbourhood on which the values of the function $G(\tau)$ lie in an angle of size $\frac{2\pi - \varepsilon}{p}$. Since $p \geq 2$ we have that $G \in A(p)$. Calculate the index of the function G. First consider the function $\varphi(e^{i\theta}) = \exp i\beta(\theta) = \exp i\alpha(\theta)\exp(-i(\frac{\pi}{2}+\theta)) = \varphi_1(\theta)\varphi_2(\theta)$. Since angular degree of simple curve equal 1 ([92], p. 84), one can see that the index of the function $\varphi_1(e^{i\theta}) = \exp i\alpha(\theta)$ in $\widetilde{\mathcal{K}}^p(\gamma)$ is equal to 1. Obviously, $\mathrm{ind}\varphi_2 = -1$. But if $\varphi_1 \in A(p)$, and φ_2 is continuous, then $\mathrm{ind}(\varphi_1\varphi_2) = \mathrm{ind}\varphi_1 + \mathrm{ind}\varphi_2$. Therefore $\mathrm{ind}\varphi = \mathrm{ind}(\varphi_1\varphi_2) = 0$. Hence it follows that $\mathrm{ind}G = 0$.

Let $X(w)$ be the factor function of G in $K^p(\Gamma)$. Then functions $\phi = CX$, where C is an arbitrary constant, is a general solution of the problem (69), (see [43], p.112–114). Since Ω is a solution of the problem (69) of the class $\widetilde{\mathcal{K}}^p(\gamma)$, we have $\Omega = C_0X$, where $C_0 \neq 0$ is a constant. But for $|w| < 1$ we have $\Omega = \sqrt[p]{z'(w)}$, and thus $\sqrt[p]{z'(w)} = C_0X(w)$. In particular, $\sqrt[p]{z'(t)} \in W^p(\gamma)$. According to Theorem 7 our assertion is proved in case I.

II. In this case the value of function G given by (70) lie in angle of size $\frac{2\pi(p-1)}{p} = \frac{2\pi}{p'}$ and since $\max(p,p') = p'$ we again have $G \in A(p)$. Hence $\sqrt[p]{z'(t)} \in W^p(\gamma)$. Consequently, by Theorem 7 the Dirichlet problem in $e^p(D)$ is uniquely solvable. □

1.4 On the Dirichlet problem in the class $\widetilde{e}^p(D)$

We consider the Dirichlet problem which is formulated as follows: Find a function $u(z)$ defined in the domain D and satisfying the following conditions:

$$\begin{cases} \Delta u = 0, \quad u \in \widetilde{e}^p(D), \quad p > 1, \\ u^+(t) = f(t), \quad f \in L^p(\Gamma). \end{cases} \tag{71}$$

It is not difficult to state that if the boundary of a simply connected domain is a curve of the class $C^{1,L}(A;\nu)$, $0 \leq \nu < p$ then any function of the class $\widetilde{e}^p(D)$ has angular boundary values for almost all $t \in \Gamma$.

Lemma 28. *Let D be internal domain bounded by the curve $\Gamma \in C^{1,L}(A;\nu)$, $0 \leq \nu \leq 2$. If $u \in \widetilde{e}^p(D)$, then the function $u^*(w) = u(z(w))$, where $z = z(w)$ is the conformal mapping of U onto D, is of the class h^β for some $\beta > 0$.*

Proof. Since $\Gamma \in C^{1,L}(A;\nu)$, for the derivative $z(w)$ of the conformal mapping of $z = z(w)$, $z(a) = A$ we have

$$z'(w) \sim (w-a)^{\nu-1} \cdot z_0(w), \tag{72}$$

where $z_0(w)$ is a continuous, different from zero function on $\overline{U}$ for $0 < \nu \leq 2$, and $[z_0(w)]^{\pm 1} \in \bigcap_{p>0} H^p$, for $\nu = 0$ (see 0.25–0.26).

We now have

$$\int_{\gamma_r} |u^*(w)|^\beta |dw| = \int_{\gamma_r} \left|\frac{u(z(w)) \cdot \sqrt[p]{z'(w)}}{\sqrt[p]{z'(w)}}\right|^\beta |dw|$$

$$\leq \Big(\int_{\gamma_r}\Big(\big|u(z(w))\sqrt[p]{z'(w)}\big|^{\beta}\Big)^{\frac{p}{\beta}}|dw|\Big)^{\beta/p}\Big(\int_{\gamma_r}\frac{dw}{|z'(w)|^{p\beta/(p-\beta)}}\Big)^{\frac{p-\beta}{p}} \tag{73}$$

Since $u(z) \in \widetilde{e}^p(D)$, the first multiplier in (73) is bounded with respect to r. If we choose β such that $\frac{p\beta}{p-\beta} < 1$, then the second multiplier will likewise be the same. □

Theorem 29. *Let D be internal domain bounded by the curve $\Gamma \in C^{1,L}(A;\nu)$, $0 \leq \nu \leq 2$. Then:*

(a) if $0 \leq \nu < p$, then the problem

$$\begin{cases} \Delta u = 0, \quad u \in \widetilde{e}^p(D), \quad p > 1 \\ u^+(t) = 0 \end{cases} \tag{74}$$

has only a trivial solution;

(b) if $1 < p \leq \nu$, then (74) has at least one non-zero solution, and such is the function

$$u_0(z) = \operatorname{Re}\frac{w(z)+w(A)}{w(z)-w(A)}. \tag{75}$$

Proof. (a) Suppose $\nu = 0$. Then, if $u(z)$ is a solution of problem (74), we can easily show that $u^*(w) = u(z(w))$ is the function of the class h^q, where q is an arbitrary positive number satisfying the condition $q < p$. But in this case it is representable by the Poisson integral of its boundary values. Taking into account the fact that by our assumption $u^+(t) \equiv 0$, we get $u(z) \equiv 0$.

If $0 < \nu < p$, then $\sqrt[p]{z'} \in W^p(\Gamma)$, and in this case $\widetilde{e}^p(D) = e^p(D)$ (see Section 1.1). Hence by Theorem 7 we conclude that homogeneous problem in $e^p(D)$ and, consequently in $\widetilde{e}^p(D)$ has only trivial solution.

(b) Let $1 < p \leq \nu$. Since $\Gamma \in C^{1,L}(A;\nu)$, according to (72) we have $|z'| \sim |1-w|^{\nu-1}$. (We put $a = 1$). To prove the Theorem, it suffices to establish that if $u^*(w) = u_0(z(w))$, then

$$I = \sup_{0<r<1}\int_0^{2\pi}|u^*(re^{i\theta})|^p(\sqrt{1+r^2-2r\cos\theta})^{\nu-1}\,d\theta = \sup_{0<r<1} I_r < \infty.$$

We have

$$I_r = \int_0^{2\pi}\Big|\frac{1-r^2}{1+r^2-2r\cos\theta}\Big|^p(1+r^2-2r\cos\theta)^{(\nu-1)/2}$$

$$\leq \int_0^{2\pi}\frac{1-r^2}{1+r^2-2r\cos\theta}\frac{(1-r^2)^{p-1}d\theta}{(1+r^2-2r\cos\theta)}^{p-1-\frac{\nu-1}{2}}$$

$$\leq 2^{p-1}\int\limits_0^{2\pi}\frac{1-r^2}{1+r^2-2r\cos\theta}\frac{(1-r)^{p-1}}{(1-r)^{\frac{2p-\nu-1}{2}}}d\theta$$

$$=2^{p-1}\int\limits_0^{2\pi}\frac{1-r^2}{1+r^2-2r\cos\theta}(1-r)^{\frac{\nu-1}{2}}d\theta<2^{p-1}\cdot 2\pi=2^p\pi. \qquad (76)$$

From (76) follows $I_r \leq 2^p\cdot\pi$. Consequently, $I\leq 2^\nu\cdot\pi$.

Thus, if $\Gamma\in C^{1,L}(A;\nu)$, $1<p\leq\nu$ then problem (74) in the class $\widetilde{e}^p(D)$ has a non-trivial solution $u_0(z)$ which is expressed by formula (75). □

In particular, if $\Gamma\in C^{1,L}(A;\nu)$, $1<\nu\leq 2$, then the homogeneous Dirichlet problem (74) in the class $\widetilde{e}^\nu(D)$ has a non-trivial solution, whereas the same problem in the class $e^\nu(D)$ has only a trivial solution (see [40], p. 165).

1.5 Notes and comments on Chapter 1

The plane boundary value problems for analytic and harmonic functions in the framework of the classical function spaces in domains with nonsmooth boundaries have been studied in [12], [13], [18], [28], [39], [40], [54], [78], [88], [89], [90], [110], etc.

The Dirichlet problem in the class $e^p(D)$ for the first time was explored in [54] (see also [52]).

Classes $\widetilde{e}^p(D)$ and the Dirichlet problem in these classes were treated in [91].

This Chapter is based on the papers [52], [54] and [91].

Chapter 2

Singular Integrals in Nonstandard Banach Function Spaces

This Chapter deals with the boundedness of the Cauchy singular integral operator in some nonstandard Banach function spaces. Namely, we establish boundedness criteria in weighted variable exponent Lebesgue spaces and generalized grand Lebesgue spaces. These results play a crucial role in the solution of boundary value problems for analytic and harmonic functions with data from the above-mentioned new function spaces.

2.1 The Cauchy singular operator in weighted variable exponent Lebesgue spaces

In this Section we aim to prove the boundedness of the Cauchy singular integral operator

$$S_\Gamma f(t) = \frac{1}{\pi i}\int\limits_\Gamma \frac{f(\tau)}{\tau - t}d\tau \tag{1}$$

in the spaces $L^{p(\cdot)}(\Gamma, \rho)$ with variable exponent $p(t)$ and power weight ρ on arbitrary Carleson curve under the assumption that $p(t)$ satisfy the log-Hölder condition on Γ. The curve Γ may be finite on infinite.

We also prove that if the singular operator is bounded in the space $L^{p(\cdot)}(\Gamma)$ then Γ is necessarily a Carleson curve. A necessary condition is also obtained for an arbitrary continuous $p(t)$.

Let $\Gamma = \{t \in \mathbb{C} : \ t = t(s), \ 0 \le s \le \ell \le \infty\}$ be a simple rectifiable curve with arc-length measure ν. In the sequel we denote

$$D(t,r) := \Gamma \cap B(t,r), \quad t \in \Gamma, \quad r > 0, \tag{2}$$

where $B(t,r)=\{z\in\mathbb{C}:|z-t|<r\}$. We also denote for brevity

$$\nu(D(t,r))=|D(t,r)|.$$

We remind that a curve is called Carleson curve (regular curve), if there exists a constant $c_0>0$ not depending on t and r, such that

$$|D(t,r)|\le c_0 r. \tag{3}$$

Let p be a measurable function on Γ such that $p:\Gamma\to(1,\infty)$. In what follows we assume that p satisfies the conditions

$$1<\underline{p}:=\operatorname*{ess\,inf}_{t\in\Gamma}p(t)\le\operatorname*{ess\,sup}_{t\in\Gamma}p(t)=:\overline{p}<\infty, \tag{4}$$

$$|p(t)-p(\tau)|\le\frac{A}{\ln\frac{1}{|t-\tau|}},\qquad t\in\Gamma,\tau\in\Gamma,\quad |t-\tau|\le\frac{1}{2}. \tag{5}$$

Observe that condition (5) may be also written in the form

$$|p(t)-p(\tau)|\le\frac{2\ell A}{\ln\frac{2\ell}{|t-\tau|}},\qquad t,\tau\in\Gamma, \tag{6}$$

where ℓ is the length of Γ.

In the case where Γ is an infinite curve, we also assume that p satisfies the following condition at infinity

$$|p(t)-p(\tau)|\le\frac{A_\infty}{\ln\frac{1}{\left|\frac{1}{t}-\frac{1}{\tau}\right|}},\qquad \left|\frac{1}{t}-\frac{1}{\tau}\right|\le\frac{1}{2},\quad |t|\ge L,|\tau|\ge L \tag{7}$$

for some $L>0$.

Variable exponent Lebesgue space is defined via the modular

$$I_\Gamma^{p(\cdot)}(f):=\int_\Gamma|f(t)|^{p(t)}d\nu$$

by the norm

$$\|f\|_{p(\cdot)}=\inf\left\{\lambda>0:\ I_\Gamma^{p(\cdot)}\left(\frac{f}{\lambda}\right)\le1\right\}.$$

Observe that

$$\|f\|_{p(\cdot)}=\|f^a\|_{\frac{p(\cdot)}{a}}^{\frac{1}{a}} \tag{8}$$

for any $0<a\le\inf p(t)$.

By $L^{p(\cdot)}(\Gamma,\omega)$ we denote the weighted Banach space of all measurable functions $f:\Gamma\to\mathbb{C}$ such that

$$\|f\|_{L^{p(\cdot)}(\Gamma,\omega)} := \|\omega f\|_{p(\cdot)} = \inf\left\{\lambda>0: \int\limits_{\Gamma}\left|\frac{\omega(t)f(t)}{\lambda}\right|^{p(t)} d\nu(t)\le 1\right\} < \infty. \tag{9}$$

We denote $p'(t)=\frac{p(t)}{p(t)-1}$. It is obvious that $L^{p(\cdot)}(\Gamma;1)=L^{p(\cdot)}(\Gamma)$.

From the Hölder inequality for the $L^{p(\cdot)}$-spaces

$$\left|\int\limits_{\Gamma} u(\tau)v(\tau)d\nu(\tau)\right| \le k\|u\|_{L^{p(\cdot)}(\Gamma)}\|v\|_{L^{p'(\cdot)}(\Gamma)}, \quad \frac{1}{p(\tau)}+\frac{1}{p'(\tau)}\equiv 1,$$

where $k=1+\frac{1}{\underline{p}}+\frac{1}{\underline{p'}}\le 2$, it follows that

$$\left|\int\limits_{\Gamma} u(t)v(t)d\nu(t)\right| \le k\|u\|_{L^{p'(\cdot)}(\Gamma,\frac{1}{\omega})}\|v\|_{L^{p(\cdot)}(\Gamma,\omega)}, \tag{10}$$

and for the conjugate space $[L^{p(\cdot)}(\Gamma,\omega)]^*$ we have

$$[L^{p(\cdot)}(\Gamma,\omega)]^* = L^{p'(\cdot)}(\Gamma,1/\omega) \tag{11}$$

which is an immediate consequence of the fact that $[L^{p(\cdot)}(\Gamma)]^* = L^{p'(\cdot)}(\Gamma)$ under conditions (4), (see [79]).

The following value

$$\frac{1}{p_D}=\frac{1}{|D|}\int\limits_{D}\frac{d\nu(\tau)}{p(\tau)}, \quad D\subset\Gamma, \quad |D|=|D(t,r)| \tag{12}$$

will be used.

By $\chi_D(\tau)=\begin{cases}1, & \tau\in D\\ 0, & \tau\in\Gamma\backslash D\end{cases}$ we denote the characteristic function of a portoin D of the curve Γ.

In the sequel we consider the power weights

$$\omega(t)=\prod_{k=1}^{n}|t-t_k|^{\beta_k}, \quad t_k\in\Gamma \tag{13}$$

in the case of finite curve and the weights

$$\omega(t)=|t-z_0|^{\beta}\prod_{k=1}^{n}|t-t_k|^{\beta_k}, \quad t_k\in\Gamma, \quad z_0\notin\Gamma \tag{14}$$

in the case of infinite curve.

Now we are ready to formulate the main statements:

Theorem 1. *Let*

i) Γ *be a simple Carleson curve;*

ii) p *satisfy conditions* (4), (5) *and also* (7) *in the case* Γ *is an infinite curve.*

Then the singular operator S_Γ *is bounded in the space* $L^{p(\cdot)}(\Gamma,\omega)$ *with weight* (13) *or* (14), *if and only if*

$$-\frac{1}{p(t_k)} < \beta_k < \frac{1}{p'(t_k)}, \quad k = 1, \dots, n, \tag{15}$$

and also

$$-\frac{1}{p_\infty} < \beta + \sum_{k=1}^{n} \beta_k < \frac{1}{p'_\infty} \tag{16}$$

in the case when Γ *is infinite.*

Let us note that from (7) it follows that exists $p_\infty = \lim_{\substack{|t|\to\infty \\ t\in\Gamma}} p(t)$ and $|p(t) - p_\infty| \le \frac{A_\infty}{\ln|t|}$, $|t| \ge \max\{L, 2\}$.

Theorem 2. *Let* Γ *be a finite rectifiable curve. Let* $p : \Gamma \to (1,\infty)$ *be a continuous function. If the singular operator* S_Γ *is bounded in the space* $L^{p(\cdot)}(\Gamma)$, *then the curve* Γ *has the property*

$$\sup_{\substack{t\in\Gamma \\ r>0}} \frac{|D(t,r)|}{r^{1-\varepsilon}} < \infty \tag{17}$$

for every $\varepsilon > 0$. *If* $p(t)$ *satisfies the log–Hölder condition* (5), *then* (17) *holds with* $\varepsilon = 0$, *that is,* Γ *is a Carleson curve.*

Theorem 3. *Let assumptions* i)–ii) *of Theorem* 1 *be satisfied, and let* $a \in C(\Gamma)$. *In the case where* Γ *is an infinite curve starting and ending at infinity, we assume that* $a \in C(\Gamma)$, *where* Γ *is the compactification of* Γ *by a single infinite point, that is,* $a(t(-\infty)) = a(t(+\infty))$. *Then under conditions* (15)–(16), *the operator*

$$(S_\Gamma aI - aS_\Gamma)f = \frac{1}{\pi i}\int_\Gamma \frac{a(\tau)-a(t)}{\tau - t} f(\tau)d\tau$$

is compact in the space $L^{p(\cdot)}(\Gamma,\omega)$ *with weight* (13)–(14).

We base ourselves on the following result for the Hardy–Littlewood maximal functions defined on Carleson curves. Let

$$Mf(t) = \sup_{r>0} \frac{1}{\nu\{D(t,r)\}} \int_{D(t,r)} |f(\tau)|d\nu(\tau) \tag{18}$$

be the maximal operator on functions defined on a curve Γ in the complex plane. The following statements are valid.

Proposition 4. *Let*

i) Γ *be a simple Carleson curve of a finite length;*

ii) p *satisfy conditions* (4)–(5).

Then the maximal operator M *is bounded in the space* $L^{p(\cdot)}(\Gamma,\omega)$ *with weight* (13), *if and only if*

$$-\frac{1}{p(t_k)} < \beta_k < \frac{1}{p'(t_k)}, \quad k = 1, \dots, n. \tag{19}$$

Proposition 5. *Let*

i) Γ *be an infinite simple Carleson curve;*

ii) p *satisfy conditions* (4)–(5) *and let there exist a circle* $B(0,R)$ *such that* $p(t) \equiv p_\infty = const$ *for* $t \in \Gamma\backslash(\Gamma \cap B(0,R))$.

Then the maximal operator M *is bounded in the space* $L^{p(\cdot)}(\Gamma,\omega)$, *with weight* (14), *if and only if*

$$-\frac{1}{p(t_k)} < \beta_k < \frac{1}{p'(t_k)} \quad \text{and} \quad -\frac{1}{p_\infty} < \beta + \sum_{k=1}^{n} \beta_k < \frac{1}{p'_\infty}. \tag{20}$$

For the proof of Proposition 4 and 5 we refer the reader to the paper [73].

We will also make use of the following Kolmogorov–type inequality.

Proposition 6. *Let* Γ *be a Carleson curve of a finite length. Then for any* $s \in (0,1)$

$$\left(\frac{1}{|\Gamma|}\int_\Gamma |S_\Gamma f(t)|^s d\nu(t)\right)^{\frac{1}{s}} \le c\frac{1}{|\Gamma|}\int_\Gamma |f(t)| d\nu(t). \tag{21}$$

Proposition 6 is a conequence of the fact that the singular operator on Carleson curves has weak (1,1)-type:

$$\nu\{t \in \Gamma : |S_\Gamma f(t)| > \lambda\} \le \frac{c}{\lambda}\int_\Gamma |f(t)| d\nu(t)$$

the latter being proved in [14].

One of main tool in the study of boundedness of the Cauchy singular integral operator is the sharp maximal function defined on Γ.

Proposition 7. *Let* $p(t)$ *satisfy condition* (4) *and the maximal operator* M *be bounded in* $L^{p(\cdot)}(\Gamma)$. *Then there exists a constant* $C > 0$ *such that*

$$\|\chi_D\|_{p(\cdot)} \le C|D|^{\frac{1}{p_D}} \quad \textit{for all} \quad D = D(t,r) \subset \Gamma \tag{22}$$

where p_D *is the mean value* (12).

Proof. Let $f(\tau)=\chi_D(\tau)|D|^{-\frac{1}{p(\tau)}}$, $D=D(t,r)$, so that $\|f\|_{p(\cdot)}=1$. For all $z\in D$ we have

$$CMf(z)\geq\frac{1}{|D|}\int\limits_D f(\tau)d\nu(\tau)=\frac{1}{|D|}\int\limits_D |D|^{-\frac{1}{p(\tau)}}d\nu(\tau)$$

$$\text{for any}\quad D=D(t,r).$$

Since the function $\Phi(x)=a^{-x}$, $x\in\mathbb{R}^1_+$, is convex for any $a>0$, by Jensen inequality

$$\Phi\bigg(\frac{1}{|D|}\int\limits_D |f(\tau)|d\nu(\tau)\bigg)\leq\frac{1}{|D|}\int\limits_D \Phi(|f(\tau)|)d\nu(\tau)$$

we obtain

$$CMf(z)\geq|D|^{-\frac{1}{|D|}\int\limits_D\frac{d\nu(\tau)}{p(\tau)}}=|D|^{-\frac{1}{p_D}},\quad z\in D.$$

Hence $\|\chi_D(z)|D|^{-\frac{1}{p_D}}\|_{p(\cdot)}\leq C\|Mf\|_{p(\cdot)}$ and by the boundedness of the maximal operator we obtain that $\|\chi_D(z)|D|^{-\frac{1}{p_D}}\|_{p(\cdot)}\leq C$, which yields (22). □

Let

$$\mathcal{M}^\sharp f(t)=\sup_{r>0}\frac{1}{|D(t,r)|}\int\limits_{D(t,r)}|f(\tau)-f_{D(t,r)}|d\nu(\tau),\quad t\in\Gamma \tag{23}$$

where $f_{D(t,r)}=\frac{1}{|D(t,r)|}\int\limits_{D(t,r)}f(\tau)d\nu(\tau)$, be the sharp maximal function on the curve Γ.

Theorem 8. *Let Γ be an infinite Carleson curve. Let $p(t)$ satisfy conditions (4)–(5) and $p(t)=p_\infty$ outside some circle $B(t_0,R)$. Let $\omega(t)=|t-t_0|^\beta$, $t_0\in\mathbb{C}$, where*

$$-\frac{1}{p(t_0)}<\beta<\frac{1}{p'(t_0)}\quad\textit{and}\quad-\frac{1}{p_\infty}<\beta<\frac{1}{p'_\infty}\quad\textit{if}\quad t_0\in\Gamma$$

and $-\frac{1}{p_\infty}<\beta<\frac{1}{p'_\infty}$ if $t_0\notin\Gamma$. Then for $f\in L^{p(\cdot)}(\Gamma,\omega)$

$$\|f\|_{L^{p(\cdot)}(\Gamma,\omega)}\leq c\|\mathcal{M}^\sharp f\|_{L^{p(\cdot)}(\Gamma,\omega)}. \tag{24}$$

Proof. As is known, $\|f\|_{L^{p(\cdot)}(\Gamma)}\sim\sup\limits_{\|g\|_{L^{p'(\cdot)}(\Gamma)}\leq 1}\left|\int\limits_\Gamma f(t)g(t)d\nu(t)\right|$ (see [79, Theorem 2.3]). Therefore,

$$\|f\omega\|_{L^{p(\cdot)}(\Gamma)}\leq c\sup_{\|g\|_{L^{p'(\cdot)}(\Gamma)}\leq 1}\left|\int\limits_\Gamma f(t)\omega(t)g(t)d\nu(t)\right|.$$

We make use of the inequality

$$\int\limits_\Gamma f(t)g(t)d\nu(t)\leq\int\limits_\Gamma M^\sharp f(t)Mg(t)d\nu(t) \tag{25}$$

where $f \in L^{p(\cdot)}(\Gamma)$, $g \in L^{p'}(\Gamma)$, which is known for the Euclidean space, see [17, Theorem 3.5] and is similarly proved for infinite Carleson curves. We obtain

$$\|f\omega\|_{L^{p(\cdot)}(\Gamma)} \leq c \sup_{\|g\|_{L^{p'(\cdot)}(\Gamma)} \leq 1} \left| \int_\Gamma \omega(t) M^\sharp f(t)[\omega(t)]^{-1} M(g\omega) d\nu(t) \right|$$

and then

$$\|f\omega\|_{L^{p(\cdot)}(\Gamma)} \leq c \sup_{\|g\|_{L^{p'(\cdot)}(\Gamma)} \leq 1} \|\omega M^\sharp f\|_{L^{p(\cdot)}(\Gamma)} \|\omega^{-1} M(g\omega)\|_{L^{p'(\cdot)}(\Gamma)}$$

by the Hölder inequality. Since $-\frac{1}{p'(t_0)} < -\beta < \frac{1}{p(t_0)}$, we may apply Proposition 5 for the space $L^{p'(\cdot)}(\Gamma)$ with β replaced by $-\beta$ and conclude that

$$\|f\omega\|_{L^{p(\cdot)}(\Gamma)} \leq c_1 \sup_{\|g\|_{L^{p'(\cdot)}(\Gamma)} \leq 1} \|\omega M^\sharp f\|_{L^{p(\cdot)}(\Gamma)} \|g\|_{L^{p'(\cdot)}(\Gamma)} \leq c_2 \|\omega M^\sharp f\|_{L^{p(\cdot)}(\Gamma)}$$

which proves (24). □

Proof of Theorem 1

General remark

Remark 9. It suffices to prove Theorem 1 for a single weight $|t - t_0|^\beta$ where $t_0 \in \Gamma$ in the case Γ is finite and t_0 may belong or not belong to Γ when Γ is infinite.

Indeed, in the case of a finite curve let $\Gamma = \bigcup_{k=1}^{n} \Gamma_k$ where Γ_k contains the point t_k in its interior and does not contain t_j, $j \neq k$ in its closure. Then

$$\|f\|_{L^{p(\cdot)}\left(\Gamma, \prod_{k=1}^{n} |t-t_k|^{\beta_k}\right)} \sim \sum_{k=1}^{n} \|f\|_{L^{p(\cdot)}(\Gamma_k, |t-t_k|^{\beta_k})} \tag{26}$$

whenever $1 < \underline{p} \leq \overline{p} < \infty$. This equivalence follows from the easily checked modular equivalence

$$I_\Gamma^{p(\cdot)}\left(f(t) \prod_{k=1}^{n} |t - t_k|^{\beta_k}\right) \sim \sum_{k=1}^{n} I_{\Gamma_k}^{p(\cdot)}(f(t)|t - t_k|^{\beta_k}),$$

since

$$\begin{aligned} c_1 \leq \|f\|_{p(\cdot)} \leq c_2 &\Longrightarrow c_3 \leq I_\Gamma^{p(\cdot)}(f) \leq c_4, \\ C_1 \leq I_\Gamma^{p(\cdot)}(f) \leq C_2 &\Longrightarrow C_3 \leq \|f\|_{p(\cdot)} \leq C_4 \end{aligned} \tag{27}$$

with $c_3=\min(c_1^{\underline{p}}, c_1^{\overline{p}})$, $c_4=\max(c_2^{\underline{p}}, c_2^{\overline{p}})$, $C_3=\min\left(C_1^{\frac{1}{\underline{p}}}, C_1^{\frac{1}{\overline{p}}}\right)$ and $C_4=\max\left(C_2^{\frac{1}{\underline{p}}}, C_2^{\frac{1}{\overline{p}}}\right)$.

Similarly, in the case of an infinite curve

$$\|f\|_{L^{p(\cdot)}\left(\Gamma,|t-z_0|^\beta\prod\limits_{k=1}^{n}|t-t_k|^{\beta_k}\right)}\sim\|f\|_{L^{p(\cdot)}(\Gamma_\infty,|t-z_0|^\beta)}+\sum_{k=1}^{n}\|f\|_{L^{p(\cdot)}(\Gamma_k,|t-t_k|^{\beta_k})}$$

where Γ_∞ is a portoin of the curve outside some large circle, so that Γ_∞ does not contain the points t_k, $k=1,\dots,n$.

Then, because of (26) the statement of Remark 9 is obtained by introduction of the standard partition of unity $1=\sum\limits_{k=1}^{n}a_k(t)$, where $a_k(t)$ are smooth functions equal to 1 in a neighborhood $D(t_k,\varepsilon)$ of the point t_k and equal to 0 outside its neighborhood $D(t_k,2\varepsilon)$ (and similarly in a neighborhood of infinity in the case Γ is infinite), so that $a_k(t)|t-t_j|^{\pm\beta_j}\equiv 0$ in a neighborhood of the point t_k, if $k\neq j$.

We have divided the proof into a sequence of Propositions.

We start with proving the following statement established by T. Alvarez and C. Pérez [2] for singular integrals in the Euclidean space.

Proposition 10. *Let Γ be a simple Carleson curve. Then the following pointwise estimate is valid*

$$\mathcal{M}^\sharp(|S_\Gamma f|^s)(t)\le c[Mf(t)]^s,\quad 0<s<1, \tag{28}$$

where the constant $c>0$ may depend on Γ and s, but does not depend on $t\in\Gamma$ and f.

To prove Proposition 10, we need the following technical lemma.

Lemma 11. *Let Γ be a simple Carleson curve, $z_0\in\Gamma$ and $D_r=D(z_0,r)$ and*

$$H_{r,z_0}(t)=\frac{1}{|D_r|^2}\int\limits_{D_r}\int\limits_{D_r}\left|\frac{1}{z-t}-\frac{1}{\tau-t}\right|d\nu(z)d\nu(\tau). \tag{29}$$

Then for any locally integrable function f the pointwise estimate holds

$$\sup_{r>0}\int\limits_{t\in\Gamma:|t-z_0|>2r}|f(t)|H_{r,z_0}(t)d\nu(t)\le CMf(z_0) \tag{30}$$

where $C>0$ does not depend on f and z_0.

Proof. We have

$$H_{r,z_0}(t)=\frac{1}{|D_r|^2}\int\limits_{D_r}\int\limits_{D_r}\frac{|\tau-z|}{|z-t|\cdot|\tau-t|}d\nu(z)d\nu(\tau).$$

For $|t - z_0| > 2r$ we have

$$|z - t| \geq |t - z_0| - |z - z_0| \geq |t - z_0| - r \geq \frac{1}{2}|t - z_0|$$

and similarly $|\tau - t| \geq \frac{1}{2}|t - z_0|$ so that

$$H_{r,z_0} \leq \frac{Cr}{|t - z_0|^2}$$

where the constant $C > 0$ depends only on the length of the curve Γ. Then

$$\sup_{r>0} \int_{t\in\Gamma:|t-z_0|>2r} |f(t)|H_{r,z_0}(t)d\nu(t) \leq c \sup_{r>0} \sum_{k=0}^{m} \int_{2^k r<|t-z_0|<2^{k+1}r} \frac{r|f(t)|}{|t - z_0|^2} d\nu(t)$$

with $m = m(r)$. Hence

$$\sup_{r>0} \int_{t\in\Gamma:|t-z_0|>2r} |f(t)|H_{r,z_0}(t)d\nu(t) \leq 2c \sup_{r>0} \sum_{k=0}^{m} \frac{1}{2^k} \frac{1}{2^{k+1}r} \int_{2^k r<|t-z_0|<2^{k+1}r} |f(t)|d\nu(t)$$

$$\leq 2cMf(z_0) \sum_{k=0}^{m} \frac{1}{2^k} \leq c_1 Mf(z_0). \qquad \square$$

We will also need the following lemma

Lemma 12. *Let f be an integrable function on Γ, $f_D = \frac{1}{|D|}\int_D f(\sigma)d\nu(\sigma)$. Then*

$$\frac{1}{|D|}\int_D |f(\tau) - f_D|d\nu(\tau) \leq \frac{2}{|D|}\int_D |f(\tau) - C|d\nu(\tau) \tag{31}$$

for any constant C on the right-hand side.

Proof. The proof is obvious:

$$\frac{1}{|D|}\int_D |f(\tau) - f_D|d\nu(\tau) \leq \frac{1}{|D|^2}\int_D\int_D |f(\tau) - f(\sigma)|d\nu(\tau)d\nu(\sigma)$$

$$\leq \frac{1}{|D|^2}\int_D\int_D (|f(\tau) - C| + |C - f(\sigma)|)d\nu(\tau)d\nu(\sigma) = \frac{2}{|D|}\int_D |f(\tau) - C|d\nu(\tau). \qquad \square$$

Proof of Proposition 10. To prove estimate (24), according to Lemma 12 it suffices to show that for any locally integrable function f and any $0 < s < 1$ there exists a positive constant A such that

$$\left(\frac{1}{|D|}\int_D \Big||S_\Gamma f(\zeta)|^s - A^s\Big| d\nu(\zeta)\right)^{\frac{1}{s}} \leq CMf(z_0), \quad D = D(z_0, r) \tag{32}$$

for almost all $z_0 \in \Gamma$, where $C > 0$ does not depend on f and z_0. We set $f = f_1 + f_2$, where $f_1 = f \cdot \chi_{D(z_0,2r)}$ and $f_2 = f \cdot \chi_{\Gamma \setminus D(z_0,2r)}$. We take

$$A = (S_\Gamma f_2)_D = \frac{1}{|D|} \int\limits_D |S_\Gamma f_2(\zeta)| d\nu(\zeta).$$

Then, taking into account that $||a|^s - |b^s|| \leq |a - b|^s$, for $0 < s < 1$, we have

$$\left(\frac{1}{|D|} \int\limits_D \Big| |S_\Gamma f(\zeta)|^s - A^s \Big| d\nu(\zeta) \right)^{\frac{1}{s}} \leq c \left(\frac{1}{|D|} \int\limits_D \Big| S_\Gamma f_1(\zeta) \Big|^s d\nu(\zeta) \right)^{\frac{1}{s}}$$

$$+ c \left(\frac{1}{|D|} \int\limits_D \Big| |S_\Gamma f_2(\zeta)| - A \Big|^s d\nu(\zeta) \right)^{\frac{1}{s}} =: c(I_1 + I_2).$$

For I_1 by (21) we obtain

$$I_1 \leq \frac{1}{|D|} \int\limits_D |f_1(\zeta)| d\nu(t) \leq \frac{1}{|D|} \int\limits_D |f(t)| d\nu(\zeta) \leq Mf(z_0). \tag{33}$$

For I_2, by Jensen inequality and Fubini theorem after easy estimations we get

$$I_2 \leq \frac{1}{|D|} \int\limits_D \left| (S_\Gamma f_2)(\zeta) - \frac{1}{|D|} \int\limits_D (S_\Gamma f_2)(\tau) d\nu(\tau) \right| d\nu(\zeta) \leq \int\limits_{\Gamma \setminus D(z_0,2r)} |f(t)| H_{r,z_0}(t) d\nu(t),$$

where $H_{r,z_0}(t)$ is the function defined in (29). Therefore, by Lemma 11 $I_2 \leq CMf(z_0)$ which comletes the proof. $\square$

Proof of Theorem 1 itself. Sufficiency part

According to Remark 9, we consider the case of a single weight $|t - t_0|^\beta$ where t_0 may be not belonging to Γ in case Γ is infinite.

I) The case of infinite curve and p constant at infinity. First we consider the case where Γ is an infinite curve and we additionally suppose at this step that $p(t) \equiv \text{const} = p_\infty$ outside some large ball $B(0, R)$.

Let $0 < s < 1$. Observe that

$$\|S_\Gamma f\|_{L^{p(\cdot)}(\Gamma,\omega)} = \||S_\Gamma f|^s\|^{\frac{1}{s}}_{L^{\frac{p(\cdot)}{s}}(\Gamma,\omega^s)}.$$

Then by Theorem 8 we have

$$\|S_\Gamma f\|_{L^{p(\cdot)}(\Gamma,\omega)} \leq C \|\mathcal{M}^\sharp(|S_\Gamma f|^s)\|^{\frac{1}{s}}_{L^{\frac{p(\cdot)}{s}}(\Gamma,\omega^s)}.$$

Indeed, Theorem 8 is applicable in this case, because $\frac{p(t)}{s}$ satisfies conditions (4)–(5) then the exponent $s\beta$ of the weight ω^s satisfies the conditions $-\frac{1}{\frac{p(t_0)}{s}} < s\beta < \frac{1}{\frac{p'(t_0)}{s}}$, required by Theorem 8. Therefore, by Proposition 10 we get

$$\|S_\Gamma f\|_{L^{p(\cdot)}(\Gamma,\omega)} \leq c\|(Mf)^s\|^{\frac{1}{s}}_{L^{\frac{p(\cdot)}{s}}(\Gamma,\omega^s)} = c\|Mf\|_{L^{p(\cdot)}(\Gamma,\omega)}.$$

It remains to apply Proposition 5 to obtain $\|S_\Gamma f\|_{L^{p(\cdot)}(\Gamma,\omega)} \leq c\|f\|_{L^{p(\cdot)}(\Gamma,\omega)}$.

II) The case of finite curve and p constant on some arc. At the next step we consider the case of finite curve under the additional assumption that there exists an arc $\gamma \subset \Gamma$ with $|\gamma| > 0$ on which $p(t) \equiv \text{const}$.

Since $|\tau'(s)| = 1$ on Carleson curves we have $\|f\tau'\|_{L^{p(\cdot)}(\Gamma,\omega)} = \|f\|_{L^{p(\cdot)}(\Gamma,\omega)}$.

The case considered now is reduced to the previous case **I**) by the change of variables. Let $z_0 \in \gamma$ be any point of γ (different from t_0 if $t_0 \in \gamma$). Without loss of generality we may assume that $z_0 = 0$. Let

$$\Gamma_* = \{t \in \mathbb{C} : t = \frac{1}{\tau},\ \tau \in \Gamma\} \quad \text{and} \quad \widetilde{p}(t) = p\left(\frac{1}{t}\right), \quad t \in \Gamma_*,$$

so that Γ_* is an infinite curve and $\widetilde{p}(z)$ is constant on Γ_* outside some large circle. By the change of variables $\frac{1}{\tau} = w$ and $\frac{1}{t} = z$ we get

$$(S_\Gamma f)(t) = -z(S_{\Gamma_*}\psi)(z), \quad z \in \Gamma_* \tag{34}$$

where $\psi(w) = \frac{1}{w} f\left(\frac{1}{w}\right)$. The following lemma is valid where the equivalence $A \sim B$ means that $c_1 A \leq B \leq c_2 A$ with $c_1 > 0$ and $c_2 > 0$.

Lemma 13. *The following modular equivalence holds*

$$I_\Gamma^{p(\cdot)}(|t - t_0|^\beta f(t)) \sim I_{\Gamma_*}^{\widetilde{p}(\cdot)}(\rho(t)\psi(t)) \tag{35}$$

where $\rho(t) = |t|^\nu |t - t_0^*|^\beta$ *with* $t_0^* = \frac{1}{t_0} \in \Gamma_*$ *and*

$$\nu = 1 - \beta - \frac{2}{\widetilde{p}(\infty)} = 1 - \beta - \frac{2}{p(0)}. \tag{36}$$

Proof. Indeed,

$$I_\Gamma^{p(\cdot)}(|t - t_0|^\beta f(t)) = \int_\Gamma |t - t_0|^{\beta p(t)} |f(t)|^{p(t)} |dt|.$$

After the change of variables $t \to \frac{1}{t}$ we get

$$I_\Gamma^{p(\cdot)}(|t - t_0|^\beta f(t)) = \int_{\Gamma_*} \left|\frac{1}{t} - t_0\right|^{\beta\widetilde{p}(t)} \left|f\left(\frac{1}{t}\right)\right|^{\widetilde{p}(t)} \frac{|dt|}{|t^2|}$$

$$= \int_{\Gamma_*} |t_0|^{\beta\widetilde{p}(t)} \frac{|t_0^* - t|^{\beta\widetilde{p}(t)}}{|t|^{\beta\widetilde{p}(t)}} \left| f\left(\frac{1}{t}\right)\right|^{\widetilde{p}(t)} \frac{|dt|}{|t^2|}.$$

Since $|t_0|^{\beta\widetilde{p}(t)} \sim$ const, we obtain

$$I_\Gamma^{p(\cdot)}(|t-t_0|^\beta f(t)) \sim \int_{\Gamma_*} \frac{|t_0^* - t|^{\beta\widetilde{p}(t)}}{|t|^{\beta\widetilde{p}(t)+2}} \left| f\left(\frac{1}{t}\right)\right|^{\widetilde{p}(t)} |dt|. \tag{37}$$

Now we have

$$f\left(\frac{1}{t}\right) = t\psi(t).$$

Therefore, from (37) we get

$$I_\Gamma^{p(\cdot)}(|t-t_0|^\beta f(t)) \sim \int_{\Gamma_*} \frac{|t_0^* - t|^{\beta\widetilde{p}(t)}}{|t|^{\beta\widetilde{p}(t)+2}} |t\psi(t)|^{\widetilde{p}(t)} |dt|.$$

Hence

$$I_\Gamma^{p(\cdot)}(|t-t_0|^\beta f(t)) \sim \int_{\Gamma_*} \left(\frac{|t_0^* - t|^{\beta}}{|t|^{(\beta-1)+\frac{2}{\widetilde{p}(t)}}} |\psi(t)|\right)^{\widetilde{p}(t)} |dt|.$$

Observe that the point $z = 0$ does not pass through the origin and therefore $|t|^{(\beta-1)+\frac{2}{\widetilde{p}(t)}} \sim |t|^{(\beta-1)+\frac{2}{\widetilde{p}(\infty)}}$. As a result we arrive at (35), (36). □

According to (27), from (35) we also have

$$\|f\|_{L^{p(\cdot)}(\Gamma,|t-t_0|^\beta)} \sim \|\psi\|_{L^{\widetilde{p}(\cdot)}(\Gamma_*,\rho(t))}$$

and

$$\|S_\Gamma f\|_{L^{p(\cdot)}(\Gamma,|t-t_0|^\beta)} \sim \|S_{\Gamma_*}\psi\|_{L^{\widetilde{p}(\cdot)}(\Gamma_*,\rho(t))} \tag{38}$$

where (34) was taken into account. Observe also that

$$-\frac{1}{p(t_0)} < \beta < \frac{1}{p'(t_0)} \Longleftrightarrow -\frac{1}{\widetilde{p}(t_0^*)} < \nu < \frac{1}{\widetilde{p}'(t_0^*)}. \tag{39}$$

Obviously, $\widetilde{p}(t)$ satisfies conditions (4)–(5). Since $\widetilde{p}(t)$ is constant at infinity, according to part I) and Remark 9, the operator S_{Γ_*} is bounded in the space $L^{\widetilde{p}(\cdot)}(\Gamma_*, \rho(t))$, the required conditions on the weight $\rho(t)$ being satisfied by (39) and by the fact that $\beta + \nu = 1 - \frac{2}{\widetilde{p}(\infty)}$ is automatically in the interval $\left(-\frac{1}{\widetilde{p}(\infty)}, \frac{1}{\widetilde{p}'(\infty)}\right)$. Then the operator S_Γ is bounded in the space $L^{p(\cdot)}(\Gamma, |t-t_0|^\beta)$ by (38).

III) The general case of finite curve. Let $\gamma_1 \subset \Gamma$ and $\gamma_2 \subset \Gamma$ be two disjoint non-empty arcs of Γ, $\overline{\gamma_1} \cap \overline{\gamma_2} = \varnothing$. According to the part II) the operator $wS_\Gamma\frac{1}{w}$ with $w(t) = |t-t_0|^\beta$, $t_0 \in \Gamma$, is bounded in the space $L^{p_1(\cdot)}(\Gamma)$, if

1 $p_1(t)$ satisfies conditions (4)–(5),

2 $p_1(t)$ is constant at γ_1,

3 $-\frac{1}{p_1(t_0)} < \beta < \frac{1}{p_1'(t_0)}$

and similarly in the space $S^{p_2(\cdot)}(\Gamma)$, if

1′ $p_2(t)$ satisfies conditions (4)–(5),

2′ $p_2(t)$ is constant at γ_2,

3′ $-\frac{1}{p_2(t_0)} < \beta < \frac{1}{p_2'(t_0)}$

Aiming to make use of the interpolation theorem, we observe that the following statement is valid.

Lemma 14. *Given a function $p(t)$ satisfying conditions (4)–(5), there exist arcs $\gamma_2 \subset \Gamma$ and $\gamma_1 \subset \Gamma$ such that $p(t)$ may be represented in the form*

$$\frac{1}{p(t)} = \frac{\theta}{p_1(t)} + \frac{1-\theta}{p_2(t)}, \qquad \theta = \frac{1}{2}, \tag{40}$$

where $p_j(t)$, $j = 1, 2,$, satisfy the above conditions 1), 2) and 1′), 2′), respectively, and

$$\frac{1}{p_j(t_0)} = \frac{\theta}{p(t_0)}, \qquad j = 1, 2,$$

so that conditions 3) and 3′) are also satisfied whenever they are satisfied for $p(t)$.

Proof. We have to prove the following. Let Γ be a Carleson curve and $a(t)$ any function on Γ, satisfying the log–Hölder condition and such that

$$0 < d \le a(t) \le D < 1 \quad \text{on} \quad \Gamma.$$

Then there exist non-intersecting non-empty arcs γ_1 and γ_2 on Γ such that

$$a(t) = \frac{b(t) + c(t)}{2} \quad \text{with} \quad b(t) \equiv 0 \quad \text{on} \quad \gamma_1 \quad \text{and} \quad c(t) \equiv 0 \quad \text{on} \quad \gamma_2$$

and $b(t)$ and $c(t)$ are log-Hölder continuous on Γ, satisfy the same condition as $a(t)$ and $b(t_0) = c(t_0) = a(t_0)$.

We will take γ_1 and γ_2 so that $t_0 \notin \overline{\gamma_1 \cup \gamma_2}$ and construct the functions $b(t)$ and $c(t)$ as follows

$$b(t) = \begin{cases} A, & t \in \gamma_1, \\ \ell(t), & t \in \Gamma \backslash (\gamma_1 \cup \gamma_2), \\ 2a(t) - B, & t \in \gamma_2, \end{cases}$$

where $A, B \in (0,1)$ are some constants. The link $\ell(t)$ between the values of $b(t)$ on γ_1 and on γ_2 may be constructed for instance in the following way: at each of the components of the set $\Gamma\backslash(\gamma_1 \cup \gamma_2)$ it is introduced as the linear interpolation between the number A and the values of $2a(t) - B$ at the endpoints of this component, if it does not belong to it, and as the piecewise linear interpolation between a the value $a(t_0)$ and the values of $2a(t) - B$ at the endpoints of this component, if it contains t_0. Then

$$c(t) = 2a(t) - b(t) = \begin{cases} 2a(t) - A, & t \in \gamma_1, \\ 2a(t) - \ell(t), & t \in \Gamma\backslash(\gamma_1 \cup \gamma_2), \\ B, & t \in \gamma_2. \end{cases}$$

Obviously, $b(t)$ and $c(t)$ are log-continuous on Γ. Checking condition which satisfies $a(t)$ for $b(t)$ and $c(t)$, we only have to verify this condition for $2a(t) - A$ on γ_1 and for $2a(t) - B$ on γ_2. To this end, we have to choose A and B so that

$$2a(t) - 1 < A < 2a(t) \quad \text{on} \quad \overline{\gamma_1}, \quad 2a(t) - 1 < B < 2a(t) \quad \text{on} \quad \overline{\gamma_2}$$

Let $a_-(\gamma_i) = \inf\limits_{t \in \gamma_i} a(t)$ and $a_+(\gamma_i) = \sup\limits_{t \in \gamma_i} a(t)$, $i = 1, 2$. It suffices to choose A and B in the intervals

$$A \in (\max\{0, 2a_+(\gamma_1) - 1\}, \min\{2a_-(\gamma_1), 1\}),$$

$$B \in (\max\{0, 2a_+(\gamma_2) - 1\}, \min\{2a_-(\gamma_2), 1\}),$$

These intervals are non-empty, if $a_+(\gamma_i) - a_-(\gamma_i) > \frac{1}{2}$, $i = 1, 2$. Obviously, γ_i may be chosen sufficiently small so that the last condition is satisfied. □

In view of Lemma 14, the boundedness of the singular operator in $L^{p(\cdot)}(\Gamma)$ with given p follows from the Riesz-Thorin interpolation theorem for the spaces $L^{p(\cdot)}(\Gamma)$ (proved in [93]).

IV). The general case of infinite curve. Obviously, the after the step III) the general case of infinite curve, that is, the case where Γ is infinite and p is not necessarily constant outside some circle, is reduced to the case of finite curve by mapping the infinite curve Γ onto a finite curve Γ_* in the same way as it was done in the step II). What is important to note is that thanks to conditions (5) and (7), the new exponent $\widetilde{p}(t)$, $t \in \Gamma_*$ is log-Hölder continuous on the curve Γ_*.

Remark 15. We emphasize the following. We had to prove the boundedness of the singular operator on an infinite curve under additional assumption that $p(t)$ is constant at infinity, then by the change of variables we could cover the case of a finite curve with $p(t)$ constant at any arc. After that we could use the interpolation theorem to get a result on boundedness on a finite curve without the assumption on $p(t)$ to be constant an arc. After that it remained to use change of variables to get the general result for an infinite curve. This order is essential. Indeed, a seeming possibility to treat first the general case of a finite curve and then cover the case of infinite curve by the change of variables, is not applicable, because the initial step in the whole proof was based on Theorem 8, which was proved for infinite curves.

Proof of the necessity part of Theorem 1

The proof of the necessity is in fact the same as in the case of smooth curves, see [69, p.153]. We dwell on the main points. Let Γ be a finite curve. From the boundedness of S_Γ in $L^{p(\cdot)}(\Gamma, \rho)$ it follows that $S_\Gamma f(t)$ exists almost everywhere for an arbitrary $f \in L^{p(\cdot)}(\Gamma, \rho)$. Thus ρ should be such that $f \in L^1(\Gamma)$ for arbitrary $f \in L^{p(\cdot)}(\Gamma, \rho)$. The function $f\rho^{-1}$ belongs to $L^1(\Gamma)$ for arbitrary $f \in L^{p(\cdot)}(\Gamma, \rho)$ if and only if $\rho^{-1} \in L^{p'(\cdot)}$. Then the function $\rho^{-1}(t) = |t - t_0|^{-\beta}$, $t_0 \in \Gamma$, belongs to $L^{p'(\cdot)}(\Gamma)$ if and only if $\beta < \frac{1}{p'(t_0)}$. Indeed, by the log-condition we have

$$|t - t_0|^{-\beta p'(t)} \sim |t - t_0|^{-\beta p'(t_0)}.$$

On the other hand, since Γ is a Carleson curve, from $|t - t_0|^{-\beta p'(t_0)} \in L^1$ we have $\beta < \frac{1}{p'(t_0)}$.

The necessity of the condition $-\frac{1}{p(t_0)} < \beta$ follows from the duality argument.

In a similar way, with slight modifications the case of infinite curve and weight fixed to infinity, is treated.

Proof of Theorem 2

We start with the following remark.

Remark 16. If the operator S_Γ is bounded in $L^{p(\cdot)}(\Gamma)$ and γ is a measurable subset of Γ, then the operator $S_\gamma = \chi_\gamma S_\Gamma \chi_\gamma$ is bounded in $L^{p(\cdot)}(\gamma)$ and $\|S_\gamma\|_{L^{p(\cdot)}(\gamma)} \leq \|S_\Gamma\|_{L^{p(\cdot)}(\Gamma)}$ (we denote the restriction of $p(\cdot)$ onto γ by the same symbol $p(\cdot)$).

Let us prove some auxiliary lemmas.

Lemma 17. *For every point $t_0 \in \Gamma$ and every $\rho \in \left(0, \frac{1}{6}\mathrm{diam}\Gamma\right)$ there exists a function $\varphi_{t_0} := \varphi_{t_0,\rho}(\tau)$ such that*

$$I^{p(\cdot)}(S_\Gamma \varphi_{t_0}) \geq m \left(\frac{|D(t_0, \rho|)}{\rho}\right)^{\underline{p}-1} I^{p(\cdot)}(\varphi_{t_0}), \tag{41}$$

where $m > 0$ is a constant not depending on t_0 and ρ.

Proof. Let us fix the point $t = t_0$ and consider circles centered at t_0 of the radii ρ, 2ρ and 3ρ and 8 rays with the angle $\frac{\pi}{4}$, one of them being parallel to the axis of abscissas. These rays split the circle $|z - t_0| < \rho$ and the annulus $2\rho < |z - t_0| < 3\rho$

into 16 parts. It suffices to treat only those partes which lie in a semiplane, for example, in the upper semiplane. We denote these parts of the circle $|z-t_0|<\rho$ by $\Gamma_k:=\Gamma_{k,t_0,\rho}$ and the parts of the annulus $2\rho<|z-t_0|<3\rho$ by $\gamma_k:=\gamma_{k,t_0,\rho}$, respectively, $k=1,2,3,4$, couting them, e. g., counter clockwise. These rays may be chosen so that there exists a pair k_0,j_0 such that

$$|\Gamma_{k_0}|\geq \frac{1}{8}|D(t_0,\rho)| \quad \text{and} \quad |\gamma_{j_0}|\geq \frac{1}{8}\rho. \tag{42}$$

Without loss of generality we may take $k_0=1$.

Let

$$\varphi_{t_0}=\varphi_{t_0,\rho}(t)=\begin{cases}1, & t\in\Gamma_1,\\ 0, & t\in\Gamma\backslash\Gamma_1.\end{cases} \tag{43}$$

We have to estimate the integral

$$I^{p(\cdot)}(S_\Gamma\varphi_{t_0,\rho})=\int_\Gamma\left|\int_\Gamma\frac{\varphi_{t_0}(\tau)}{\tau-t}d\nu(\tau)\right|^{p(t)}d\nu(t). \tag{44}$$

Let $\tau-t=|\tau-t|e^{i\alpha(\tau,t)}$. We have

$$I^{p(\cdot)}(S_\Gamma\varphi_{t_0})\geq\int_{\gamma_{j0}}\left|\int_{\Gamma_1}\frac{\cos\alpha(\tau,t)-i\sin\alpha(\tau,t)}{|\tau-t|}d\nu(\tau)\right|^{p(t)}d\nu(t). \tag{45}$$

Let first $j_0=1$. We put $M_1=(\rho,0)$ and $M_2=\left(2\rho\cos\frac{\pi}{4},2\rho\sin\frac{\pi}{4}\right)$. It is easily seen that $\max|\alpha(\tau,t)|\leq\beta_1<\frac{\pi}{2}$, where β_1 is the angle between the vector $\overline{M_2M_2}$ and the axis of abscissas. Similarly it can be seen that

$$\begin{aligned}&\text{if}\quad \tau\in\Gamma_1,\ t\in\gamma_2,\ \text{then}\ \frac{\pi}{4}\leq\alpha(\tau,t)\leq\pi-\beta_2,\\ &\text{if}\quad \tau\in\Gamma_1,\ t\in\gamma_3,\ \text{then}\ \frac{\pi}{2}\leq\alpha(\tau,t)\leq\pi-\beta_3,\\ &\text{if}\quad \tau\in\Gamma_1,\ t\in\gamma_4,\ \text{then}\ \frac{3\pi}{4}\leq\alpha(\tau,t)\leq\pi+\beta_4\end{aligned}$$

where $\beta_2=\mathrm{arctg}2$, $\beta_3=\mathrm{arctg}\frac{1}{3}$ and $\beta_4=\mathrm{arctg}\frac{2\sqrt{2}-1}{7}$. Therefore, when $\tau\in\Gamma_1$ and $t\in\gamma_{j_0}$, $j_0=1,2,3,4$, then

$$\text{either}\quad |\cos\alpha(\tau,t)|\geq m_0>0,\quad \text{or}\quad |\sin\alpha(\tau,t)|\geq m_0>0.$$

Moreover, when $\tau\in\Gamma_1$ and $t\in\gamma_2$ or $t\in\gamma_4$, then $\cos\alpha(\tau,t)$ preserves the sign and when $\tau\in\Gamma_1$ and $t\in\gamma_2$ or $t\in\gamma_3$, then $\sin\alpha(\tau,t)$ preserves the sign. Consequently, from (45) we get

$$I^{p(\cdot)}(S_\Gamma\varphi_{t_0})\geq\int_{\gamma_{j0}}\max\left(\left|\mathrm{Re}\int_{\Gamma_1}\frac{\varphi_{t_0}(\tau)d\nu(\tau)}{\tau-t}\right|^{p(t)},\ \left|\mathrm{Im}\int_{\Gamma_1}\frac{\varphi_{t_0}(\tau)d\nu(\tau)}{\tau-t}\right|^{p(t)}\right)|dt|.$$

Hence

$$I^{p(\cdot)}(S_\Gamma \varphi_{t_0}) \geq \int\limits_{\gamma_{j0}} \left| \int\limits_{\Gamma_1} \frac{m_0}{3\rho} d\nu(\tau) \right|^{p(t)} d\nu(t) \geq \left(\frac{m_0}{3}\right)^{\overline{p}} \int\limits_{\gamma_{j0}} \left(\frac{|\Gamma_1|}{\rho}\right)^{p(t)} d\nu(t).$$

Then by (42)

$$I^{p(\cdot)}(S_\Gamma \varphi_{t_0}) \geq \left(\frac{m_0}{3 \cdot 8}\right)^{\overline{p}} \int\limits_{\gamma_{j0}} \left(\frac{|D(t_0,\rho)|}{\rho}\right)^{p(t)} d\nu(t) \geq m_1 \left(\frac{|D(t_0,\rho)|}{\rho}\right)^{\underline{p}} |\gamma_{j_0}|.$$

Since $|D(t_0,\rho)| \geq I^{p(\cdot)}(\varphi_{t_0}) = |\Gamma_1|$ and $|\gamma_{j_0}| \geq \frac{\rho}{8}$, we obtain

$$I^{p(\cdot)}(S_\Gamma \varphi_{t_0}) \geq \frac{m_1}{8} \left(\frac{|D(t_0,\rho)|}{\rho}\right)^{\underline{p}} \frac{\rho}{|D(t_0,\rho)|} |\Gamma_1| = m \left(\frac{|D(t_0,\rho)|}{\rho}\right)^{\underline{p}-1} I^{p(\cdot)}(\varphi_{t_0})$$

which proves (41) with (43). □

We denote for brevity

$$\alpha(f) = \alpha_\gamma(f) = \begin{cases} \overline{p}, & \text{if } \|f\|_{p(\cdot)} \geq 1, \\ \underline{p}, & \text{if } \|f\|_{p(\cdot)} < 1, \end{cases}$$

and

$$\beta(f) = \beta_\gamma(f) = \begin{cases} \underline{p}, & \text{if } \|f\|_{p(\cdot)} \geq 1, \\ \overline{p}, & \text{if } \|f\|_{p(\cdot)} < 1, \end{cases}$$

so that $\alpha(f) + \beta(f) \equiv \overline{p} + \underline{p}$ and

$$\|f\|_{p(\cdot)}^{\beta(f)} \leq I^{p(\cdot)}(f) \leq \|f\|_{p(\cdot)}^{\alpha(f)}. \tag{46}$$

Lemma 18. *If the operator S_Γ is bounded in the space $L^{p(\cdot)}(\Gamma)$, then for every $t_0 \in \Gamma$ the estimate holds*

$$\frac{|D(t_0,\rho)|}{\rho} \leq C_\Gamma |D(t_0,\rho)|^{\delta_\Gamma(t_0)} \tag{47}$$

where $\delta_\Gamma(t_0) = \frac{1}{\underline{p}-1}\left(\frac{\alpha(S_\Gamma \varphi_{t_0})}{\beta(\varphi_{t_0})} - 1\right)$, $C_\Gamma = \left(\frac{8\|S_\Gamma\|^{\overline{p}}_{L^{p(\cdot)}(\Gamma)}}{m}\right)^{\frac{1}{\underline{p}-1}}$, φ_{t_0} is function indicated in Lemma 17 and the constant m is number from (41).

Proof. Let $K = \|S_\Gamma\|_{L^{p(\cdot)}(\Gamma)}$ for brevity. By the boundedness $\|S_\Gamma f\|_{L^{p(\cdot)}(\Gamma)} \leq K\|f\|_{L^{p(\cdot)}}$ and property (46) we have

$$I^{p(\cdot)}(S_\Gamma f) \leq K^{\alpha(S_\Gamma f)}\|f\|^{\alpha(S_\Gamma f)}_{L^{p(\cdot)}(\Gamma)} \leq K^{\alpha(S_\Gamma f)}[I^{p(\cdot)}(f)]^{\frac{\alpha(S_\Gamma f)}{\beta(f)}}.$$

We choose $f = \varphi_{t_0}$ with φ_{t_0} from Lemma 17 and take (43) into account, which yields

$$K^{\alpha(S_\Gamma f)}[I^{p(\cdot)}(\varphi_{t_0})]^{\frac{\alpha(S_\Gamma \varphi_{t_0})}{\beta(\varphi_{t_0})}} \geq I^{p(\cdot)}(S_\Gamma \varphi_{t_0}) \geq m\left(\frac{|D(t_0,\rho)|}{\rho}\right)^{\underline{p}-1} I^{p(\cdot)}(\varphi_{t_0})$$
$$\geq \frac{m}{8}\left(\frac{|D(t_0,\rho)|}{\rho}\right)^{\underline{p}-1}|D(t_0,\rho)|. \qquad (48)$$

We observe that in the first term in this chain of inequalities we have $I^{p(\cdot)}(\varphi_{t_0}) \leq |D(t_0,\rho)|$ and then (48) yields (47). □

Proof of Theorem 2 itself

Let $\gamma = D(\xi, 3\rho) = \Gamma \cap \{z : \ |z-\xi| < 3\rho\}$. According to Remark 16, the operator S_γ is bounded in $L^{p(\cdot)}(\gamma)$. Then by Lemma 18 we obtain

$$\frac{|D(\xi,\rho)|}{\rho} \leq C_\gamma |D(\xi,\rho)|^{\delta_\gamma(\xi)} \leq C_\Gamma |D(\xi,\rho)|^{\delta_\gamma(\xi)}, \quad \xi \in \gamma, \qquad (49)$$

where C_Γ is the same as in Lemma 18 and

$$\delta_\gamma(\xi) = \frac{1}{\underline{p}(\gamma)-1}\left(\frac{\alpha(S_\gamma \varphi_\xi)}{\beta(\varphi_\xi)} - 1\right)$$

with $\underline{p}(\gamma) = \min\limits_{\tau\in\gamma} p(\tau)$. Depending on the values $\|S_\gamma \varphi_\xi\|_{L^{p(\cdot)}(\gamma)}$ and $\|\varphi_\xi\|_{L^{p(\cdot)}(\gamma)}$, the exponent $\delta_\gamma(\xi)$ may take only three values 0, $-\delta_1$ and δ_2, where

$$\delta_1 = \frac{\overline{p}(\gamma) - \underline{p}(\gamma)}{\overline{p}(\gamma)}\frac{1}{\underline{p}(\gamma)-1}, \qquad \delta_2 = \frac{\overline{p}(\gamma) - \underline{p}(\gamma)}{\underline{p}(\gamma)}\frac{1}{\underline{p}(\gamma)-1}$$

Therefore, when ρ is small, $|\delta_\gamma(\xi)|$ also has small values:

$$|\delta_\gamma(\xi)| \leq \lambda\omega(p, 6\rho), \quad \lambda = \frac{1}{(\underline{p}(\Gamma)-1)\underline{p}(\Gamma)}. \qquad (50)$$

where $\omega(p,h)$ is the continuity modulus of the function p, since $p(t)$ is continuous on the compact set Γ and consequently is uniformly continuous.

Let $\rho_1 < 1$ be sufficiently small such that $\lambda\omega(p, 6\rho_1) < \varepsilon$.

From (49) we have $|D(\xi,\rho)|^{1-\delta_\gamma(\xi)} \le C\rho$ and then

$$|D(\xi,\rho)| < C_\Gamma^{\frac{1}{1-\delta(\xi)}} \rho^{\frac{1}{1-\delta_\gamma(\xi)}} < C_\Gamma^{\frac{1}{1-\varepsilon}} \rho^{\frac{1}{1+\varepsilon}} \quad \text{for} \quad \rho < \rho_1 \tag{51}$$

(where we took into account that $C_\Gamma > 1$ and $\rho \le \rho_1 < 1$). Thus (17) has been proved.

Let now $p(t)$ satisfy the log–Hölder condition (5). For the function

$$\psi_\xi(\rho) = |D(\xi,\rho)|^{\delta_\gamma(\xi)}$$

by (51) we have

$$|\ln \psi_\xi(\rho)| = |\delta_\gamma(\xi) \ln |D(\xi,\rho)|| \le \lambda\omega(p, 6\rho) \left(\frac{\ln C_\Gamma}{1-\varepsilon} + \frac{|\ln \rho|}{1+\varepsilon}\right).$$

In view of (50) and (6) we then obtain

$$|\ln \psi_\xi(\rho)| \le \frac{\lambda A}{1-\varepsilon} \frac{\ln \frac{C_\Gamma}{\rho}}{\ln \frac{\ell}{3\rho}}, \quad \rho < \min\left\{\rho_1, \frac{\ell}{6}\right\}. \tag{52}$$

It is easy to see that $\frac{\ln \frac{C_\Gamma}{\rho}}{\ln \frac{\ell}{3\rho}}$ is bounded for small ρ, so that $|\ln \psi_\xi(\rho)| \le C < \infty$. Since $\frac{|D(\xi,\rho)|}{\rho} \le C_\Gamma \psi_\xi(\rho)$ by (49), we get $\frac{|D(\xi,\rho)|}{\rho} \le C_\Gamma e^C$, which means that Γ is a Carleson curve.

Proof of Theorem 3

Theorem 3 is derived from Theorem 1, which is standard. Indeed, it is known that any function $a(t)$ continuous on Γ may be approximated in $C(\Gamma)$ by a rational function $r(t)$, whatsoever Jordan curve Γ we have, as is known from the Mergelyan's Theorem. Therefore, since the singular operator S_Γ is bounded in $L^{p(\cdot)}(\Gamma,\omega)$ by Theorem 1, we obtain that the commutator $aS_\Gamma - S_\Gamma aI$ is approximated in the operator norm in $L^{p(\cdot)}(\Gamma,\omega)$ by the commutator $rS_\Gamma - S_\Gamma rI$ which is finite-dimensional operator, and consequently compact in $L^{p(\cdot)}(\Gamma,\omega)$. Therefore, $aS_\Gamma - S_\Gamma aI$ is compact.

2.2 The generalized grand Lebesgue spaces

The grand Lebesgue spaces $L^{p)}$ where introduced by T. Iwaniec and C. Sbordone [35] when they studied the problem of the integrability of Jakobian under minimal hypothesis. Later the generalized grand Lebesgue spaces $L^{p),\theta}$ appeared in the paper by L. Greco, T. Iwaniec and C. Sbordone [29]. It was realized the necessity of the study of these spaces, because of their rather essential role in various fields. It turns out that in the theory of PDEs they are right spaces in which some nonlinear equations have been considered.

Let Ω be a bounded subset of $\mathbb{R}^n$ and let ω be a weight (i.e. a. e. positive integrable function) on Ω. The generalized weighted grand Lebesgue space $L_\omega^{p),\theta}(\Omega)$ $(1 < p < \infty)$ is a Banach space defined by the norm

$$\|f\|_{L_\omega^{p),\theta}(\Omega)} = \sup_{0<\varepsilon<p-1} \left(\frac{\varepsilon^\theta}{|\Omega|} \int_\Omega |f(y)|^{p-\varepsilon} \omega(y) dy \right)^{1/(p-\varepsilon)}, \qquad 1 < p < \infty.$$

In the case $\omega \equiv$ *const* the space $L_\omega^{p),\theta}(\Omega)$, denoted by $L^{p),\theta}(\Omega)$, is rearrangement invariant and non–reflexive. If $\theta = 1$, then the space $L^{p),\theta}(\Omega)$ is the grand Lebesgue space which is denoted by $L^{p)}(\Omega)$.

It is worth mentioning that, in general, the weighted space $L_\omega^{p),\theta}(\Omega)$ is not rearrangement–invariant.

It is easy to check that the following continuous embeddings hold:

$$L_\omega^p(\Omega) \subset L_\omega^{p),\theta_1}(\Omega) \subset L_\omega^{p),\theta_2}(\Omega) \subset L_\omega^{p-\varepsilon}(\Omega),$$

where $0 < \varepsilon < p - 1$ and $\theta_1 < \theta_2$.

Note that the extension of grand Lebesgue spaces to weighted case has a relevant difference from the case of classical Lebesgue spaces. In the latter case, for a weight w, we have that f belongs to the weighted Lebesgue space if and only if $f\omega^{1/p} \in L^p$. This equivalence, however, fails in grand Lebesgue spaces as can be seen from the following example, taken from the paper [23]:

Let $\omega(x) = x^\alpha$, $\alpha > 0$. Let $f(x) = x^\beta$, $\beta > -\frac{\alpha}{p} - 1$. It is evident that $f \in L_\omega^p(0,1)$. On the other hand, for ε, $0 < \varepsilon < p - 1$ we have $\left(f\omega^{1/p}\right)^{p-\varepsilon} = x^{(\beta+\alpha/p)(p-\varepsilon)}$, so that if we choose $\beta < -1 - \alpha/p$, then the exponent of x becomes less than -1. Therefore $\left(f\omega^{1/p}\right)^{p-\varepsilon} \notin L(0,1)$ and so $f\omega^{1/p} \neq L^{p)}(0,1)$.

Along the space $L_\omega^{p),\theta}(\Omega)$ in Section 1.4 we treat also the Banach function space $L^{p),\theta}(\Omega, \omega)$ defined as

$$L^{p),\theta}(\Omega, \omega) = \{f : \|f\omega\|_{L^{p),\theta}(\Omega)} < \infty\}.$$

We will use the fact that boundedness some operator T in $L^{p),\theta}(\Omega, \omega)$ is equivalent to the boundedness of weighted operator $\omega T\left(\frac{f}{w}\right)$ in the space $L^{p),\theta}$.

2.3 The Cauchy singular integrals and maximal functions defined on Carleson curves

In this Section mapping properties of the Cauchy singular integrals and maximal functions defined on Carleson curves in generalized grand Lebesgue spaces are studied. We establish that an operator of this type is bounded in $L^{p),\theta}(\Omega, \omega)$ if and only if the weight ω satisfies the Muckenhoupt condition on the curve Γ. From this result it follows that the Cauchy singular integral operator is bounded in generalized Lebesgue space if and only if Γ is a Carleson curve.

The weighted grand Lebesgue space $L_{\omega}^{p),\theta}(\Gamma)$ $(1 < p < \infty, \theta > 0)$ is a Banach function space defined by the norm

$$\|f\|_{L_{\omega}^{p),\theta}(\Gamma)} = \sup_{0<\varepsilon<p-1} \left(\frac{\varepsilon^{\theta}}{\nu\Gamma} \int\limits_{\Gamma} |f(t)|^{p-\varepsilon} \omega(t) d\nu \right)^{\frac{1}{p-\varepsilon}}$$

where ω is an almost everywhere positive integrable function on Γ (i. e. weight).

Our goal is to give a complete characterization of that weight functions ω which govern one-weighted norm inequality for the following two operators: the Cauchy singular integral

$$(S_{\Gamma} f)(t) = \frac{1}{\pi i} \int\limits_{\Gamma} \frac{f(\tau)}{\tau - t} d\tau$$

and the Hardy-Littlewood maximal function

$$(M_{\Gamma} f)(t) = \sup_{r>0} \frac{1}{r} \int\limits_{D(t,r)} |f(\tau)| d\nu$$

defined on Carleson curves.

G. David's [14] well-known theorem states that for the boundedness of S_{Γ} in $L^p(\Gamma)$, $p > 1$, it is necessary and sufficient that Γ is a Carleson curve.

Our main result in this section is the following statement:

Theorem 19. *Let Γ be a Carleson curve and $\omega \in A_p(\Gamma)$, $(1 < p < \infty)$ i. e.*

$$\sup \left(\frac{1}{r} \int\limits_{D(z,r)} \omega(\tau) d\nu \right) \left(\frac{1}{r} \int\limits_{D(z,r)} \omega^{1-p'}(\tau) d\nu \right)^{p-1} < \infty, \tag{53}$$

where the supremum is taken over all $z \in \Gamma$ and r, $0 < r < diam\,\Gamma$. Let $\theta > 0$. Then we have

$$\text{(i)} \quad S_{\Gamma} \text{ is bounded in } L_{\omega}^{p),\theta}(\Gamma); \tag{54}$$

$$\text{(ii)} \quad M_{\Gamma} \text{ is bounded in } L_{\omega}^{p),\theta}(\Gamma). \tag{55}$$

If for some rectifiable curve Γ and a weight function ω the operator S_{Γ} (resp. M_{Γ}) is bounded in $L_{\omega}^{p),\theta}(\Gamma)$, $p > 1, \theta > 0$, then condition (53) is fulfilled.

For the real line, condition (53) coincides with the well-known B. Muckenhoupt's $A_p(I)$ condition.

To prove a sufficiency part of Theorem 19 we need the following

Theorem 20. *Let some sublinear operator T is bounded in $L^p_\omega(\Gamma)$ $1<p<\infty$ and in $L^{p-\sigma}_\omega(\Gamma)$ for some σ, $0<\sigma<p-1$ and weight function ω. Then T is bounded in $L^{p),\theta}(\Gamma)$, $0<\theta<\infty$.*

Proof. By the Riesz-Thorin Theorem we conclude that

$$\|Tf\|_{L^{p-\varepsilon}_\omega(\Gamma)} \le c\|f\|_{L^{p-\varepsilon}_\omega(\Gamma)}, \quad \varepsilon\in(0,\sigma], \tag{56}$$

for some constant c independent of f and ε.

Fix $\varepsilon\in(\sigma,p-1)$. Then $\frac{p-\sigma}{p-\varepsilon}>1$. By using Hölder's inequality with respect to the exponent $(p-\sigma)/(p-\varepsilon)$ and observing that $\left(\frac{p-\sigma}{p-\varepsilon}\right)'=\frac{p-\sigma}{\varepsilon-\sigma}$ we have

$$\|Tf\|_{L^{p-\varepsilon}_\omega(\Gamma)} \le \left(\int_\Gamma |Tf(x)|^{p-\sigma}\omega(x)dx\right)^{1/(p-\sigma)} \omega(\Gamma)^{(\varepsilon-\sigma)/[(p-\sigma)(p-\varepsilon)]}. \tag{57}$$

Further, the conditions $\sigma<p-1$ and $\varepsilon\in(\sigma,p-1)$ yield

$$0<\frac{\varepsilon-\sigma}{(p-\sigma)(p-\varepsilon)}<\frac{p-1-\sigma}{p-\sigma} \quad \text{and} \quad (p-1)\sigma^{-1/(p-\sigma)}>1.$$

Applying (57) we find that

$$\|Tf\|_{L^{p),\theta}_\omega(\Gamma)} = \max\left\{\sup_{0<\varepsilon\le\sigma}\varepsilon^{\frac{\theta}{p-\varepsilon}}\|Tf\|_{L^{p-\varepsilon}_\omega(\Gamma)},\ \sup_{\sigma<\varepsilon<p-1}\varepsilon^{\frac{\theta}{p-\varepsilon}}\|Tf\|_{L^{p-\varepsilon}_\omega(\Gamma)}\right\}$$

$$\le \max\left\{\sup_{0<\varepsilon\le\sigma}\varepsilon^{\frac{\theta}{p-\varepsilon}}\|Tf\|_{L^{p-\varepsilon}_\omega(\Gamma)},\ \sup_{\sigma<\varepsilon<p-1}\varepsilon^{\frac{\theta}{p-\varepsilon}}\|Tf\|_{L^{p-\sigma}_\omega(\Gamma)}\omega(\Gamma)^{\frac{\varepsilon-\sigma}{(p-\sigma)(p-\varepsilon)}}\right\}$$

$$=\max\left\{\sup_{0<\varepsilon\le\sigma}\varepsilon^{\frac{\theta}{p-\varepsilon}}\|Tf\|_{L^{p-\varepsilon}_\omega(\Gamma)},\ \sup_{\sigma<\varepsilon<p-1}\varepsilon^{\frac{\theta}{p-\varepsilon}}\sigma^{-\frac{\theta}{p-\sigma}}\sigma^{\frac{\theta}{p-\sigma}}\|Tf\|_{L^{p-\sigma}_\omega(\Gamma)}\omega(\Gamma)^{\frac{\varepsilon-\sigma}{(p-\sigma)(p-\varepsilon)}}\right\}$$

$$\le \max\left\{\sup_{0<\varepsilon\le\sigma}\varepsilon^{\frac{\theta}{p-\varepsilon}}\|Tf\|_{L^{p-\varepsilon}_\omega(\Gamma)},\right.$$

$$\left.\left(\sup_{\sigma<\varepsilon<p-1}\varepsilon^{\frac{\theta}{p-\varepsilon}}\sigma^{-\frac{\theta}{p-\sigma}}\omega(\Gamma)^{\frac{\varepsilon-\sigma}{(p-\sigma)(p-\varepsilon)}}\right)\left(\sup_{0<\varepsilon\le\sigma}\varepsilon^{\frac{\theta}{p-\varepsilon}}\|Tf\|_{L^{p-\varepsilon}_\omega(\Gamma)}\right)\right\}$$

Using the notation:

$$H_1 := \sup_{0<\varepsilon\le\sigma}\varepsilon^{\frac{\theta}{p-\varepsilon}}\|Tf\|_{L^{p-\varepsilon}_\omega(\Gamma)}, \quad H_2 := \sup_{\sigma<\varepsilon<p-1}\varepsilon^{\frac{\theta}{p-\varepsilon}}\sigma^{-\frac{\theta}{p-\sigma}}\omega(\Gamma)^{\frac{\varepsilon-\sigma}{(p-\sigma)(p-\varepsilon)}}$$

we have that

$$\|Tf\|_{L^{p),\theta}_\omega(\Gamma)} \le \max\{1,H_2\}\cdot H_1$$

$$\le \max\left\{1,\ \sup_{\sigma<\varepsilon<p-1}\varepsilon^{\frac{\theta}{p-\varepsilon}}\sigma^{-\frac{\theta}{p-\sigma}}(1+\omega(\Gamma))^{\frac{\varepsilon-\sigma}{(p-\sigma)(p-\varepsilon)}}\right)\cdot H_1.$$

Denoting $h(\varepsilon) := \varepsilon^{\frac{\theta}{p-\varepsilon}}$, $g(\varepsilon) := (1+\omega(\Gamma))^{\frac{\varepsilon-\sigma}{(p-\sigma)(p-\varepsilon)}}$. We, see that h and g are increasing in $(\sigma, p-1)$.

Therefore, by (56) we conclude that

$$\|Tf\|_{L^{p),\theta}_\omega(\Gamma)} \le c\max\Big\{1, (p-1)^\theta \sigma^{-\frac{\theta}{p-\sigma}}\Big(1+\omega(\Gamma)\Big)^{\frac{p-1-\sigma}{p-\sigma}}\Big\} \sup_{0<\varepsilon\le\sigma} \varepsilon^{\frac{\theta}{p-\varepsilon}}\|f\|_{L^{p-\varepsilon}_\omega(\Gamma)}$$

$$\le c(p-1)^\theta \sigma^{-\frac{\theta}{p-\sigma}}\Big(1+\omega(\Gamma)\Big)^{\frac{p-1-\sigma}{p-\sigma}}\|f\|_{L^{p),\theta}_\omega(\Gamma)}. \qquad \square$$

We need also the following

Lemma 21. *Let $1<p<\infty$ and $\theta>0$. Then for all $f \in L^{p),\theta}_\omega(\Gamma)$ and all subsets $J \subset \Gamma$, the inequality*

$$\|f\chi_J\|_{L^{p),\theta}_\omega(\Gamma)} \le \omega(J)^{-1/p}\|f\chi_J\|_{L^p_\omega(\Gamma)}\|\chi_J\|_{L^{p),\theta}_\omega(\Gamma)}$$

holds.

Proof. Let $f \ge 0$. By Hölder's inequality we have

$$\|f\chi_J\|_{L^{p),\theta}_\omega(\Gamma)} = \sup_{0<\varepsilon<p-1}\left(\frac{\varepsilon^\theta}{\nu\Gamma}\int_J f^{p-\varepsilon}\omega\right)^{\frac{1}{p-\varepsilon}} = \sup_{0<\varepsilon<p-1}\left(\frac{\varepsilon^\theta}{\nu\Gamma}\int_J f^{p-\varepsilon}\omega^{\frac{p-\varepsilon}{p}}\omega^{\frac{\varepsilon}{p}}\right)^{\frac{1}{p-\varepsilon}}$$

$$\le \sup_{0<\varepsilon<p-1}\left(\frac{\varepsilon^\theta}{\nu\Gamma}\right)^{\frac{1}{p-\varepsilon}}\left(\int_J\left(f^{p-\varepsilon}\omega^{\frac{p-\varepsilon}{p}}\right)^{\frac{p}{p-\varepsilon}}\right)^{\frac{1}{p}}\cdot\left(\int_J\left(\omega^{\frac{\varepsilon}{p}}\right)^{\frac{p}{\varepsilon}}\right)^{\frac{\varepsilon}{p(p-\varepsilon)}}$$

$$= \sup_{0<\varepsilon<p-1}\left(\frac{\varepsilon^\theta}{\nu\Gamma}\right)^{\frac{1}{p-\varepsilon}}\left(\int_J f^p\omega\right)^{\frac{1}{p}}\left(\int_J\omega\right)^{\frac{\varepsilon}{p(p-\varepsilon)}}$$

$$= \sup_{0<\varepsilon<p-1}\left(\frac{\varepsilon^\theta}{\nu\Gamma}\right)^{\frac{1}{p-\varepsilon}}\|f\chi_J\|_{L^p_\omega(\Gamma)}\cdot\big(\omega(J)\big)^{-\frac{1}{p}}\big(\omega(J)\big)^{\frac{1}{p-\varepsilon}}$$

$$= \big(\omega(J)\big)^{-\frac{1}{p}}\|f\chi_J\|_{L^p_\omega(\Gamma)}\sup_{0<\varepsilon<p-1}\left(\frac{\varepsilon^\theta}{\nu\Gamma}\int_\Gamma(\chi_J)^{p-\varepsilon}\omega\right)^{\frac{1}{p-\varepsilon}}$$

$$= \big(\omega(J)\big)^{-\frac{1}{p}}\|f\chi_J\|_{L^p_\omega(\Gamma)}\|\chi_J\|_{L^{p),\theta}_\omega(\Gamma)}. \qquad \square$$

Proof of Theorem 19. Sufficiency follows from Theorem 20, openess of $A_p(\Gamma)$ class, and from the boundedness of S_Γ and M_Γ in $L^p_\omega(\Gamma)$ with $\omega \in A_p(\Gamma)$, $1<p<\infty$.

Our aim now is to prove that from the boundedness of S_Γ in $L_\omega^{p),\theta}(\Gamma)$ it follows that

$$\sup_{\substack{t\in\Gamma\\ 0<r<\frac{d}{3}}} \left(\frac{1}{r}\int_{D(t,r)} \omega(\tau)d\nu\right)\left(\frac{1}{r}\int_{D(t,r)} \omega^{1-p'}d\nu\right)^{p-1} < \infty$$

where $d = \min_{t\in\Gamma}\max_{\tau\in\Gamma}|t-\tau|$. Fix $t \in \Gamma$ and let $x \in \Gamma$ be such that $|x-t| = 3r$. Given a nonnegative function $f \in L_\omega^{p),\theta}(\Gamma)$, we define a function $g \in L_\omega^{p),\theta}(\Gamma)$ as follows $g(t) = f(\tau)\frac{d\tau}{|d\tau|}e^{i\arg(t-x)} \cdot \chi_{D(t,r)}$ (see [5, p. 128]).

It is easy to see that for almost $z \in D(x,r)$

$$(S_\Gamma g)(z) = \frac{1}{\pi i}\int_{D(t,r)} \frac{g(\tau)}{\tau - z}d\tau = \frac{1}{\pi i}\int_{D(t,r)} \frac{f(\tau)}{|\tau - z|}e^{i\alpha(\tau,z)}d\nu$$

where $\alpha(\tau,z) := \arg(t-x) - \arg(\tau - z)$. Since $|\tau - z| \le |t-x| + 2r = 5r$ and

$$|\sin\alpha(\tau,z)| \le \frac{r}{\left(\frac{3}{2}\right)r} = \frac{2}{3}, \quad \cos\alpha(\tau,z) \ge \left(1 - \left(\frac{2}{3}\right)^2\right)^{\frac{1}{2}} = \frac{\sqrt{5}}{3}$$

for $\tau \in D(t,r)$ and $z \in D(x,r)$, we see that such τ and z,

$$\begin{aligned} |S_\Gamma g(z)| &\ge \frac{1}{\pi}\int_{D(t,r)} \frac{f(\tau)}{|\tau - z|}\cos\alpha(\tau,z)d\nu \\ &\ge \frac{\sqrt{5}}{3\pi}\int_{D(t,r)} \frac{f(\tau)}{|\tau - z|}d\nu \ge \frac{c}{r}\int_{D(t,r)} f(\tau)d\nu \end{aligned} \tag{58}$$

for almost all $z \in D(x,r)$.

From (2.3) we infer that

$$\|S_\Gamma g(z)\|_{L_\omega^{p),\theta}(D(x,r))} = \sup_{0<\delta\le p-1}\left(\frac{\delta^\theta}{\nu D(x,r)}\int_{D(x,r)} |S_\Gamma g(z)|^{p-\delta}\omega(z)d\nu\right)^{\frac{1}{p-\delta}}$$

$$\ge \left(\frac{c}{r}\int_{D(t,r)} f(\tau)d\nu\right)\sup_{0<\delta\le p-1}\left(\frac{\delta^\theta}{\nu D(x,r)}\int_{D(x,r)} \omega(z)d\nu\right)^{\frac{1}{p-\delta}}.$$

Thus we have

$$\|S_\Gamma g(z)\|_{L_\omega^{p),\theta}(D(x,r))} \ge \frac{c}{r}\int_{D(t,r)} f(\tau)d\nu \|\chi_{D(x,r)}\|_{L_\omega^{p),\theta}(\Gamma)}. \tag{59}$$

From the boundedness inequality and (59) we derive that

$$\|g(z)\|_{L^{p),\theta}_{\omega}(D(x,r))} \geq \frac{c}{r} \int\limits_{D(t,r)} f(\tau)d\nu \|\chi_{D(x,r)}\|_{L^{p),\theta}_{\omega}(\Gamma)}. \tag{60}$$

But

$$\|g(z)\|_{L^{p),\theta}_{\omega}(D(x,r))} \leq \|f\|_{L^{p),\theta}_{\omega}(D(t,r))}.$$

Now (60) yields that

$$\frac{1}{r} \int\limits_{D(t,r)} f(\tau)d\nu \|\chi_{D(x,r)}\|_{L^{p),\theta}_{\omega}(\Gamma)} \leq c\|f\|_{L^{p),\theta}_{\omega}(D(t,r))}$$

for arbitrary $f \geq 0$.

For $f \equiv 1$ the last inequality implies that

$$\frac{1}{r}\nu D(t,r)\|\chi_{D(x,r)}\|_{L^{p),\theta}_{\omega}(\Gamma)} \leq c\|\chi_{\Gamma(t,r)}\|_{L^{p),\theta}_{\omega}(\Gamma)}. \tag{61}$$

Hence, since Γ is rectifiable we have that

$$\|\chi_{D(x,r)}\|_{L^{p),\theta}_{\omega}(\Gamma)} \leq c\|\chi_{D(t,r)}\|_{L^{p),\theta}_{\omega}(\Gamma)}.$$

Changing the roles of $D(x,r)$ and $D(t,r)$ we come to the inequality

$$\|\chi_{D(t,r)}\|_{L^{p),\theta}_{\omega}(\Gamma)} \leq c\|\chi_{D(x,r)}\|_{L^{p),\theta}_{\omega}(\Gamma)} \tag{62}$$

for arbitrary t, z and x such that $|t-x| = 3r$, $z \in \Gamma(x,r)$ and $0 < r \leq \frac{d}{3}$.

Let us now take

$$g(\tau) = \omega^{1-p'}(\tau)\chi_{D(t,r)}$$

in the definition of f. According to (2.3) we have that

$$|S_{\Gamma}g(z)| \geq \frac{c}{r} \int\limits_{D(t,r)} \omega^{1-p'}(\tau)d\nu \tag{63}$$

for almost all $z \in D(x,r)$.

Thus

$$\int\limits_{D(t,r)} \omega^{1-p'}(\tau)d\nu < +\infty$$

for arbitrary $D(t,r)$, $t \in \Gamma$ and $0 < r < d$.

The boundedness of S_{Γ} and the inequality (63) imply that

$$\|g(z)\|_{L^{p),\theta}_{\omega}(\Gamma)} \geq c\|S_{\Gamma}g\|_{L^{p),\theta}_{\omega}(D(x,r)}$$

$$\geq \frac{c}{r}\int\limits_{D(t,r)} \omega^{1-p'}(\tau)d\nu \cdot \sup_{0<\delta\leq p-1}\left(\frac{\delta^\theta}{\nu D(x,r)}\int\limits_{D(x,r)} \omega(z)d\nu\right)^{\frac{1}{p-\delta}}$$

$$= \frac{c}{r}\int\limits_{D(t,r)} \omega^{1-p'}(\tau)d\nu\|\chi_{D(x,r)}\|_{L^{p),\theta}_\omega(D(x,r)}. \tag{64}$$

On the other hand,

$$\|g\|_{L^{p),\theta}_\omega(\Gamma)} \leq \|\omega^{1-p'}\chi_{D(t,r)}\|_{L^{p),\theta}_\omega(\Gamma)}.$$

From the last inequality and (64) we deduce that

$$\frac{1}{r}\int\limits_{D(t,r)} \omega^{1-p'}(\tau)d\nu\,\|\chi_{D(x,r)}\|_{L^{p),\theta}_\omega(D(x,r))} \leq c\|\omega^{1-p'}\chi_{D(t,r)}\|_{L^{p),\theta}_\omega(\Gamma)}. \tag{65}$$

Now we need the inequality

$$\|f\chi_{D(t,r)}\|_{L^{p),\theta}_\omega(\Gamma)} \leq c\,(\omega D(t,r))^{-\frac{1}{p}}\,\|f\chi_{D(t,r)}\|_{L^p_\omega(\Gamma)}\cdot\|\chi_{D(t,r)}\|_{L^{p),\theta}_\omega(\Gamma)} \tag{66}$$

for arbitrary f, proof of which follows in the same way as in the case of a finite interval (see Lemma 21) and therefore is omitted.

Applying (66), from (65) we derive that

$$\frac{1}{r}\int\limits_{D(t,r)} \omega^{1-p'}(\tau)d\nu\|\chi_{D(x,r)}\|_{L^{p),\theta}_\omega(D(x,r))}$$

$$\leq c(\omega D(t,r))^{-\frac{1}{p}}\left(\int\limits_{D(t,r)} \omega^{1-p'}(\tau)d\nu\right)^{\frac{1}{p}}\|\chi_{D(t,r)}\|_{L^{p),\theta}_\omega(\Gamma)}.$$

Taking (62) into account this inequality implies that

$$\frac{1}{r}\int\limits_{D(t,r)} \omega^{1-p'}(\tau)d\nu\|\chi_{\Gamma(x,r)}\|_{L^{p),\theta}_\omega(\Gamma)}$$

$$\leq c(\omega D(t,r))^{-\frac{1}{p}}\left(\int\limits_{D(t,r)} \omega^{1-p'}(\tau)d\nu\right)^{\frac{1}{p}}\|\chi_{D(x,r)}\|_{L^{p),\theta}_\omega(\Gamma)}.$$

Hence,

$$\frac{1}{r}(\omega D(t,r))^{\frac{1}{p}}\left(\int\limits_{D(t,r)} \omega^{1-p'}(\tau)d\nu\right)^{\frac{1}{p'}} \leq c.$$

Thus, $\omega \in A_p(\Gamma)$. □

2.4 Boundedness of weighted singular integral operator in $L^{p),\theta}(\Gamma)$

In this Section we establish that weighted Cauchy singular integral defined on a Carleson curve generates a bounded operator in generalized grand Lebesgue spaces.

Let S_Γ be again the Cauchy singular operator when Γ is a Carleson curve. As it was mentioned in Section 1.3 boundedness of operator S_Γ in $L^{p),\theta}(\Gamma, w)$ is equivalent to the boundedness of weighted operator $\omega S_\Gamma\left(\frac{f}{\omega}\right)$ in $L^{p),\theta}(\Gamma)$. Our goal is to prove the following

Theorem 22. *Let Γ be a Carleson curve. Let $1 < p < \infty$, $0 < \theta < \infty$ and $\omega^p \in A_p(\Gamma)$. Then weighted operator*

$$T_\omega f = \omega S_\Gamma\left(\frac{f}{\omega}\right)$$

is bounded in $L^{p),\theta}(\Gamma)$.

Proof. Since the openess of $A_p(\Gamma)$ there exists $\sigma \in (0, p-1)$ such that $\omega^p \in A_{p-\sigma}(\Gamma)$. On the other hand, $\omega^{p\alpha} \in A_{p-\sigma}(\Gamma)$ for arbitrary α, $0 < \alpha < 1$. Let now $\alpha = \frac{p-\sigma}{p}$. Then $0 < \alpha < 1$ and $\omega^{p-\sigma} \in A_{p-\sigma}(\Gamma)$. According to the boundedness of the Cauchy singular operator in weighted Lebesgue space we have:

$$\int_\Gamma |S_\Gamma f(t)|^p \omega^p(t) d\nu \le c_1 \int_\Gamma |f(t)|^p \omega^p(t) d\nu$$

and

$$\int_\Gamma |S_\Gamma f(t)|^{p-\sigma} \omega^{p-\sigma}(t) d\nu \le c_2 \int_\Gamma |f(t)|^{p-\sigma} \omega^{p-\sigma}(t) d\nu$$

Thus for given ω we have boundedness of operator T_ω at the same time in $L^p(\Gamma)$ and in $L^{p-\sigma}(\Gamma)$.

Applying Theorem 20 we conclude that T_ω is bounded in $L^{p),\theta}(\Gamma)$ with $\omega^p \in A_p(\Gamma)$. □

Corollary 23. *Let φ be a real continuous function given on Γ. Assume that $t_k \in \Gamma$, $k = 1, 2, \ldots, n$ such that $t_j \neq t_j$ when $i \neq j$.*

Let

$$\rho(t) = \prod_{k=1}^n |t - t_k|^{\alpha_k} \left| e^{\int_\Gamma \frac{\varphi(\tau)d\tau}{t-\tau}} \right|, \quad -\frac{1}{p} < \alpha_k < \frac{1}{p'} \tag{67}$$

then T_ρ is bounded in $L^{p),\theta}(\Gamma)$ i. e. S_Γ is bounded in $L^{p),\theta}(\Gamma, \rho)$.

Indeed, as it is shown in [40, p. 153] $\rho^p \in A_p(\Gamma)$. Applying Theorem 22 we deduce Corollary 23.

Note that in [76] it is shown that if T_ρ is bounded in $L^{p)}(\Gamma;\rho)$ with

$$\rho(t)=\prod_{k=1}^{n}|t-t_k|^{\alpha_k}$$

then

$$-\frac{1}{p}<\alpha_k<\frac{1}{p'}.$$

Remark 24. A different proof of Corollary 23 is given in Chapter 4, Theorem 50.

2.5 Notes and comments on Chapter 2

The variable exponent Lebesgue spaces were introduced and studied for the first time by W. Orlicz [100]. Then H. Nakano ([99]) considered spaces $L^{p(\cdot)}$ as an example of his modular spaces. The theory of modular spaces and, in particular, generalized Orlicz spaces generated by Young functions with a parameter (Musielak-Orlicz spaces) were explored in [93], [94]. The variable exponent Lebesgue spaces are a special case of Musielak-Orlicz spaces. The bases of the variable exponent Lebesgue and Sobolev spaces were developed by J. Musielak [93], I. I. Sharapudinov [117], O. Kováčik and J. Rákosník [79], S. Samko (see survays [115], [116] and references therein).

The problem of the boundedness of the Hardy-Littlewood maximal operator in variable exponent Lebesgue spaces has been extensively studied in [15], [16], [10]. Criteria of weighted inequalities for the operator mentioned above when weight is a power type function have been given in [70]. The necessary and sufficient condition for the weights which governs one-weighted inequality for maximal functions was obtained very recently [9] (see also [16]). The maximal functions defined on Carleson curves and, in general, on quasi-metric measure space in point of view of their boundedness were studied in [72], [73], [74], [75].

The boundedness of Calderón-Zygmund singular operator in unweighted variable exponent Lebesgue spaces was established in [17]. The mapping properties of the Cauchy singular integrals defined on Lyapunov or curves of bounded rotation (Radon curves) without cusps were explored in [69]. The similar problem for the Carleson curves was solved in [68]. Later on, more general results were established in [66]. Namely, necessary and sufficient conditions both for weights and curves were established which govern the boundedness of the Cauchy singular operator in weighted variable exponent Lebesgue spaces.

The grand Lebesgue spaces for the first time appeared in [35]. In generalized form these spaces were introduced in [29]. The one-weighted problem is grand Lebesgue spaces was studied in [23], where the necessary and sufficient condition for the Hardy-Littlewood function was established. The similar problem for the Hilbert transform is investigated in [51]. For more general results on the boundedness of singular integral operator in weighted grand Lebesgue spaces we refer to [47], [48], [50]. The boundedness of weighted singular integral operators was explored in [76].

Section 1.1 is based on the paper [66]. The results of Sections 1.3 and 1.4 are presented in [50] and [76] respectively.

Chapter 3

Variable Exponent Hardy and Smirnov Classes of Analytic Functions

In this Chapter various type of weighted variable exponent Hardy and Smirnov classes of analytic functions in simple and doubly-connected domains are introduced and studied. In particular, a wide class of those domains are revealed in which the functions from the above-mentioned classes are representable by the Cauchy type integrals with densities of weighted variable exponent Lebesgue spaces.

3.1 Classes $E^{p(t)}(D;\omega)$ and $e^{p(\cdot)}(D;\omega)$

Let D be the internal domain bounded by a simple rectifiable curve Γ, and $p(t)$ is a positive measurable function on Γ, and $\omega(z)$ is measurable, a.e. different from zero, function on D.

Let us suppose that $z = z_1(w)$ be some conformal mapping of U on D, and $\Phi(z)$ be an analytic function in D such that

$$I_1(\Phi;\omega;p) = \sup_{0<r<1} \int_0^{2\pi} |\Phi(z_1(re^{i\vartheta}))\omega(z_1(re^{i\vartheta}))|^{p(z_1(e^{i\vartheta}))} |z_1'(re^{i\vartheta})|\, d\vartheta < \infty.$$

We will show that if $z = z_2(w)$ is any other function conformal mapping of U on D, then for this function we have

$$I_2(\Phi;\omega;p) = \sup_{0<\rho<1} \int_0^{2\pi} |\Phi(z_2(\rho e^{i\vartheta}))\omega(z_2(\rho e^{i\vartheta}))|^{p(z_2(e^{i\vartheta}))} |z_2'(\rho e^{i\vartheta})|\, d\vartheta < \infty.$$

A function $re^{i\vartheta} \to z_2^{-1}[z_1(re^{i\vartheta})]$, where z_2^{-1} is the inverse function to z_2, is a one-to-one mapping of U on D. Assume that $\rho e^{i\varphi} = z_2^{-1}[z_1(re^{i\vartheta})]$. Then $z_2(re^{i\varphi}) =$

$z_1(\rho e^{i\vartheta})$, i.e. $z_2(\eta) = z_1(\zeta)$, $\eta = \rho e^{i\varphi}$, $\zeta = re^{i\vartheta}$. Hence $z_2'(\eta) = z_1'(\zeta)\,\frac{d\zeta}{d\eta}$, i.e.

$$\begin{aligned} z_2'(\eta)\rho\, d\varphi &= z_1'(\zeta)\,\frac{d\zeta}{d\eta}\,\rho\, d\varphi = z_1'(\zeta)\,\frac{d\zeta}{d\vartheta} : \frac{d\eta}{d\varphi}\,\rho\, d\vartheta \\ &= z_1'(\zeta)\,\frac{ire^{i\vartheta}}{i\rho e^{i\varphi}}\,\rho\, d\vartheta = z_1'(\zeta)e^{i(\vartheta-\varphi)}r\, d\vartheta. \end{aligned}$$

Therefore $|z_2'(\eta)|\rho\, d\varphi = |z_1'(\zeta)|r\, d\vartheta$. Moreover, $z_2(\rho e^{i\varphi}) = z_1(re^{i\vartheta})$ and thus $\omega(z_2(\rho e^{i\vartheta})) = \omega(z_1(re^{i\vartheta}))$ and $z_2(e^{i\varphi}) = z_1(e^{i\vartheta})$. Hence

$$\begin{aligned} &\int_0^{2\pi} |\Phi(z_1(re^{i\vartheta}))\omega(z_1(re^{i\vartheta}))|^{p(z_1(e^{i\vartheta}))}|z_1'(re^{i\vartheta})|\, d\vartheta \\ &= \int_0^{2\pi} |\Phi(z_2(\rho e^{i\vartheta}))\omega(z_2(\rho e^{i\vartheta}))|^{p(z_2(e^{i\vartheta}))}|z_2'(\rho e^{i\vartheta})|\, d\vartheta, \end{aligned}$$

i.e. $I_1(\Phi;\omega;p) = I_2(\Phi;\omega;p)$.

Now we can formulate

Definition 1. An analytic function $\Phi(z)$ in the domain D belongs to the class $E^{p(\cdot)}(D,\omega)$ if

$$\sup_{0<r<1} \int_0^{2\pi} |\Phi(z(re^{i\vartheta}))\omega(z(re^{i\vartheta}))|^{p(z(e^{i\vartheta}))}|z'(re^{i\vartheta})|\, r\, d\vartheta < \infty,$$

where $z = z(re^{i\vartheta})$ as an arbitrary conformal mapping of U on D.

Analogously, we denote by $h^{p(\cdot)}$ the set of harmonic functions $u(z)$ in U for which

$$\sup_{0<r<1} \int_0^{2\pi} |u(re^{i\vartheta})|^{p(\vartheta)} d\vartheta < \infty.$$

We assume that $E^{p(\cdot)}(D) = E^{p(\cdot)}(D;1)$, $H^{p(\cdot)}(\omega) = E^{p(\cdot)}(U;\omega)$, $H^{p(\cdot)} = E^{p(\cdot)}(U)$, $e^{p(\cdot)}(D;\omega) = \{u :\ u = \operatorname{Re}\Phi,\ \Phi \in E^{p(\cdot)}(D;\omega)\}$, $e^{p(\cdot)}(D) = e^{p(\cdot)}(D;1)$, $h^{p(\cdot)}(\omega) = \{u : u = \operatorname{Re}\Phi,\ \Phi \in H^{p(\cdot)}(\omega)\}$, $h^{p(\cdot)} = h^{p(\cdot)}(1)$.

If $p = \text{const}$ then classes $H^{p(\cdot)}$ and $E^{p(\cdot)}(D)$ coincide with Hardy and Smirnov classes.

3.2 On the existence of boundary values for the functions in $H^{p(\cdot)}(\omega)$

Lemma 2. *If* $\Phi \in H^{p(\cdot)}(\omega)$, $\inf p(\vartheta) = \underline{p} > 0$, *and* ω *is a function such that for some positive* a

$$C_\omega = \sup_r \int_0^{2\pi} |\omega(re^{i\vartheta})|^{-a} d\vartheta < \infty$$

and a.e. on $[0, 2\pi]$ *there exists the limit*

$$\lim_{r\to 1} \omega(re^{i\vartheta}) = \omega^+(e^{i\vartheta}) \equiv \omega(e^{i\vartheta}),$$

then $\Phi \in H^\eta$, $\eta \in \left(0, \frac{\underline{p}a}{\underline{p}+a}\right)$, *and* $\Phi^+ \in L^{p(\cdot)}(\gamma; \omega)$.

Proof. Since

$$\int_0^{2\pi} |(\Phi\omega)(re^{i\vartheta})|^{\underline{p}} d\vartheta \le \int_{|\Phi\omega|\le 1} |(\Phi\omega)(re^{i\vartheta})|^{\underline{p}} d\vartheta + \int_{|\Phi\omega|>1} |(\Phi\omega)(re^{i\vartheta})|^{\underline{p}} d\vartheta$$
$$\le 2\pi + I(\Phi; \omega; p)$$

and $\Phi \in H^{p(\cdot)}(\omega)$, we conclude that $\Phi\omega \in H^{\underline{p}}$. Furthermore we have

$$\int_0^{2\pi} |\Phi(re^{i\vartheta})|^\eta d\vartheta \le \int_0^{2\pi} |(\Phi\omega)(re^{i\vartheta})|^\eta |\omega(re^{i\vartheta})|^{-\eta} d\vartheta$$
$$\le \left(\int_0^{2\pi} |(\Phi\omega)(re^{i\vartheta})|^{\underline{p}} d\vartheta\right)^{\frac{\eta}{\underline{p}}} \left(\int_0^{2\pi} |\omega(re^{i\vartheta})|^{\frac{-\underline{p}\eta}{\underline{p}-\eta}} d\vartheta\right)^{\frac{\underline{p}-\eta}{\underline{p}}}.$$

Since $0 < \frac{\underline{p}\eta}{\underline{p}-\eta} < a$, we obtain

$$\sup_r \int_0^{2\pi} |\Phi(re^{i\vartheta})|^\eta d\vartheta \le I(\Phi; \omega; \underline{p}) \left[\int_{|\omega|\le 1} |\omega(re^{i\vartheta})|^{-a} d\vartheta + 2\pi\right]^{\frac{\underline{p}-\eta}{\underline{p}}}$$
$$\le I(\Phi; \omega; \underline{p})(2\pi + C_\omega) < \infty.$$

Therefore $\Phi \in H^\eta$. Moreover, since a.e. on γ there exists the angular limit $\Phi^+(t) = \Phi^+(e^{i\vartheta})$, we obtain

$$\lim_{r\to 1} |\Phi(re^{i\vartheta})\omega(re^{i\vartheta})|^{p(\vartheta)} = |\Phi^+(e^{i\vartheta})\omega(e^{i\vartheta})|^{p(\vartheta)}.$$

Applying the Fatou lemma and taking into account that $\Phi \in H^{p(\cdot)}(\omega)$, we conclude that $\Phi^+ \in L^{p(\cdot)}(\gamma; \omega)$. □

3.3 Some properties of the Cauchy type integrals with density from $L^{p(\cdot)}(\gamma)$

Definition 3. Let Γ be an arbitrary plane curve. We will say that a function p belongs to the class $\mathcal{P}(\Gamma)$ if

1) there exists a constant A such that

$$|p(t_1) - p(t_2)| < \frac{A}{|\ln|t_1 - t_2||}, \quad t_1, t_2 \in \Gamma;$$

2) $\underline{p} > 1$.

Proposition 4. *Let $p \in \mathcal{P}(\Gamma)$ and Γ a closed Jordan curve. Then the set of rational functions with a unique pole inside of Γ is dense in $L^{p(\cdot)}(\Gamma)$.*

The validity of this statement follows from the denseness in $L^{p(\cdot)}(\Gamma)$ of the set of continuous functions and the fact that any continuous function may be approximated in $C(\Gamma)$ by rational functions, whatsoever Jordan curve Γ we have according to the Theorem 7 of Chapter 2 from [132].

Let f be a periodic function with period 2π and $f \in L[0, 2\pi]$. Consider the Poisson integral

$$u_f(r, \vartheta) = \int_0^{2\pi} f(e^{i\sigma}) P(r, \sigma - \vartheta)\, d\sigma. \tag{1}$$

Proposition 5. *Let f be a 2π-periodic function and $f \in L^{p(\cdot)}(T)$, $T = [0, 2\pi]$. If $p \in \mathcal{P}(\gamma)$, then for a harmonic function u_f defined by integral (1) the equalities*

$$\lim_{r\to 1} \int_0^{2\pi} |u_f(re^{i\vartheta}) - u^+(e^{i\vartheta})|^{p(\vartheta)} d\vartheta = \lim_{r\to 1} \int_0^{2\pi} |u_f(re^{i\vartheta}) - f(e^{i\vartheta})|^{p(\vartheta)} d\vartheta = 0 \tag{2}$$

and

$$\lim_{r\to 1} \|u_f(re^{i\vartheta}) - u^+(e^{i\vartheta})\|_{p(\cdot)} = 0$$

are valid and therefore $u_f \in h^{p(\cdot)}$.

Proof. Let us denote $u_f(re^{i\vartheta})$ by $u_f(r, \vartheta)$.

As is known (see [136, Ch. IV]),

$$|u_f(r, \vartheta)| \le M(f)(\vartheta), \quad f \in L^1(T),$$

where $M(f)$ is a maximal function of f. If $p \in \mathcal{P}(\gamma)$ then the operator $M : f \to Mf$ is continuous in $L^{p(\cdot)}(T)$ [15]. Hence $Mf \in L^{p(\cdot)}(T)$ and

$$\|u_f(r, \cdot)\|_{p(\cdot)} \le c\|f\|_{p(\cdot)} .$$

Since $p \in \mathcal{P}(\gamma)$ we have $f \in L^{\underline{p}}(\gamma)$, $\underline{p} = \inf_{t\in\gamma} p(t)$. Therefore

$$\lim_{r\to 1} u_f(r,\vartheta) = f(\vartheta)$$

almost everywhere. On other hand we have

$$|u_f(r,\vartheta) - f(\vartheta)|^{p(\vartheta)} \leq 2^{\overline{p}-1}\left(|u_f(r,\vartheta)|^{p(\vartheta)} + |f(\vartheta)|^{p(\vartheta)}\right), \quad \overline{p} = \sup_{t\in\gamma} p(t).$$

Consequently,

$$|u_f(r,\vartheta) - f(\vartheta)|^{p(\vartheta)} \leq 2^{\overline{p}-1}\left(|Mf(\vartheta)|^{p(\vartheta)} + |f(\vartheta)|^{p(\vartheta)}\right).$$

The right side of latter inequality is summable. Applying the Lebesgue convergence theorem we deduce (2). Then the sequence of norms of operators $u_{r_n} : f \to u_f(r_n, \cdot)$ is bounded. On the other hand, the set of all continuous functions φ is everywhere dense in $L^{p(\cdot)}(T)$ (see Proposition 4) and uniform on T

$$\lim_{r\to 1} u_\varphi(r,\vartheta) = \varphi(\vartheta).$$

Therefore for any continuous function φ on T we have

$$\lim \|u_\varphi(r,\cdot) - \varphi\|_{p(\cdot)} = 0\,.$$

Now the assertion of the theorem follows from the Banach–Steinhaus theorem. □

Corollary 6. *If $\Phi \in H^{p(\cdot)}$, $p \in \mathcal{P}(\gamma)$, then*

$$\lim_{r\to 1} \int_0^{2\pi} |\Phi(re^{i\vartheta}) - \Phi^+(e^{i\vartheta})|^{p(\vartheta)} d\vartheta = 0, \tag{3}$$

and therefore $\Phi^+ \in L^{p(\cdot)}(\gamma)$.

Indeed, since in the considered case $\Phi \in H^1$, $\Phi^+ \in L^{p(\cdot)}(\gamma)$, we have $\Phi = u + iv$, where u and v are Poisson integrals with densities u^+ and v^+ (see [108, p. 97]) and, moreover, u^+ and v^+ belong to $L^{p(\cdot)}(\gamma)$. Thereby (3) is a corollary of (2).

Now we are able to obtain an analogy of the following statement belonging to Fichtenholtz ([108, p. 97]): each of the classes of analytic functions Φ in U which are representable by the formula

$$\Phi(z) = \Phi(re^{i\vartheta}) = \int_0^{2\pi} f(e^{i\sigma}) P(r, \sigma - \vartheta)\, d\sigma \tag{4}$$

or by

$$\Phi(z) = \frac{1}{2\pi i} \int_\Gamma \frac{\Phi^+(t)\, dt}{t - z} \tag{5}$$

coincides with the class H^1.

Theorem 7. *For an analytic function* Φ *in* U *to be representable by formula* (4) (*or by* (5)) *with a density from* $L^{p(\cdot)}(\gamma)$, $p \in \mathcal{P}(\gamma)$, *it is necessary and sufficient that* Φ *belong to the class* $H^{p(\cdot)}$.

Proof. *Necessity.* In both considered cases the necessity follows from the above statement taking into account that $H^{p(\cdot)} \subset H^1$.

Sufficiency. We first assume that (4) is valid with f from $L^{p(\cdot)}(\gamma)$, $p \in \mathcal{P}(\gamma)$. Then $\Phi = u + iv$, where u and v are Poisson integrals with a density from $L^{p(\cdot)}(\gamma)$. According to Proposition 5 they belong to $h^{p(\cdot)}$ and thereby $\Phi \in H^{p(\cdot)}$.

Let now (5) hold and $\Phi^+ \in L^{p(\cdot)}(\gamma)$. Then $\Phi \in H^{\underline{p}}$, $\underline{p} > 1$ and, by Fichtenholtz' theorem, $\Phi(z)$ is representable by formula (4) where $f(e^{i\sigma}) = \Phi^+(e^{i\vartheta}) \in L^{p(\cdot)}(\gamma)$. Consequently we have $\Phi \in H^{p(\cdot)}$. □

In addition to Theorem 7 we will prove a stronger statement about Cauchy type integrals.

Theorem 8. *If* $f \in L^{p(\cdot)}(\gamma)$, $p \in \mathcal{P}(\gamma)$, *then a Cauchy type integral*

$$\Phi(z) = (K_\gamma f)(z) = \frac{1}{2\pi i} \int_\gamma \frac{f(t)\,dt}{t - z}$$

belongs to the class $H^{p(\cdot)}$.

Proof. By virtue of Theorem 7 it is sufficient to show that $\Phi^+ \in L^{p(\cdot)}(\gamma)$. Using the Plemelj–Sokhotskiĭ formula we obtain $\Phi^+(t) = \frac{1}{2} f(t) + \frac{1}{2}(S_\gamma f)(t)$, $t \in \gamma$. In the assumptions made with regard to $p(t)$ the operator $S_\gamma : f \to S_\gamma f$ is continuous in the space $L^{p(\cdot)}(\gamma)$. Therefore $\Phi^+ \in L^{p(\cdot)}(\gamma)$ and thereby $\Phi \in H^{p(\cdot)}$. □

Theorem 9. *If* $\Phi \in H^{p(\cdot)}$, $\underline{p} > 0$, *and* $\Phi^+ \in L^{p_1(\cdot)}(\gamma)$, $p_1 \in \mathcal{P}(\gamma)$, *then* Φ *belongs to the class* $H^{\widetilde{p}(\cdot)}$, *where* $\widetilde{p}(t) = \max(p(t), p_1(t))$, $t \in \gamma$.

Proof. We have $\Phi \in H^{\underline{p}}$ and $\Phi^+ \in L^{p_1(\cdot)}(\gamma) \subset L(\gamma)$. Apply the Smirnov theorem we conclude $\Phi \in H^1$. Therefore Φ is representable by a Cauchy integral with density Φ^+. The latter belongs, to $L^{p_1(\cdot)}(\gamma)$, $p_1 \in \mathcal{P}(\gamma)$. By virtue of Theorem 8 we have $\Phi \in H^{p_1(\cdot)}$. Since Φ belongs also to $H^{p(\cdot)}$, we conclude that $\Phi \in H^{\widetilde{p}(\cdot)}$. □

Lemma 10. *If* $f \in H^{p(\cdot)}$, $p \in H(\gamma)$, $\underline{p} > 0$, *then there exist constants* K_1, K_2 *not depending on* r *and* f *and such that*

$$\int_0^{2\pi} |f(re^{i\vartheta})|^{p(\vartheta)} d\vartheta \le K_1 \Big(\int_0^{2\pi} |f(e^{i\vartheta})|^{p(\vartheta)} d\vartheta + 2\pi \Big) \exp\Big(K_2 \int_0^{2\pi} |f(e^{i\vartheta})|^{\underline{p}} d\vartheta \Big). \quad (6)$$

Proof. Since $f \in H^{\underline{p}}$, we have

$$f(z) = b(z)\sigma(z)(\exp i\lambda)D(z),$$

where $b(z)$ is the Blashke product, $|\sigma(z)| \leq 1$, $\sigma(z) \neq 0$, $z \in U$, and

$$D(z) = \exp \frac{1}{2\pi} \int_0^{2\pi} \ln |f(e^{i\varphi})| \frac{e^{i\varphi} + z}{e^{i\varphi} - z} \, d\varphi, \quad |z| < 1$$

([108, p. 110]). Hence it follows that

$$|f(z)| \leq |D(z)| = \exp \frac{1}{2\pi} \int_{|f|\leq 1} \ln |f(e^{i\varphi})| \operatorname{Re} \frac{e^{i\varphi} + z}{e^{i\varphi} - z} \, d\varphi$$

$$\times \exp \int_{|f|>1} \ln |f(e^{i\varphi})| \operatorname{Re} \frac{e^{i\varphi} + z}{e^{i\varphi} - z} \, d\varphi = D_1(z) D_2(z).$$

Since

$$\operatorname{Re} \frac{e^{i\varphi} + z}{e^{i\varphi} - z} = \operatorname{Re} \frac{e^{i\varphi} + re^{i\vartheta}}{e^{i\varphi} - re^{i\vartheta}} = P(r, \vartheta - \varphi) > 0,$$

we have $|D_1| \leq 1$ and therefore

$$|f(z)| \leq |D_2(z)| = \exp \frac{1}{2\pi} \int_0^{2\pi} \ln \widetilde{f}(e^{i\varphi}) P(r, \vartheta - \varphi) \, d\varphi,$$

$$\widetilde{f}(e^{i\varphi}) = \begin{cases} |f(e^{i\varphi})| & \text{if } |f(e^{i\varphi})| > 1, \\ 1 & \text{if } |f(e^{i\varphi})| \leq 1. \end{cases}$$

Now we have

$$|f(re^{i\vartheta})|^{p(\vartheta)} \leq \exp \frac{1}{2\pi} \int_0^{2\pi} p(\varphi) \ln \widetilde{f}(e^{i\varphi}) P(r, \vartheta - \varphi) \, d\varphi$$

$$\times \exp \frac{1}{2\pi} \int_0^{2\pi} (p(\vartheta) - p(\varphi)) \ln \widetilde{f}(e^{i\varphi}) P(r, \vartheta - \varphi) \, d\varphi = A(r, \vartheta) B(r, \vartheta). \quad (7)$$

Let us estimate the function $B(r, \vartheta)$. We have

$$|B(r, \vartheta)| = \left| \exp \frac{1}{2\pi} \int_0^{2\pi} \ln \widetilde{f}(e^{i\varphi}) \frac{(1 - r^2)(p(\vartheta) - p(\varphi))}{1 + r^2 - 2r\cos(\vartheta - \varphi)} \, d\varphi \right|$$

$$\leq \exp \frac{1}{2\pi} \int_0^{2\pi} \ln \widetilde{f}(e^{i\varphi}) \frac{(1-r^2)|p(\vartheta)-p(\varphi)|}{1+r^2-2r\cos(\vartheta-\varphi)} d\varphi$$

$$\leq \exp \frac{1}{2\pi} \int_0^{2\pi} \ln \widetilde{f}(e^{i\varphi}) |p(\vartheta)-p(\varphi)| \left| \frac{e^{i\varphi}+re^{i\vartheta}}{e^{i\varphi}-re^{i\vartheta}} \right| d\varphi. \tag{8}$$

But

$$\left| \frac{e^{i\varphi}+re^{i\vartheta}}{e^{i\varphi}-re^{i\vartheta}} \right| \leq \frac{2}{\sqrt{1+r^2-2r\cos(\vartheta-\varphi)}} = \frac{2}{\sqrt{(1-r)^2+4r\sin^2\frac{\vartheta-\varphi}{2}}}$$

$$\leq \begin{cases} 4 & \text{if } 0 \leq r \leq \frac{1}{2}, \\ \frac{\sqrt{2}}{|\sin\frac{\vartheta-\varphi}{2}|} & \text{if } \frac{1}{2} \leq r \leq 1. \end{cases} \tag{9}$$

Since $p \in H(\gamma)$, there exist constants M and α, $0 < \alpha \leq 1$, such that

$$|p(\vartheta)-p(\varphi)| = |p(e^{i\vartheta})-p(e^{i\varphi})| \leq M|e^{i\vartheta}-e^{i\varphi}|^{\alpha}.$$

As $|e^{i\vartheta}-e^{i\varphi}| = |2\sin\frac{\vartheta-\varphi}{2}|$, from (8) we obtain by virtue of (9)

$$B(r,\vartheta) \leq \exp \frac{4M}{\pi} \int_{\vartheta-\pi}^{\vartheta+\pi} \ln \widetilde{f}(e^{i\varphi}) \frac{d\varphi}{|e^{i\varphi}-e^{i\vartheta}|^{1-\alpha}}. \tag{10}$$

Further, for ε from $(0,\infty)$ and for all x from $[1,\infty)$ we have $\ln x < x^{\varepsilon}(\varepsilon e)^{-1}$. Assuming that $\varepsilon = \frac{\alpha}{2}\underline{p}$ and applying the Hölder inequality with exponent $\frac{2}{\alpha}$, we obtain

$$\int_{\vartheta-\pi}^{\vartheta+\pi} |\widetilde{f}(e^{i\varphi})|^{\varepsilon} \frac{d\varphi}{|e^{i\varphi}-e^{i\vartheta}|^{1-\alpha}} \leq \left(\int_{\vartheta-\pi}^{\vartheta+\pi} |\widetilde{f}(e^{i\varphi})|^{\underline{p}} d\varphi \right)^{\frac{\alpha}{2}} \left(\int_{\vartheta-\pi}^{\vartheta+\pi} \frac{d\varphi}{|e^{i\varphi}-e^{i\vartheta}|^{\beta}} \right)^{\frac{2-\alpha}{2}}$$

$$\leq M_1 \left(\int_0^{2\pi} |\widetilde{f}(e^{i\varphi})|^{\underline{p}} d\varphi \right)^{\frac{\alpha}{2}} \leq M_1 \left(\int_0^{2\pi} |f(e^{i\varphi})|^{\underline{p}} d\varphi + 2\pi \right), \tag{11}$$

$$M_1 = \max_{\vartheta} \left(\int_{\vartheta-\pi}^{\vartheta+\pi} \frac{d\varphi}{|e^{i\vartheta}-e^{i\varphi}|^{\beta}} \right)^{\frac{2-\alpha}{2}}, \quad \beta = \frac{(1-\alpha)2}{2-\alpha} < 1.$$

For $\varphi \in [\vartheta-\pi, \vartheta+\pi]$ we have $|\frac{\vartheta-\varphi}{2}| \leq \frac{\pi}{2}$ and since for $|x| \leq \frac{\pi}{2}$ we have $|\sin x| \geq \frac{2}{\pi}|x|$, we obtain the inequality $|e^{i\vartheta}-e^{i\varphi}| \geq \frac{2}{\pi}|\vartheta-\varphi|$. Therefore

$$M_1 \leq \left(\frac{\pi}{2}\right)^{1-\alpha} \left(\int_{\vartheta-\pi}^{\vartheta+\pi} \frac{d\varphi}{|\vartheta-\varphi|^{\beta}} \right)^{\frac{2-\alpha}{2}} \leq \frac{\pi}{2} \cdot 2 \left(\frac{2\pi^{1-\beta}}{1-\beta} \right)^{\frac{2-\alpha}{2}}$$

$$\leq 2\pi \cdot \pi^{\frac{\alpha}{2}} \left(\frac{2-\alpha}{\alpha}\right)^{\frac{2-\alpha}{2}} < \frac{2\pi^2 \cdot 2}{\alpha} = \frac{4\pi^2}{\alpha}.$$

Since $(\varepsilon e)^{-1} < \alpha^{-1}$, with regard to (11), inequality (10) can be written in the form

$$|B(r,\vartheta)| \leq \exp \frac{4M}{\pi} \cdot \frac{4\pi^2}{\alpha^2} \Big(\int_0^{2\pi} |f(e^{i\varphi})|^{\underline{p}} \, d\varphi + 2\pi \Big)$$

$$= \exp \frac{16M\pi}{\alpha^2} \Big(\int_0^{2\pi} |f(e^{i\varphi})^{\underline{p}} d\varphi + 2\pi \Big) = K_1 \exp K_2 \int_0^{2\pi} |f(e^{i\varphi})|^{\underline{p}} d\varphi,$$

where

$$K_1 = \exp \frac{32M\pi^2}{\alpha^2}, \qquad K_2 = \frac{16M\pi}{\alpha^2}. \tag{12}$$

Now (7) implies

$$\int_0^{2\pi} |f(re^{i\vartheta})|^{p(\vartheta)} \, d\vartheta \leq K_1 \Big(\exp K_2 \int_0^{2\pi} |f(e^{i\varphi})|^{\underline{p}} d\varphi \Big) \cdot \int_0^{2\pi} A(r,\vartheta) \, d\vartheta. \tag{13}$$

Since the mean geometric value is not greater than the mean arithmetic value,

$$\int_0^{2\pi} A(r,\vartheta) \, d\vartheta = \int_0^{2\pi} \exp \frac{1}{2\pi} \int_0^{2\pi} \ln[\widetilde{f}(e^{i\varphi})]^{p(\varphi)} P(r, \vartheta - \varphi) \, d\varphi d\vartheta$$

$$\leq \int_0^{2\pi} \frac{1}{2\pi} \int_0^{2\pi} |\widetilde{f}(e^{i\varphi})|^{p(\varphi)} P(r, \vartheta - \varphi) \, d\varphi \, d\vartheta = \int_0^{2\pi} |\widetilde{f}(e^{i\varphi})|^{p(\varphi)} d\varphi$$

$$< \int_0^{2\pi} |f(e^{i\varphi})|^{p(\varphi)} d\varphi + 2\pi.$$

Now from (13) we obtain inequality (6), where K_1 and K_2 are defined by (12). □

3.4 Convergence of sequences in $H^{p(\cdot)}$

In investigating the questions of the convergence of sequences from $H^{p(\cdot)}$, we focus attention mainly on the generalization of Tumarkin's theorem ([108, p. 268]) which in the case of a circle is formulated as follows:

If the sequence $\{f_n(\zeta)\}$ of boundary values of functions $f_n(z)$ from the class H^p converges in measure on the set of a positive Lebesgue measure $\ell \subset \gamma$ $(m\ell > 0))$ and $\int_\gamma |f_n(\zeta)|^p |d\zeta| < C$, where C does not depend on n, then the sequence $\{f_n(z)\}$ uniformly converges in the circle U to a function $f(z)$ of the class H^p, while the sequence $\{f_n(\zeta)\}$ converges in measure on ℓ to a function $f^+(\zeta)$.

Theorem 11. *Let $\{f_n(\zeta)\}$ be the sequence of boundary values of functions $f_n(z)$ of the class $H^{p(\cdot)}$ and*

$$\int_\gamma |f_n(\zeta)|^{p(\zeta)}|d\zeta| = \int_0^{2\pi} |f_n(e^{i\vartheta})|^{p(e^{i\vartheta})} d\vartheta < C,$$

where C does not depend on n. If, in addition, we have that

I. *$p \in \mathcal{P}(\gamma)$ and $\{f_n(\zeta)\}$ converges in measure on γ or*

II. *$p \in H(\gamma)$, $\underline{p} > 0$, and $\{f_n(\zeta)\}$ converges in measure on $\ell \in \gamma$, $m\ell > 0$,*

then $f_n(z)$ uniformly converges in U to a function $f(z)$ of the class $H^{p(\cdot)}$, while $\{f_n(\zeta)\}$ converges in measure to a function $f^+(\zeta)$ on γ in case I *and on ℓ in case* II.

Proof. I. Since the functions $f_n(z)$ belong to $H^{\underline{p}}$ and

$$\int_\gamma |f_n(\zeta)|^{\underline{p}}|d\zeta| < 1 + \int_\gamma |f_n(\zeta)|^{p(\zeta)}|d\zeta| < 1 + C,$$

by Tumarkin's theorem we conclude that the sequence $\{f_n(z)\}$ converges in the circle U to a function $f(z) \in H^{\underline{p}} \subset H^1$, while $\{f_n(\zeta)\}$ converges in measure on γ to $f^+(\zeta)$. Let us show that $f^+ \in L^{p(\cdot)}(\gamma)$. From $\{f_n(\zeta)\}$ we can choose the sequence $\{f_{n_k}(\zeta)\}$ converging almost everywhere on γ. Then $|f_{n_k}(e^{i\vartheta})|^{p(\vartheta)}$ $(p(\vartheta) = p(e^{i\vartheta}))$ converges to $|f^+(e^{i\vartheta})|^{p(\vartheta)}$ almost everywhere on $[0, 2\pi]$. By virtue of the Fatou lemma

$$\int_0^{2\pi} |f^+(e^{i\vartheta})|^{p(\vartheta)} d\vartheta \le \varliminf_{k\to\infty} \int_0^{2\pi} |f_{n_k}(e^{i\vartheta})|^{p(\vartheta)}\, d\vartheta < C.$$

Thus $f(z)$, being a function from H^1, is representable by a Cauchy integral with density $f^+ \in L^{p(\cdot)}(\gamma)$. By Theorem 7 we conclude that $f \in H^{p(\cdot)}$.

II. Again applying Tumarkin's theorem, we find that $\{f_n(z)\}$ converges to a function $f \in H^{\underline{p}}$, while $\{f_n(\zeta)\}$ converges in measure to f^+ on ℓ. Since $p \in H(\gamma)$, we can apply the lemma 10 by which

$$\int_0^{2\pi} |f_n(re^{i\vartheta})|^{p(\vartheta)} d\vartheta \le K_1\left[\exp K_2 \int_0^{2\pi} |f_n(e^{i\vartheta})|^{\underline{p}} d\vartheta\right]\left(2\pi + \int_0^{2\pi} |f_n(e^{i\vartheta})|^{p(\vartheta)} d\vartheta\right)$$
$$< K_1\left(\exp(K_2 + K_2 C)\right)(2\pi + C) < C_1.$$

Passing here to the limit, we establish that $f \in H^{p(\cdot)}$. □

Theorem 11 enables us to formulate one characteristic property of functions from $H^{p(\cdot)}$ (cf. [108, p. 138]).

Theorem 12. *Let $p \in H(\gamma)$, $\underline{p} > 0$. The necessary and sufficient condition that a measurable function $\varphi(t) = \varphi(e^{i\vartheta})$ given on a set $E \subset \gamma$, $mE > 0$, coincide almost everywhere on E with the boundary value $f(e^{i\vartheta})$ of a function $f(z)$ of the class $H^{p(\cdot)}$ is the existence of polynomials $\mathcal{P}_n(z)$ such that*

1) *$\{\mathcal{P}_n(e^{i\vartheta})\}$ converges to $\varphi(e^{i\vartheta})$ almost everywhere on E;*
2) $\overline{\lim\limits_{n\to\infty}} \int\limits_0^{2\pi} |\mathcal{P}_n(e^{i\vartheta})|^{p(\vartheta)} d\vartheta < \infty$.

Proof. *Necessity.* Let $f(e^{i\vartheta})$ be an angular boundary value of a function $f \in H^{p(\cdot)}$. Let us consider the functions $f_n(z) = f(r_n z)$, $|z| \leq 1$, $r_n \in (0,1)$, $r_n \to 1$. The function $f_n(z)$ is analytic in $\overline{U}$ and therefore for a given number ε_n there exists, by virtue of the Runge theorem, a polynomial $\mathcal{P}_n(z)$ such that

$$|f(z) - \mathcal{P}_n(z)| < \varepsilon_n, \quad |z| \leq 1.$$

But

$$\lim_{n\to\infty} f_n(e^{i\vartheta}) = \lim_{n\to\infty} f(r_n e^{i\vartheta}) = f(e^{i\vartheta}) \text{ a.e. on } E.$$

Taking $\varepsilon_n \to 0$, we conclude that $\lim\limits_{n\to\infty} \mathcal{P}_n(e^{i\vartheta}) = f(e^{i\vartheta})$ almost everywhere on E. Moreover, by Theorem 11 we conclude that

$$\int\limits_0^{2\pi} |\mathcal{P}_n(e^{i\vartheta})|^{p(\vartheta)} d\vartheta \leq \int\limits_0^{2\pi} |f(r_n e^{i\vartheta})|^{p(\vartheta)} d\vartheta + \varepsilon_n \cdot 2\pi < \widetilde{C}.$$

Thus conditions 1)–2) are fulfilled.

Sufficiency. If $p \in H(\gamma)$, $\underline{p} > 0$, and conditions 1)–2) are fulfilled, then to the sequence $\{\mathcal{P}_n(z)\}$ we can apply Theorem 11 by which $\{\mathcal{P}_n(z)\}$ converges to a function $f(z)$ from $H^{p(\cdot)}$. □

3.5 Invariance of the classes $\mathcal{P}(\Gamma)$ under a conformal mapping

Lemma 13. *If $\Gamma \in C^1(A_1, \dots, A_i; \nu_1, \dots, \nu_i)$, $0 < \nu_k \leq 2$, $k = \overline{1,i}$, is the simple closed curve bounding the domain D, $p \in \mathcal{P}(\Gamma)$, and $z = z(w)$ is the conformal mapping of U onto D, then the functions $\ell(\tau) = p(z(\tau))$ and $\frac{1}{\ell(\tau)}$ belong to $\mathcal{P}(\gamma)$.*

Proof. Since $p \in \mathcal{P}(\Gamma)$, we have

$$|\ell(\tau_1) - \ell(\tau_2)| = |p(z(\tau_1)) - p(z(\tau_2))| < \frac{A}{|\ln|z(\tau_1) - z(\tau_2)||}. \tag{14}$$

Assuming $\tau_i = \exp \sigma_i$, $\sigma_i \in [0, 2\pi]$, $\sigma_1 < \sigma_2$, we have

$$|z(\tau_1) - z(\tau_2)| = \left| \int\limits_{\tau_1}^{\tau_2} z'(\tau)\, d\tau \right| \leq \int\limits_{\sigma_1}^{\sigma_2} |z'(\tau)|\, |d\tau|.$$

Let us now apply the following result (see 0.25): if the domain D is bounded by a curve $\Gamma \in C^1(A_1,\dots,A_i;\nu_1,\dots,\nu_i)$, $0 \le \nu_k \le 2$, and $z(a_k) = A_k$, then

$$z'(w) \sim \prod_{k=1}^{i}(w-a_k)^{\nu_k-1}\exp\left(\frac{1}{\pi}\int_{\gamma}\frac{\beta(\zeta)\,d\zeta}{\zeta-w}\right),$$

$\beta(\zeta)$ is a real continuous function on γ.

Since, by assumption, $0 < \nu_k \le 2$, there exists a number $\delta_1 > 1$ such that $\prod_{k=1}^{i}(\tau-\tau_k)^{\nu_k-1} \in L^{\delta_1}(\gamma)$. The function $\exp\left(\frac{1}{\pi}\int_{\gamma}\frac{\beta(\zeta)\,d\zeta}{\zeta-w}\right)$ belongs to the set $\bigcap_{\eta>0} L^{\eta}(\gamma)$ (see [123]). Therefore $z' \in L^{\delta}(\gamma)$ for some $\delta > 1$. Now, we have

$$|z(\tau_1)-z(\tau_2)| \le \left(\int_{\sigma_1}^{\sigma_2}|z'(\tau)|^{\delta}\,|d\tau|\right)^{\frac{1}{\delta}}|\sigma_2-\sigma_1|^{\frac{1}{\delta'}}, \quad \delta' = \frac{\delta}{\delta-1}.$$

As $|\sigma_1-\sigma_2| < \frac{\pi}{2}|\tau_1-\tau_2|$, we conclude that there exists $M > 1$ such that

$$|\ell(\tau_1)-\ell(\tau_2)| < M|\tau_1-\tau_2|^{\mu}, \quad \mu = \frac{1}{\delta'}.$$

Assuming $|\tau_1-\tau_2| < M^{-\frac{1}{\mu}} = \eta_1$, from inequality (14) we obtain

$$|\ell(\tau_1)-\ell(\tau_2)| < \frac{A}{\mu|\ln|\tau_1-\tau_2||-\ln M}.$$

Let us take $|\tau_1-\tau_2|$ so small that $\ln M < \frac{\mu}{2}|\ln|\tau_1-\tau_2||$, i.e., we assume $|\tau_1-\tau_2| < M^{-\frac{2}{\mu}} = \eta\ (<\eta_1)$. Then for $|\tau_1-\tau_2| < \eta$ the latter inequality implies

$$|\ell(\tau_1)-\ell(\tau_2)| < \frac{2A}{\mu|\ln|\tau_1-\tau_2||}.$$

For $|\tau_1-\tau_2| \ge \eta$ we have

$$|\ell(\tau_1)-\ell(\tau_2)| < \frac{\max p(t)\cdot\ln\frac{1}{\eta}}{|\ln|\tau_1-\tau_2||}.$$

Thus the inequality

$$|\ell(\tau_1)-\ell(\tau_2)| < \frac{B}{|\ln|\tau_1-\tau_2||}, \quad B = \max\left(\frac{2A}{\mu};\left(\ln\frac{1}{\eta}\right)\max_{t\in\Gamma}p(t)\right),$$

holds for any $\tau_1,\tau_2 \in \gamma$, and therefore $\ell \in \mathcal{P}(\gamma)$.

For the function $[\ell(\tau)]^{-1}$ we have

$$\left|[\ell(\tau_1)]^{-1}-[\ell(\tau_2)]^{-1}\right| \le |\ell(\tau_1)\cdot\ell(\tau_2)|^{-1}|\ell(\tau_1)-\ell(\tau_2)| \le \underline{p}^{-2}|\ell(\tau_1)-\ell(\tau_2)|,$$

i.e., $[\ell(\tau)]^{-1} \in \mathcal{P}(\gamma)$. □

3.6 On Smirnov classes in domains with piecewise Lyapunov boundary

In studying the properties of functions from classes $E^p(D)$, the fact that the inclusion of a function $\Phi \in E^p(D)$ is equivalent to the belonging of the function $\Psi(w) = \Phi(z(w))[z'(w)]^{\frac{1}{p}}$ to the class H^p plays an important role. For variable p, the function $\Psi(w)$ is not even analytic. When the domain D is bounded by a simple piecewise Lyapunov curve, we succeed in obtaining a certain analogy of the above given statement.

Proposition 14. *Let the domain D be bounded by a given curve $\Gamma \in C^{1,L}(A_1, \ldots, A_n; \nu_1, \ldots, \nu_n)$, $0 < \nu_k \leq 2$. The function $\Phi(z)$ belongs to the class $E^{p(\cdot)}(D)$ if and only if the function $\Phi(z(w))\rho(w) = \Phi(z(w)) \prod\limits_{k=1}^{n} (w - a_k)^{\frac{\nu_k - 1}{\ell(\beta_k)}}$, $a_k = z^{-1}(A_k)$, $a_k = \exp i\beta_k$, belongs to $H^{\ell(\cdot)}$, $\ell(\tau) = p(z(\tau))$, $\ell(\vartheta) \equiv \ell(z(e^{i\vartheta}))$.*

Proof. Let $\omega(w) = |z'(re^{i\vartheta})|^{\frac{1}{\ell(\vartheta)}}$. To prove the proposition it is sufficient to establish that

$$0 < \inf_{|w|<1} \omega(w)|\rho(w)|^{-1} \leq \sup_{|w|<1} \omega(w)|\rho(w)|^{-1} < \infty, \quad \rho(w) = \prod_{k=1}^{n} (w - a_k)^{\frac{\nu_k - 1}{\ell(\beta_k)}}.$$

Assuming that $\alpha_k = \nu_k - 1$, we have

$$\omega(w)|\rho(w)|^{-1} = \prod_{k=1}^{n} |w - a_k|^{\alpha_k\left(\frac{1}{\ell(\vartheta)} - \frac{1}{\ell(\beta_k)}\right)} |z_0(w)|^{\frac{1}{\ell(\vartheta)}}.$$

Since $|z_0(w)|^{\pm\frac{1}{\ell(\vartheta)}}$ are bounded functions, we should establish that the functions $[\varphi(w)]^{\pm 1}$, where

$$\varphi(w) = \prod_{k=1}^{n} |w - a_k|^{\alpha_k\left(\frac{1}{\ell(\vartheta)} - \frac{1}{\ell(\beta_k)}\right)},$$

are bounded in $\overline{U}$. We have

$$\begin{aligned}
[\varphi(w)]^{\pm 1} &= \prod_{k=1}^{n} |w - a_k|^{\pm\left(\frac{\alpha_k}{\ell(\vartheta)} - \frac{\alpha_k}{\ell(\beta_k)}\right)} \\
&= \prod_{k=1}^{n} \exp\left[\frac{\pm\alpha_k(\ell(\vartheta) - \ell(\beta_k))}{\ell(\vartheta)\ell(\beta_k)} \ln\left|re^{i\vartheta} - e^{i\beta_k}\right|\right].
\end{aligned}$$

In the assumptions made for Γ and p by Lemma 13 we have that the function $\ell(\tau)$ belongs to $\mathcal{P}(\gamma)$. Taking this fact into account, for $|re^{i\vartheta} - e^{i\beta_k}| < \frac{1}{2}$ we obtain

$$\delta_k = \left|\frac{\pm\alpha_k(\ell(\vartheta) - \ell(\beta_k))}{\ell(\vartheta)\ell(\beta_k)} \ln\left|re^{i\vartheta} - e^{i\beta_k}\right|\right|$$

$$\leq \frac{A|\alpha_k|}{\underline{\ell}^2}\left|\frac{\ln|re^{i\vartheta}-e^{i\beta_k}|}{\ln|e^{i\vartheta}-e^{i\beta_k}|}\right| = \frac{A}{\underline{\ell}^2}\left|\frac{\ln((1-r)^2+4\sin^2\frac{\vartheta-\beta_k}{2})}{\ln 4\sin^2\frac{\vartheta-\beta_k}{2}}\right|$$

$$\leq \frac{A}{\underline{\ell}^2}\left|\frac{\ln 4\sin^2\frac{\vartheta-\beta_k}{2}}{\ln 4\sin^2\frac{\vartheta-\beta_k}{2}}\right| \leq \frac{A}{\underline{l}^2}.$$

When $|re^{i\vartheta}-e^{i\beta_k}| \geq \frac{1}{2}$, keeping in mind that $|re^{i\vartheta}-e^{i\beta_k}| \leq 2$ we obtain $\delta_k \leq \frac{A}{\underline{l}^2}\ln 2$.

The obtained estimates imply the boundedness of the functions $[\varphi(w)]^{\pm 1}$ and thereby the validity of Proposition 14. □

3.7 Belonging of Cauchy type integrals to $E^{p(\cdot)}(D)$

Now we are able to prove that a Cauchy type integral with a density from $L^{p(\cdot)}(\Gamma)$ belongs to $E^{p(\cdot)}(D)$.

Theorem 15. *If D is an inner domain bounded by a simple close curve Γ from $C^{1,L}(A_1,\dots,A_n;\nu_1,\dots,\nu_n)$, $0<\nu_k\leq 2$, $p\in\mathcal{P}(\Gamma)$ and $f\in L^{p(\cdot)}(\Gamma)$, then a Cauchy type integral, $\Phi(z)=(K_\Gamma f)(z)$, $z\in D$, belongs to $E^{p(\cdot)}(D)$.*

Proof. Since $f\in L^{p(\cdot)}(\Gamma)\subset L^{\underline{p}}(\Gamma)$ and $p\in\mathcal{P}(\Gamma)$ and piecewise Lyapunov curves are regular the function $\Phi(z)\in E^{\underline{p}}(D)$. According to Proposition 14 the function $\Psi(w)=\Phi(z(w))\rho(w)$ belongs to $H^{\underline{p}}$. Let us show that $\Psi^+\in L^{\ell(\cdot)}(\gamma)$. Since $p\in\mathcal{P}(\Gamma)$, a Cauchy operator S_Γ is continuous in $L^{p(\cdot)}(\Gamma)$. Thus the function $\Phi^+=\frac{1}{2}(f+S_\Gamma f)$ belongs to $L^{p(\cdot)}(\Gamma)$, i.e. $|\Phi(z(\tau))|\,|z'(\tau)|^{\frac{1}{\ell(\tau)}}\in L^{\ell(\cdot)}(\gamma)$, $\ell(\tau)=p(z(\tau))$. By virtue of Proposition 14, this is equivalent to $\Psi^+(\tau)=\Phi(z(\tau))\rho(\tau)\in L^{\ell(\cdot)}(\gamma)$. Therefore $\Psi\in H^{\underline{p}}$ and $\Psi^+\in L^{\ell(\cdot)}(\gamma)$. Applying Theorem 9, we conclude that $\Psi\in H^{\ell(\cdot)}$. But then, by virtue of Proposition 14, $\Phi\in E^{p(\cdot)}(D)$. □

Corollary 16. *Let* 1) *D be an inner domain bounded by a curve Γ from $C^{1,L}(A_1,\dots,A_n;\nu_1,\dots,\nu_n)$, $0<\nu_k\leq 2$;* 2) *$p(t)\in\mathcal{P}(\Gamma)$;* 3) *the sequence $\{f_n(t)\}$ ($f_n(z)\in E^{p(\cdot)}(D)$) converges in measure on Γ;* 4) *$\int_\Gamma |f_n(t)|^{p(t)}|dt|<C$, where C does not depend on n.*

Then the sequence $\{f_n(z)\}$ converges in D to a function $f(z)\in E^{p(\cdot)}(D)$, while $\{f_n(t)\}$ converges in measure on Γ to a function $f^+(t)$.

Proof. By virtue of Tumarkin's theorem, $\{f_n(z)\}$ converges to a function $f(z)\in E^{\underline{p}}(D)$ and $\{f_n(t)\}$ converges in measure on Γ to $f^+(t)$. It remains to show that $f\in E^{p(\cdot)}(D)$. Since $\underline{p}>1$, $f(z)$ is representable by a Cauchy integral with density f^+. Choosing from $\{f_n(t)\}$ a sequence $\{f_{n_k}(t)\}$ converging to $f^+(t)$ almost everywhere on Γ, by the Fatou lemma and assumption 4) we conclude that $f^+\in L^{p(\cdot)}(\Gamma)$. Thus $f(z)=(K_\Gamma f)(z)$, $f\in L^{p(\cdot)}(\Gamma)$; hence by Theorem 15 we have $f\in E^{p(\cdot)}(D)$. □

Remark 17. If $\Gamma\in C^{1,L}(A_1,\dots,A_n;\nu_1,\dots,\nu_n)$ and D is an outer domain bounded by Γ, the assertion of the theorem remains valid if it is assumed that $0\leq\nu_k<2$.

Now we are able to generalize the Theorem 9.

Proposition 18. *Let D be an inner domain bounded by a simple curve $\Gamma \in C^{1,L}(A_1,\dots,A_n;\nu_1,\dots,\nu_n)$, $0<\nu_k\le 2$, $k=\overline{1,n}$, $p\in\mathcal{P}(\Gamma)$. If $\Phi\in E^{p(\cdot)}(D)$ and $\Phi^+\in L^{p_1(\cdot)}(\Gamma)$, $p_1\in\mathcal{P}(\Gamma)$, then $\Phi\in E^{\widetilde{p}(\cdot)}(D)$, $\widetilde{p}(t)=\max(p(t),p_1(t))$.*

Proof. Since $\Phi\in E^{\underline{p}}(D)$ and $\underline{p}>1$, we have $\Phi\in E^1(D)$ (see [108, p. 264]). Therefore Φ is representable by the Cauchy integral with a density from $L^{p_1(\cdot)}(\Gamma)$, $p_1\in\mathcal{P}(\Gamma)$. By virtue of Theorem 15 we have $\Phi\in E^{p_1(\cdot)}(D)$, and since $\Phi\in E^{p(\cdot)}(D)$, we conclude $\Phi\in E^{\widetilde{P}(\cdot)}(D)$. □

We have a special interest in functions $\omega(re^{i\vartheta})$ for which there exists a boundary function $\omega(e^{i\vartheta})=\lim\limits_{r\to 1}\omega(re^{i\vartheta})$ with a good weight property with respect to the operator S_γ.

3.8 Class $W^{p(\cdot)}(U)$

It is assumed that

$$W^{p(\cdot)}(U)=\left\{\omega:\ [\omega(z)]^{\pm 1}\in\bigcup_{\delta>0}H^\delta,\ \ \omega^+(e^{i\vartheta})\in W^{p(\cdot)}(\gamma)\right\}.$$

Examples: 1) Let $p\in\mathcal{P}(\gamma)$, $a_k\in\gamma$, $k=\overline{1,n}$. Make cuts ℓ_k which lie outside U and connect the point a_k with the point $z=\infty$. Fix an arbitrary branch of the analytic function $(z-a_k)^{\alpha_k}$, $\alpha_k\in\mathbb{R}$, cut along γ. The function

$$\rho(z)=\prod_{k=1}^{n}(z-a_k)^{\alpha_k},\qquad -\frac{1}{p(a_k)}<\alpha_k<\frac{1}{p'(a_k)},\tag{15}$$

belongs to $W^{p(\cdot)}(U)$ since it is obvious that $[\rho(z)]^{\pm 1}\in\bigcup\limits_{\delta>0}H^\delta$, and the function $\rho^+(t)=\prod\limits_{k=1}^{n}(t-a_k)^{\alpha_k}$ belongs to $W^{p(\cdot)}(\gamma)$ according to the conditions on α_k.

2) Let $z=z(w)$ be the conformal mapping of the circle U onto the domain D bounded by a simple piecewise smooth curve with angular points t_k at which angle values with respect to D are equal to $\pi\nu_k$, where $0<\nu_k<p(t_k)$. Then $z'(w)\in W^{p(\cdot)}(U)$. Indeed, in the considered case

$$z'(w)\sim\prod_{k=1}^{n}(w-t_k)^{\nu_k-1}\exp\left\{\frac{1}{\pi}\int_\gamma\frac{\varphi(\tau)\,d\tau}{\tau-w}\right\},$$

(see 0.26). Therefore $z'\in\bigcup\limits_{\delta>1}H^\delta$. Furthermore, since $\frac{\nu_k-1}{p(t_k)}\in\left(-\frac{1}{p(t_k)},\frac{1}{p'(t_k)}\right)$, $\rho_0(t)=\prod\limits_{k=1}^{n}(t-t_k)^{\nu_k-1}$ belongs to $W^{p(\cdot)}(\gamma)$ and it is obvious that $\frac{1}{\rho(t)}\in L^{p'(\cdot)+\varepsilon}(\gamma)$

for some $\varepsilon > 0$. But a function of the form $\rho_0(t)\exp\left(\int_\gamma \frac{\varphi(\tau)\,d\tau}{\tau-t}\right)$, where φ is a real continuous function, belongs to $W^{p(\cdot)}(\gamma)$ (see Chapter 4 Corollary 11).

Theorem 19. *For an analytic function Φ in U to belong to the class $H^{p(\cdot)}(\omega)$, where $p \in \mathcal{P}(\gamma)$ and $\omega \in W^{p(\cdot)}(U)$, it is necessary and sufficient that it would be representable in the form*

$$\Phi(z) = (K_\gamma\varphi)(z) = \frac{1}{2\pi i}\int\limits_\gamma \frac{\varphi(t)\,dt}{t-z}, \quad z \in U, \tag{16}$$

where $\varphi \in L^{p(\cdot)}(\gamma;\omega)$.

Proof. *Necessity.* Let $\Phi \in H^{p(\cdot)}(\omega)$, $p \in \mathcal{P}(\gamma)$, $\omega \in W^{p(\cdot)}(U)$. Since $\omega^+(t) = \omega(t) \in W^{p(\cdot)}(\gamma)$, we have $[w(t)]^{-1} \in L^{p'(\cdot)}(\gamma)$. But $[\omega(z)]^{-1} \in \bigcup\limits_{\delta>0} H^\delta$, hence by virtue of Smirnov's Theorem we conclude that $[\omega(z)]^{-1} \in H^{p'(\cdot)}$. Since $\Phi(z) = [\Phi(z)\omega(z)]\,[\omega(z)]^{-1}$, we have $\Phi \in H^1$. That is, $\Phi(z) = (K_\gamma\Phi^+)(z)$. But, by virtue of Lemma 2 we deduce $\Phi^+ \in L^{p(\cdot)}(\gamma;\omega)$. Therefore the equality (16), where $\varphi(t) = \Phi^+(t) \in L^{p(\cdot)}(\gamma;\omega)$, is valid.

Sufficiency. Let (16) be fulfilled with $\varphi \in L^{p(\cdot)}(\gamma;\omega)$. Then $\varphi(t) = \psi(t)[\omega(t)]^{-1}$, where $\psi \in L^{p(\cdot)}(\gamma)$.

Let us show that $\Phi(z)\omega(z) \in H^{p(\cdot)}$.

From the representation (16) it follows that $\Phi(z) \in H^{\delta_0}$, $\delta_0 \in (0,1)$ (see, for example, [19, p. 39]). Besides, by assumption, $\omega \in \bigcup\limits_{\delta>0} H^\delta$. Therefore $\Phi(z)\omega(z) \in H^\eta$ for some $\eta > 0$. Furthermore,

$$(\Phi\omega)^+ = \omega\left(\frac{1}{2}\varphi + \frac{1}{2}\,S_\gamma\varphi\right) = \frac{\omega}{2}\left(\frac{\psi}{\omega} + S_\gamma\,\frac{\psi}{\omega}\right) = \psi + \omega S_\gamma\,\frac{\psi}{\omega}\,.$$

Since $\psi \in L^{p(\cdot)}(\gamma)$ and $\omega \in W^{p(\cdot)}(\gamma)$, we have $\left(\psi + \omega S_\gamma \frac{\psi}{\omega}\right) \in L^{p(\cdot)}(\gamma)$. Thus $\Phi(z)\omega(z) \in H^\eta$ and $[\Phi(z)\omega(z)]^+ \in L^{p(\cdot)}(\gamma)$. By Theorem 9 we conclude that $[\Phi(z)\omega(z)] \in H^{p(\cdot)}$, which means that $\Phi \in H^{p(\cdot)}(\omega)$. □

3.9 Classes $H^{p(z)}(\omega)$ and $\mathcal{P}(G)$

Let $p(z) = p(r,\vartheta)$ be a measurable positive function, and $\omega(z)$ be a measurable function, a.e. different from zero in U.

An analytic function Φ in U belongs to the class $H^{p(z)}(\omega)$ if

$$\sup_{r<1}\int\limits_0^{2\pi} |\Phi(re^{i\vartheta})\omega(re^{i\vartheta})|^{p(r,\vartheta)} r\,d\vartheta < \infty.$$

Assume $H^{p(z)}(1) = H^{p(z)}$.

Below when we write $p(z)$, we will mean that z varies in U, whereas in the notation $p(t)$ it will be assumed that t varies on γ. Instead of $H^{p(z)}(\omega)$ we will sometimes write $H^{p(r,\vartheta)}(\omega)$.

Let G be the set of the complex plane, which is a simple rectifiable curve or a simply- or doubly- connected domain bounded by such curves.

We denote by $\mathcal{P}(G)$ the set of measurable positive functions $p(z)$ on G, for which the following conditions are fulfilled:

a) there exists a constant C (depending on p) such that for any $z_1, z_2 \in G$ we have

$$|p(z_1) - p(z_2)| < \frac{C}{|\ln|z_1 - z_2||},$$

b) $\inf\limits_{z\in G} p(z) = \underline{p} > 1$.

Analogously to the proof of Lemma 2 we prove

Proposition 20. *If $\Phi \in H^{p(z)}(\omega)$, where $\inf\limits_{z\in U} p(z) = \underline{p} > 0$, for almost all $t \in \gamma$ the limit $\lim\limits_{z\to t} p(z) = p(t)$ exists and the conditions of Lemma 2 are fulfilled for the function $\omega(z)$, then $\Phi \in H^{\eta}$, $\eta \in \left(0, \frac{\underline{p}a}{\underline{p}+a}\right)$ and $\Phi^+ \in L^{p(t)}(\gamma;\omega)$. Moreover, if $\Phi \in H^{p(z)}$, then $\Phi \in H^{\underline{p}}$ and $\Phi^+ \in L^{p(t)}(\gamma)$.*

Theorem 21. *If $\Phi \in H^{p(z)}$, $p(z) \in \mathcal{P}(\overline{U})$ and $\min\limits_{t\in\gamma} p(t) > 1$, then $p(t) \in \mathcal{P}(\gamma)$ and $\Phi \in H^{p(t)}$.*

Proof. By Proposition 20 we have $\Phi \in H^{\underline{p}}$, $\Phi^+ \in L^{p(t)}(\gamma)$. Since $p(z) \in \mathcal{P}(\overline{U})$,

$$\left|p(e^{i\vartheta_1}) - p(e^{i\vartheta_2})\right| < \frac{C}{|\ln|e^{i\vartheta_1} - e^{i\vartheta_2}||}$$

for any $\vartheta_1, \vartheta_2 \in [0, 2\pi]$. Together with the condition $\min\limits_{t\in\gamma} p(t) > 1$, this shows that $p \in \mathcal{P}(\gamma)$. Moreover, since $\Phi^+ \in L^{p(t)}(\gamma) \subset L^1(\gamma)$, by Smirnov's Theorem $\Phi \in H^1$. Therefore $\Phi(z)$ is representable by the Cauchy integral. Since $p \in \mathcal{P}(\gamma)$, by virtue of Theorem 8 we conclude that $\Phi \in H^{p(t)}$. □

Lemma 22. *If $\Phi \in H^{p(t)}$, $p \in \mathcal{P}(\gamma)$ and $q(z)$ is an arbitrary function from $\mathcal{P}(\overline{U})$, for which $q(t) = p(t)$ on γ, we have $\Phi \in H^{q(z)-\varepsilon}$ for $\varepsilon \in (0, \underline{q})$, $\underline{q} = \min\limits_{z\in\overline{U}} q(z)$.*

Proof. Assuming $q(z) = q(r,\vartheta)$, $q(1, e^{i\vartheta}) = q(\vartheta)$, we obtain

$$\int\limits_0^{2\pi} |\Phi(re^{i\vartheta})|^{q(r,\vartheta)-\varepsilon} d\vartheta = \int\limits_{|\Phi|\leq 1} |\Phi(re^{i\vartheta})|^{q(r,\vartheta)-\varepsilon} d\vartheta + \int\limits_{|\Phi|>1} |\Phi(re^{i\vartheta})|^{q(r,\vartheta)-\varepsilon} d\vartheta$$

$$< 2\pi + \int\limits_{|\Phi|>1} |\Phi(re^{i\vartheta})|^{q(r,\vartheta)-q(1,\vartheta)-\varepsilon} |\Phi(re^{i\vartheta})|^{p(\vartheta)} d\vartheta. \quad (17)$$

Since $q \in \mathcal{P}(\overline{U})$, we have

$$|q(r,\vartheta) - q(1,\vartheta)| < \frac{C}{|\ln|re^{i\vartheta} - e^{i\vartheta}||} = \frac{C}{|\ln(1-r)|}.$$

Therefore, for $r > \widetilde{r} = 1 - \exp\left(-\frac{2C}{\varepsilon}\right)$ we have $\frac{C}{|\ln(1-r)|} < \frac{\varepsilon}{2}$ and then $-\frac{3\varepsilon}{2} < q(re^{i\vartheta}) - q(1,\vartheta) < -\frac{\varepsilon}{2}$. For such r, when $|\Phi(re^{i\vartheta})| > 1$ we have $|\Phi(re^{i\vartheta})|^{q(r,\vartheta)-q(1,\vartheta)-\varepsilon} < 1$. Now, for $\varepsilon \in (0, \underline{q})$, from (17) we obtain

$$\int\limits_0^{2\pi} |\Phi(re^{i\vartheta})|^{q(r,\vartheta)-\varepsilon} d\vartheta$$

$$< 2\pi + \sup_{r<\widetilde{r}} \int\limits_0^{2\pi} |\Phi(re^{i\vartheta})|^{q(r,\vartheta)-\varepsilon} d\vartheta + \sup_{\widetilde{r}\le r<1} \int\limits_{|\Phi|>1} |\Phi(re^{i\vartheta})|^{p(\vartheta)} d\vartheta$$

$$\le 2\pi + 2\pi \left(\sup_{r<\widetilde{r}} |\Phi(re^{i\vartheta})|^{\max q(r,\vartheta)} + 1 \right) + \sup_{0\le r<1} \int\limits_0^{2\pi} |\Phi(re^{i\vartheta})|^{p(\vartheta)} d\vartheta.$$

Since $\Phi \in H^{p(t)}$, $p \in \mathcal{P}(\gamma)$, this implies that $\Phi \in H^{q(r,\vartheta)-\varepsilon}$. □

3.10 Variable exponent Hardy classes in a circular ring

Classes $H^{p(\cdot)}(K;\omega)$

Assume that $\gamma = \gamma_1 = \{z : |z| = 1\}$, $\gamma_2 = \{z : |z| < \rho,\ 0 < \rho < 1\}$, $K = \{z : \rho < |z| < 1\}$, and $\omega(z)$ is a function different from zero in K.

We will say that an analytic function Φ in K belongs to the Hardy class $H^p(K;\omega)$, $p > 0$, if

$$\sup_{\rho<r<1} \int\limits_0^{2\pi} |\Phi(re^{i\vartheta})\omega(re^{i\vartheta})|^p d\vartheta < \infty,$$

(see [126]).

If ω is an analytic function, then the condition $\Phi \in H^p(K;\omega)$ is equivalent to the condition $\omega\Phi \in H^p(K)$. Let $H^p(K) = H^p(K;1)$.

Class of weight functions $W^p(K)$

Let $[\omega(z)]^{\pm 1} \in \bigcup\limits_{\delta>0} H^\delta(K)$. Assume that $\omega_1(\vartheta) = \omega^+(e^{i\vartheta})$, $\omega_2(\vartheta) = \omega^+(\rho e^{i\vartheta})$. We will say that $\omega(z)$ belongs to $W^p(K)$ if $\omega_i(\vartheta) \in W^p(\gamma_i)$. Thus

$$W^p(K) = \Big\{\omega(z): \ [\omega(z)]^{\pm 1} \in \bigcup_{\delta>0} H^\delta(K), \ \omega_i(\vartheta) \in W^p(\gamma_i)\Big\}.$$

Denote by K_i, $i = 1, 2$, the domain bounded by the curve γ_i and containing the ring K. For an analytic function in K_i assume

$$H^p(K_1; \omega) = \Big\{\Phi: \ \sup_{r<1} \int_0^{2\pi} |\omega(re^{i\vartheta})\Phi(re^{i\vartheta})|^p r \, d\vartheta < \infty\Big\},$$

$$H^p(K_2; \omega) = \Big\{\Phi: \ \sup_{r>\rho} \int_0^{2\pi} |\omega(re^{i\vartheta})\Phi(re^{i\vartheta})|^p r \, d\vartheta < \infty\Big\}.$$

If $\Phi \in H^p(K; \omega)$, then

$$\Phi(z) = \Phi_1(z) + \Phi_2(z), \quad \Phi_i \in H^p(K_i; \omega_i). \tag{18}$$

Conversely, if (18) holds for Φ, then $\Phi \in H^p(K; \omega)$ (see [126], [39, Theorem 1]).

These classes turned out useful when investigating the Dirichlet problem for harmonic functions that are the real parts of functions from $H^p(K)$ [39]. In that case, the boundary data on the curves γ_i, $i = 1, 2$, must be integrable to one and the same degree.

If the boundary data belong to $L^{p_1}(\gamma_1)$ and $L^{p_2}(\gamma_2)$, respectively, and if $p_1 < p_2$, then the problem can be solved only in a wide class $\operatorname{Re} H^{p_1}(K)$. We will therefore know of the solution that near γ_2 it is integrable to the degree p_1. It is certainly desirable to have another, more narrow, class of Hardy type, in which the Dirichlet problem is solvable and its solution near γ_2 is integrable to the degree p_2.

Classes $H^{p_1(\cdot),p_2(\cdot)}(K; \omega)$ and $W^{p_1(\cdot),p_2(\cdot)}(K)$

Let $p_i(\vartheta)$, $i = 1, 2$, be positive measurable functions on $[0, 2\pi]$.

An analytic function Φ in K belongs to the class $H^{p_1(\cdot),p_2(\cdot)}(K; \omega)$ if

$$\left[\varlimsup_{r\to 1-} \int_0^{2\pi} |(\Phi\omega)(re^{i\vartheta})|^{p_1(\vartheta)} d\vartheta + \varlimsup_{r\to \rho+} \int_0^{2\pi} |(\Phi\omega)(re^{i\vartheta})|^{p_2(\vartheta)} d\vartheta\right] < \infty.$$

It is not difficult to verify that the class $H^{p_1(\cdot),p_2(\cdot)}(K;\omega)$ coincides with the class of those analytic functions Φ in K, for which we have

$$\left[\sup_{r_0\le r<1}\int_0^{2\pi}|(\Phi\omega)(re^{i\vartheta})|^{p_1(\vartheta)}d\vartheta+\sup_{\rho<r<r_0}\int_0^{2\pi}|(\Phi\omega)(re^{i\vartheta})|^{p_2(\vartheta)}d\vartheta\right]<\infty \tag{19}$$

for any $r_0\in(\rho,1)$. Hence it follows that if $\Phi\in H^{p_1(\cdot),p_2(\cdot)}(K;\omega)$, then

$$\Phi\in H^{p_1(\cdot),p_2(\cdot)}(K(r_0;1);\omega)\quad\text{and}\quad\Phi\in H^{p_1(\cdot),p_2(\cdot)}(K(\rho;r_0);\omega),$$

where $K(r_0,1)=\{z:r_0<|z|<1\}$, $K(\rho,r_0)=\{z:\rho<|z|<r_0\}$.

Assume $H^{p_1(\cdot),p_2(\cdot)}(K)=H^{p_1(\cdot),p_2(\cdot)}(K;1)$, $h^{p_1(\cdot),p_2(\cdot)}(K;\omega)=\{u:\ \mathrm{Re}\Phi,\ \Phi\in H^{p_1(\cdot),p_2(\cdot)}(K;\omega)\}$.

$$W^{p_1(\cdot),p_2(\cdot)}(K)=\Big\{\omega:\omega^{\pm1}\in\bigcup_{\delta>0}H^{\delta}(K),$$
$$\omega_1=\omega^+(e^{i\vartheta})\in W^{p_1(\cdot)}(\gamma_1),\ \ \omega_2=\omega^+(\rho e^{i\vartheta})\in W^{p_2(\cdot)}(\gamma_2)\Big\}.$$

Lemma 23. *If $\omega\in W^{p_1(\cdot),p_2(\cdot)}(K)$, then there exists $\eta>0$ such that*

$$H^{p_1(\cdot),p_2(\cdot)}(K;\omega)\subset H^{\eta}(K) \tag{20}$$

and for any function Φ from $H^{p_1(\cdot),p_2(\cdot)}(K;\omega)$ we have

$$\Phi^+(t)\in L^{p_i(\cdot)}(\gamma_i;\omega_i). \tag{21}$$

Furthermore, if $p_i\in\mathcal{P}(\gamma_i)$, then

$$\Phi(z)=\Phi_1(z)+\Phi_2(z) \tag{22}$$

when

$$\Phi_i(z)=\frac{1}{2\pi i}\int_{\gamma_i}\frac{\varphi_i(t)\,dt}{t-z},\quad \varphi_i\in L^{p_i(\cdot)}(\gamma_i;\omega_i). \tag{23}$$

Conversely, if Φ is representable in the form (22), where Φ_i given by (23), $p_i\in\mathcal{P}(\gamma_i)$ and $\omega_i(t)=\omega^+(t)$, $t\in\gamma_i$, where $\omega\in W^{p_1(\cdot),p_2(\cdot)}(K)$, then $\Phi\in H^{p_1(\cdot),p_2(\cdot)}(K;\omega)$.

Proof. Let $p_0=\min(\underline{p_1},\underline{p_2})$. Then

$$\sup_{\rho<r<1}\int_0^{2\pi}|(\Phi\omega)(re^{i\vartheta})|^{p_0}d\vartheta$$

$$\leq 2\pi + \sup_{r_0 \leq r < 1} \int_0^{2\pi} |(\Phi\omega)(re^{i\vartheta})|^{p_1(\vartheta)} d\vartheta + \sup_{\rho < r < r_0} \int_0^{2\pi} |(\Phi\omega)(re^{i\vartheta})|^{p_2(\vartheta)} d\vartheta.$$

Therefore $\Phi\omega \in H^{p_0}(K)$. Moreover, $1/\omega_1 \in H^{p_1'(\cdot)}(K(r_0;1)$, $1/\omega_2 \in H^{p_2'(\cdot)}(K(\rho;r_0)$, therefore the equality $\Phi = (\Phi\omega)\omega^{-1}$ readily implies the validity of the inclusion (20), whereas (21) is obtained by applying the Fatou lemma.

Since $\Phi \in H^{\eta}(K)$, from (18) it follows that

$$\Phi = \Phi_1 + \Phi_2, \quad \Phi_i \in H^{\eta}(K_i).$$

If now $p_i \in \mathcal{P}(\gamma_i)$, then by Smirnov's Theorem we conclude that $\Phi_i \in H^{p_i(\cdot)}(K_i;\omega_i)$, $i = 1,2$ holds. Hence the equality (23) follows by virtue of Theorem 7.

Let now (23) be fulfilled. Put

$$\Phi_i(z) = \frac{1}{2\pi i} \int_{\gamma_i} \frac{\Phi_i^+ \, dt}{t - z}, \qquad z \in K_i.$$

Hence we see that $\Phi_1(z)$ is bounded for $|z| < 1 - \varepsilon$, $\varepsilon > 0$, whereas $\Phi_2(z)$ is bounded for $|z| > \rho + \varepsilon$. Thus by (18) we conclude that $\Phi \in H^{p_1(\cdot),p_2(\cdot)}(K;\omega)$. □

Lemma 24. *If $\omega \in W^{p_1(\cdot),p_2(\cdot)}(K)$ and $f_i \in L^{p_i(\cdot)}(\gamma_i;\omega_i)$, where $p_i \in \mathcal{P}(\gamma_i)$, $\omega_i(t) = \omega^+(t)$, $t \in \gamma_i$, then the Cauchy type integral*

$$\Phi(z) = \frac{1}{2\pi i} \int_{\gamma_1} \frac{f_1(t)\, dt}{t - z} + \frac{1}{2\pi i} \int_{\gamma_2} \frac{f_2(t)\, dt}{t - z} = (K_{\gamma_1} f_1)(z) + (K_{\gamma_2} f_2)(z), \quad z \in K,$$

belongs to the class $H^{p_1(\cdot),p_2(\cdot)}(K;\omega)$.

Proof. The functions $\Phi_i(z) = (K_{\gamma_i} f_i)(z)$ belong to $\bigcap_{\delta<1} H^{\delta}$ (see, for example, [19, p. 39]). Along with this, $\Phi_i^+ = \frac{1}{2}(f_i + S_{\gamma_i} f_i)$. But $f_i = \varphi_i \omega_i^{-1}$, $\varphi_i \in L^{p_i(\cdot)}(\gamma_i)$ and therefore $\Phi_i^+ = \frac{1}{2}\left(\frac{\varphi_i}{\omega_i} + S_{\gamma_i} \frac{\varphi_i}{\omega_i}\right)$. Hence it follows that $\omega_i \Phi_i^+ = \frac{1}{2}\left(\varphi_i + \omega_i S_{\gamma_i} \frac{\varphi_i}{\omega_i}\right)$. Since $\omega_i \in W^{p_i(\cdot)}(\gamma_i)$, we have $\omega_i \Phi_i^+ \in L^{p_i(\cdot)}(\gamma_i)$. But $\omega_i^{-1} \in L^{p_i'(\cdot)}(\gamma_i)$ and hence we obtain $\Phi_i \in H^1(U_i)$ ($U_1 = U$, $U_2 = \{z : |z| < \rho\}$). Thus $\Phi_i = K_{\gamma_i}(\Phi_i^+)$ and since $\Phi_i^+ \in L^{p_i(\cdot)}(\gamma_i;\omega_i)$, by virtue of Lemma 23 $\Phi \in H^{p_1(\cdot),p_2(\cdot)}(K;\omega)$. □

From Lemmas 23 and 24 follows

Theorem 25. *Let $p_i \in \mathcal{P}(\gamma_i)$, $\omega \in W^{p_1(\cdot),p_2(\cdot)}(K)$. For an analytic function Φ in K to belong to the class $H^{p_1(\cdot),p_2(\cdot)}(K;\omega)$ it is necessary and sufficient that Φ would be representable as*

$$\Phi(z) = \frac{1}{2\pi i} \int_{\gamma_1} \frac{\varphi_1(t)\, dt}{t - z} + \frac{1}{2\pi i} \int_{\gamma_2} \frac{\varphi_2(t)\, dt}{t - z}, \qquad z \in K,$$

where $\varphi_i \in L^{p_i(\cdot)}(\gamma_i;\omega_i)$, $\omega_i(t) = \omega^+(t)$, $t \in \gamma_i$.

Classes $H^{p(z)}(K;\omega)$

We will say that an analytic function Φ in K belongs to the class $H^{p(z)}(K;\omega)$ if

$$\sup_{\rho<r<1}\int_0^{2\pi}|\Phi(re^{i\vartheta})\omega(re^{i\vartheta})|^{p(re^{i\vartheta})}d\vartheta<\infty.$$

Assume $H^{p(z)}(K)=H^{p(z)}(K;1)$.

Analogously to the proof of Lemma 2 we prove

Lemma 2′. *Let* $\inf\limits_{z\in K}p(z)=\underline{p}>0$ *and a.e. on* $[0,2\pi]$

$$\lim_{r\to1}p(re^{i\vartheta})=p_1(\vartheta),\quad \lim_{r\to\rho}p(re^{i\vartheta})=p_2(\vartheta).$$

If $\Phi\in H^{p(z)}(K;\omega)$, *where* ω *is a function such that for some* $a>0$ *we have*

$$\sup_{\rho<r<1}\int_0^{2\pi}|\omega(re^{i\vartheta})|^{-a}d\vartheta<\infty\quad\text{and}\quad \lim_{r\to1-}\omega(re^{i\vartheta})=\omega_1(\vartheta),\quad \lim_{r\to\rho+}\omega(re^{i\vartheta})=\omega_2(\vartheta),$$

then $\Phi\in H^{\eta}(K)$, $\eta\in\left(0;\frac{\underline{p}a}{\underline{p}+a}\right)$, *and* $\Phi^+|_{\gamma_i}\in L^{p_i(\vartheta)}(\gamma_i;\omega_i)$.

Theorem 26. *Let* $p(z)\in\mathcal{P}(\overline{K})$ *and* $p_1(\vartheta)=p(1,\vartheta)$, $p_2(\vartheta)=p(\rho,\vartheta)$, *and the conditions of Lemma 2′ be fulfilled for* ω. *Then: for* $\Phi\in H^{p(z)}(K;\omega)$ *we have* $\Phi(z)=\Phi_1(z)+\Phi_2(z)$, *where* $\Phi_i(z)\in H^{\eta}(K_i)$, $\eta\in\left(0;\frac{p_0a}{p_0+a}\right)$, *and* $\Phi|_{\gamma_i}\in L^{p_i(\cdot)}(\gamma_i;\omega_i)$; *for* $\omega\equiv1$ *we have* $\Phi_i\in H^{p_i(\cdot)}(K_i)$.

Proof. The first part of the theorem follows from Lemma 2′ and the representation (18) for functions from $H^{\eta}(K)$.

Assume now that $\omega\equiv1$. Since $p_i\in\mathcal{P}(\gamma)$ and therefore in the representation (18) we have $\Phi|_{\gamma_i}\in L^{p_i(\cdot)}(\gamma_i)\subset L^1(\gamma_i)$. Thus $\Phi_i\in H^1$ and therefore Φ_i is representable by the Cauchy integral with a density from $L^{p_i(\cdot)}(\gamma_i)$. By Theorem 8 we have $\Phi_i\in H^{p_i(\cdot)}(K_i)$. □

Lemma 27. *If* $\Phi\in H^{p_1(\cdot),p_2(\cdot)}(K)$ *and* $q(r,\vartheta)$ *is an arbitrary function from* $\mathcal{P}(\overline{K})$, *for which* $q(1,\vartheta)=p_1(\vartheta)$ *and* $q(\rho,\vartheta)=p_2(\vartheta)$, $\min p_i>0$, $i=1,2$, *then* $\Phi\in H^{q(r,\vartheta)-\varepsilon}(K)$ *for arbitrary* $\varepsilon\in(0,\underline{q})$.

Proof. The representation (22) holds for Φ by virtue of Lemma 23. Following the reasoning of Lemma 23 it can be easily proved that

$$\sup_{r_0\le r<1}\int_0^{2\pi}|\Phi_1(re^{i\vartheta})|^{q(r,\vartheta)-\varepsilon}d\vartheta<\infty,\qquad \sup_{r_0\le r<1}\int_0^{2\pi}|\Phi_2(re^{i\vartheta})|^{q(r,\vartheta)-\varepsilon}d\vartheta<\infty.$$

But

$$\sup_{0<r<1} \int_0^{2\pi} |\Phi_1(re^{i\vartheta})|^{q(r,\vartheta)-\varepsilon} d\vartheta$$

$$\le \left(\max_{0<r\le r_0} |\Phi_1(re^{i\vartheta})| \right)^{\max q(r,\vartheta)} + 2\pi + \sup_{r_0\le r<1} \int_0^{2\pi} |\Phi_1(re^{i\vartheta})|^{p_1(\vartheta)} d\vartheta < \infty.$$

Analogously, it is shown that

$$\sup_{\rho<r\le r_0} \int_0^{2\pi} |\Phi_2(re^{i\vartheta})|^{q(r,\vartheta)-\varepsilon} d\vartheta < \infty.$$

Therefore $\Phi \in H^{q(r,\vartheta)-\varepsilon}(K)$. □

3.11 Smirnov classes in doubly-connected domains

Classes $E^{p_1(\cdot),p_2(\cdot)}(D)$

Let Γ_1 and Γ_2 be simple rectifiable curves, Γ_2 lying in a bounded domain with boundary Γ_1. Denote by D a doubly-connected domain with boundary $\Gamma_1 \cup \Gamma_2$. This domain can be conformally mapped onto some ring $K = \{w : \rho < |w| < 1\}$ (see, for example, [27, p. 207]). Assume that $w = w(z)$ gives this conformal mapping, and $z = z(w)$ is its inverse.

Definition 28. Let $p_i = p_i(t)$ be a positive measurable function on Γ_i. Assume $p_1(\vartheta) = p(z(e^{i\vartheta}))$, $p_2(\vartheta) = p(z(\rho e^{i\vartheta}))$. We will say that an analytic function Φ in D belongs to the class $E^{p_1(\cdot),p_2(\cdot)}(D)$ if for any $r_0 \in (\rho, 1)$ we have

$$\left[\sup_{r_0<r<1} \int_0^{2\pi} |\Phi(z(re^{i\vartheta}))|^{p_1(\vartheta)} |z'(re^{i\vartheta})|\, d\vartheta \right.$$
$$\left. + \sup_{\rho<r\le r_0} \int_0^{2\pi} |\Phi(z(re^{i\vartheta}))|^{p_2(\vartheta)} |z'(re^{i\vartheta})|\, d\vartheta \right] < \infty. \quad (24)$$

Suppose

$$e^{p_1(\cdot),p_2(\cdot)}(D) = \{u : \ u = \operatorname{Re}\Phi, \ \Phi \in E^{p_1(\cdot),p_2(\cdot)}(D)\}.$$

It is not difficult to verify that if (24) is fulfilled for some $r_0 = \widetilde{r}_0$, then (24) will also hold for any $r_0 \in (\rho; 1)$. Like in Section 3.1 it is proved that if (24) is fulfilled for some function $z = \widetilde{z}(w)$ conformal mapping of K onto D, then (24) will also hold for any such function $z = z(w)$.

Analogously to the proof of Lemma 23 we prove

Lemma 29. *If* $\Phi \in E^{p_1(\cdot),p_2(\cdot)}(D)$, $\inf p_i = \underline{p_i} > 0$ *and* $p_0 = \min \underline{p_i}$, *then* $E^{p_1(\cdot),p_2(\cdot)}(D) \subset E^{p_0}(D)$ *and* $\Phi^+|_{\Gamma_i} \in L^{p_i(\cdot)}(\Gamma_i)$. *Furthermore, if* $p_i \in \mathcal{P}(\Gamma_i)$, *then*

$$\Phi(z) = \Phi_1(z) + \Phi_2(z), \quad \Phi_i \in E^{p_i(\cdot)}(D_i), \tag{25}$$

where D_i *is the domain bounded by the curve* Γ_i *and containing* D.

Conversely, if the equality (25) *holds for* Φ, *then* $\Phi \in E^{p_1(\cdot),p_2(\cdot)}(D)$.

3.12 On the belonging of a Cauchy type integral to the class $E^{p_1(\cdot),p_2(\cdot)}(D)$

From (25) it follows that in the assumptions of Lemma 29 we have

$$\Phi(z) = (K_{\Gamma_1} f_1)(z) + (K_{\Gamma_2} f_2)(z), \quad f_i \in L^{p_i(\cdot)}(\Gamma_i). \tag{26}$$

To establish an analogue of Lemma 24, i.e., to prove that the functions representable by the formula (26) belong to the class $E^{p_1(\cdot),p_2(\cdot)}(D)$, we have to impose additional conditions on the domain boundary.

Theorem 30. *Let* D *be a doubly-connected domain bounded by the curves* $\Gamma_i \in C^{1,L}(A_1^{(i)}, \dots, A_{n_i}^{(i)}; \nu_1^{(i)}, \dots, \nu_{n_i}^{(i)})$, $0 < \nu_k^{(i)} \le 2$, $k = \overline{1, n_i}$, *and the functions* $p_i(t)$ *belong to* $\mathcal{P}(\Gamma_i)$. *Then for* $f_i \in L^{p_i(\cdot)}(\Gamma_i)$ *the function* Φ *given by* (26) *belongs to the class* $E^{p_1(\cdot),p_2(\cdot)}(D)$.

Proof. Since $\Gamma_1 \in C^{1,L}(A_1^{(1)}, \dots, A_{n_1}^{(1)}; \nu_1^{(1)}, \dots, \nu_{n_1}^{(1)})$, $0 < \nu_k^{(1)} \le 2$, $k = \overline{1, n_1}$, by Theorem 15 the function $(K_{\Gamma_1} f_1)(z)$ belongs to $E^{p_1(\cdot)}(D_1)$.

By substitution of the variables $t = \frac{1}{\tau}$, $z = \frac{1}{\zeta}$ in the Cauchy type integral $(K_{\Gamma_2} f_2)(z)$ it is not difficult to verify that the Theorem 15 also remains valid in the case of external domains bounded by piecewise Liapunov curves. In our case Γ_2 is exactly such a curve and therefore $(K_{\Gamma_2} f_2)(z) \in E^{p_2(\cdot)}(D_2)$.

Thus by formula (26) the representation (25) holds for the Cauchy type integral and therefore by virtue of Lemma 29 we conclude that $\Phi \in E^{p_1(\cdot),p_2(\cdot)}(D)$. □

On Domains D for which $p(z(w)) \in \mathcal{P}(\overline{K})$ when $p(z) \in \mathcal{P}(\overline{D})$

When investigating the Dirichlet problem in simply-connected domains for harmonic functions from Smirnov classes with a variable exponent, it was important to take into consideration the fact that in the case of piecewise Lyapunov curves Γ when $p(t) \in \mathcal{P}(\Gamma)$, the function $p(z(t))$ belongs to $\mathcal{P}(\gamma)$ (see Lemma 13). Now we give the generalization of this result and keeping in mind possible applications, we will focus attention on finding the doubly-connected domains for which $p(z(w))$ belongs to $\mathcal{P}(\overline{U})$ when $p(z)$ belongs to $\mathcal{P}(\overline{D})$.

Theorem 31. *Let* I. $p(z) \in \mathcal{P}(\overline{D})$ *and* $z = z(w)$ *be the conformal mapping of the ring* K *onto* D. *If* $z(w)$ *is a function such that for some* $a > 0$ *we have*

$$\sup_{w_1,w_2\in\overline{K}} \frac{|z(w_1) - z(w_2)|}{|w_1 - w_2|^a} = M < \infty, \tag{27}$$

then $p(z(w)) \in \mathcal{P}(\overline{K})$.

or II. $p(t) \in \mathcal{P}(\Gamma)$ *and*

$$\sup_{w_1,w_2\in\gamma} \frac{|z(w_1) - z(w_2)|}{|w_1 - w_2|^a} < \infty, \tag{28}$$

then $p(z(\tau)) \in \mathcal{P}(\gamma)$.

Proof. I. Since $p(z) \in \mathcal{P}(\overline{D})$, for $w_1, w_2 \in \overline{K}$ we have

$$|p(z(w_1)) - p(z(w_2))| < \frac{C}{|\ln|z(w_1) - z(w_2)||} = \frac{C}{\left|\ln\left|\frac{z(w_1)-z(w_2)}{|w_1-w_2|^a}\right| + a\ln|w_1 - w_2|\right|}.$$

From (27) it follows that for $|a\ln|w_1 - w_2|| > 2M$, i.e., when $|w_1 - w_2| < \exp\left(-\frac{2M}{a}\right)$, we have

$$|p(z(w_1)) - p(z(w_2))| < \frac{2C}{|\ln|w_1 - w_2||}.$$

If $|w_1 - w_2| \geq \exp\left(-\frac{2M}{a}\right)$, then

$$\begin{aligned} |p(z(w_1)) - p(z(w_2))| &< \frac{2C\max|p(z)|}{|\ln|w_1 - w_2||}|\ln|w_1 - w_2|| \\ &= \frac{2C\max|p(z)|\max\left(\frac{2M}{a}, \ln 2\right)}{|\ln|w_1 - w_2||} = \frac{C'}{|\ln|w_1 - w_2||}. \end{aligned} \tag{29}$$

These estimate imply that $p(z(w)) \in \mathcal{P}(\overline{K})$.

II. The proof is carried out exactly in the same manner as above for the assumption (27). □

Theorem 32. I. *If* $z'(w) \in H^{1+\varepsilon}(K)$ *for some* $\varepsilon > 0$ *and* $p(z) \in \mathcal{P}(\overline{D})$, *then* $p(z(w)) \in \mathcal{P}(\overline{U})$. II. *If* Γ *bounded the simply-connected domain* D *for which* $z'(w) \in H^{1+\varepsilon}$ *and* $p(t) \in \mathcal{P}(\Gamma)$, *then* $p(z(\tau)) \in \mathcal{P}(\gamma)$.

Proof. I. To prove the first claim it suffices due to Theorem 31 to show that the inequality (27) holds for some $a > 0$.

Since $z' \in H^{1+\varepsilon}(K)$, we have [126, p. 66]

$$z'(w) = \widetilde{z}_1(w) + \widetilde{z}_2(w), \quad \widetilde{z}_i \in H^{1+\varepsilon}(K_i).$$

Then $z(w) = z_1(w) + z_2(w)$, where $z_i' \in H^{1+\varepsilon}(K_i)$.

Let us first show that the inequality (27) holds for z_1. We have

$$|z_1(w_1) - z_1(w_2)| = \left| \int\limits_{\ell(w_1,w_2)} z_1'(w)\,dw \right| \le \int\limits_{\ell(w_1,w_2)} |z_1'(w)|\,|dw|, \tag{30}$$

where $\ell(w_1, w_2)$ is an arbitrary rectifiable curve lying in U and connecting the points w_1 and w_2.

Let $w_i = \rho_i \exp \vartheta_i$ and $\rho_1 < \rho_2$, $\vartheta_1 \ne \vartheta_2$. Let further $\ell_1(w_1, w)$ be a small arc of a circle of radius ρ_1 with ends w_1 and $w = \rho_1 \exp \vartheta_2$, and $\ell_2(w, w_2)$ be a segment of the straight line with ends w and w_2. We take the union of $\ell_1(w_1, w)$ and $\ell_2(w, w_2)$ as the curve $\ell(w_1, w_2)$. Then the inequality (30) implies

$$\begin{aligned}|z_1(w_1) - z_1(w_2)| &\le \Bigg(\int\limits_{\ell_1} |z_1'|^{1+\varepsilon} |dw| \Bigg)^{\frac{1}{1+\varepsilon}} |\ell_1|^{\frac{\varepsilon}{1+\varepsilon}} + \Bigg(\int\limits_{\ell_2} |z_1'|^{1+\varepsilon} |dw| \Bigg)^{\frac{1}{1+\varepsilon}} |\ell_2|^{\frac{\varepsilon}{1+\varepsilon}} \\ &\le \Bigg(\int\limits_0^{2\pi} |z_1'(\rho_1 e^{i\vartheta})|^{1+\varepsilon} d\vartheta \Bigg)^{\frac{1}{1+\varepsilon}} |\rho_1(\vartheta_1 - \vartheta_2)|^{\frac{\varepsilon}{1+\varepsilon}} \\ &\quad + \Bigg(\int\limits_0^{1} |z_1'(re^{i\vartheta})|^{1+\varepsilon} dr \Bigg)^{\frac{1}{1+\varepsilon}} |\rho_2 - \rho_1|^{\frac{\varepsilon}{1+\varepsilon}}. \end{aligned} \tag{31}$$

Since $z_1' \in H^{1+\varepsilon}(K_1) = H^{1+\varepsilon}(U)$, the inequality

$$\int\limits_{-1}^{1} |z_1'(re^{i\vartheta})|^{1+\varepsilon} dr \le \frac{1}{2} \int\limits_0^{2\pi} |z_1'(e^{i\vartheta})|^{1+\varepsilon} d\vartheta \tag{32}$$

holds for any $\vartheta \in [0, 2\pi]$ and $r \in (0, 1)$ (see [19, p. 46]). Therefore the inequality (31) implies

$$\begin{aligned}&|z_1(w_1) - z_1(w_2)| \\ &< \Bigg(\int\limits_0^{2\pi} |z_1'(e^{i\vartheta})|^{1+\varepsilon} d\vartheta \Bigg)^{\frac{1}{1+\varepsilon}} \left(|\rho_1(\vartheta_1 - \vartheta_2)|^a + |\rho_1 - \rho_2|^a \right), \quad a = \frac{\varepsilon}{1+\varepsilon} > 0. \end{aligned} \tag{33}$$

Let us consider the triangle ABC with the vertices at the points w_1, w and w_2, respectively. Assume that $\gamma = |\vartheta_1 - \vartheta_2|$ and $\beta = \frac{\pi - \gamma}{2} = \angle OAB = \angle OBA$, where $O(0, 0)$ is the origin. Then $\alpha = \angle ABC = \gamma + \beta$. But since $2\beta + \gamma = \pi$, we have $\alpha = \gamma + \beta > \frac{\pi}{2}$ and therefore the angle α in the triangle ABC is the largest one. Hence it follows that $|AB| < |AC| = |w_2 - w_1|$, $|BC| < |AC| = |w_2 - w_1|$. Since

$$\left| \exp \rho_1 e^{i\vartheta_1} - \exp \rho_1 e^{i\vartheta_2} \right| = 2\rho_1 \sin \frac{\gamma}{2} = 2\rho_1 \frac{\sin \frac{|\vartheta_1 - \vartheta_2|}{2}}{|\vartheta_1 - \vartheta_2|} |\vartheta_1 - \vartheta_2| \ge m\rho_1 |\vartheta_1 - \vartheta_2|,$$

we have $|\rho_1(\vartheta_1 - \vartheta_2)| \leq M_1|w_1 - w_2|$. Finally, for $\vartheta_2 \neq \vartheta_1$,

$$|\rho_1(\vartheta_1 - \vartheta_2| \leq M_1|w_1 - w_2|, \quad |\rho_2 - \rho_1| \leq |w_1 - w_2|. \tag{34}$$

For $\vartheta_2 = \vartheta_1$ these inequalities are obvious.

Taking the inequalities (34) and (33) into account, we obtain

$$|z_1(w_1) - z_1(w_2)| < 4M_1|w_1 - w_2|^a, \quad a = \frac{\varepsilon}{1+\varepsilon}.$$

Thus the inequality (27) with z_1 is fulfilled for $a = \frac{\varepsilon}{1+\varepsilon}$.

Now let us consider the difference $|z_2(w_1) - z_2(w_2)|$. Since $z_2\left(\frac{1}{\zeta}\right)\zeta^{-\frac{2}{p}} \in H^{1+\varepsilon}(K_1)$, relying on (32) we obtain

$$\int_0^\infty |z_2'(re^{i\vartheta})|^{1+\varepsilon} dr \leq \frac{1}{2}\int_0^{2\pi} |z_2'(e^{i\vartheta})|^{1+\varepsilon} d\vartheta.$$

Following the reasoning we used when investigating $z_1(w)$, the latter inequality allows us to establish the validity of (27) for the function z_2. Hence (27) holds for $z = z_1 + z_2$, too, and therefore $p(z(w)) \in \mathcal{P}(\overline{U})$.

II. To prove the second claim, it suffices by virtue of Theorem 31 to show the fulfillment of (28). Let ℓ be an arc of γ with ends $w_i = e^{i\vartheta_i}$, $i = 1, 2$. Then

$$\begin{aligned}|z(w_1) - z(w_2)| &\leq \int_{\vartheta_1}^{\vartheta_2} |z'(e^{i\vartheta})| d\vartheta \leq \left(\int_{\vartheta_1}^{\vartheta_2} |z'(e^{i\vartheta})|^{1+\varepsilon} d\vartheta\right)^{\frac{1}{1+\varepsilon}} |\vartheta_1 - \vartheta_2|^{\frac{\varepsilon}{1+\varepsilon}} \\ &\leq M|e^{i\vartheta_1} - e^{i\vartheta_2}|^{\frac{\varepsilon}{1+\varepsilon}}.\end{aligned}$$

Therefore the inequality (28) holds for $a = \frac{\varepsilon}{1+\varepsilon}$. □

Corollary 33. *I. If the domain D is bounded by piecewise smooth curves without zero angles and $p(z) \in \mathcal{P}(\overline{D})$, then $p(z(w)) \in \mathcal{P}(\overline{U})$. II. If the simply connected domain D bounded by piecewise smooth curve Γ without zero angles and $p(t) \in \mathcal{P}(\Gamma)$ then $p(z(\tau)) \in \mathcal{P}(\gamma)$.*

Indeed, in that case we have [39, Statement 9]

$$z'(w) \sim \prod_{i=1}^{2}\sum_{k=1}^{n_i}(w - a_{k_i})^{n_k^{(i)}-1} z_0(w), \quad a_k^{(i)} = w(A_k^{(i)}), \quad z_0^{\pm 1} \in \bigcap_{\delta>1} H^\delta(K).$$

If $\eta = \min(\nu_k^{(i)} - 1) \geq 0$, then $z' \in \bigcap_{\delta>0} H^\delta(K)$. If $\eta < 0$, then $\eta \in (-1, 0)$ and therefore for $\varepsilon \in \left(0; \frac{1}{|\eta|} - 1\right)$ we have $z' \in H^{1+\varepsilon}(K)$. Thus the condition of Theorem 32 is fulfilled and thus claim of the corollary is valid.

3.13 Notes and comments on Chapter 3

The well-known Hardy classes of analytic functions has been extended for simply-connected domains by V. I. Smirnov [122], [123]. Different, but equivalent definition of these classes was introduced by M. A. Lavrentiev and M. V. Keldysh. For the definition of analytic functions of Smirnov classes in multiply-connected domains and their properties we refer the reader to the paper [126].

Variable exponent Hardy and Smirnov classes of analytic functions for the first time were introduced and studied in [55], [57]. Further various approaches to the above-mentioned classes were developed in [62]. In that paper various type weighted Hardy and Smirnov classes of analytic functions both in simply and double-connected domains were introduced. The relations between different definitions are treated. In the papers mentioned above we were focused on the representability problems by Cauchy type integrals of analytic functions from the above mentioned classes.

The results related to the convergence of sequences of functions from variable exponent Hardy and Smirnov classes extend G. Tumarkin's theorem (see, for example, [108, p. 268].

In Chapter 3 the results of paper [55], [57], and [62] are exposed.

Chapter 4

Boundary Value Problems for Analytic Functions

This Chapter is devoted to the solution of the Riemann boundary value problem for analytic functions with data in boundary condition from various nonstandard Banach function spaces. We mainly are focusing on the case of piecewise continuous conjugacy coefficients. This Chapter contain three parts. In Part I the Riemann boundary value problem is investigated when the boundary function belongs to the variable exponent Lebesgue space. In Part II the problem is studied in more general setting. Namely, we treat with so called Riemann boundary value problem with shift which in literature is known also as Haseman problem. Part III deals with the problem of linear conjugation with data from the grand Lebesgue spaces. In all above-mentioned cases the classes of sought functions are sets of Cauchy type integrals with densities from the same classes as data.

We establish the solvability conditions and construct solutions in explicit form.

Let Γ be an oriented rectifiable simple closed curve bounding the domains D^+ and D^-. Further, let $A(\Gamma)$ be some class of functions analytic in $\mathbb{C}\backslash\Gamma$. Riemann's boundary value problem is called the following one: find an analytic function Φ from $A(\Gamma)$ whose boundary values satisfy the conjugacy condition

$$\Phi^+(t) = G(t)\Phi^-(t) + g(t), \quad t \in \Gamma, \tag{1}$$

where G and g are the given functions on Γ and Φ^+ and Φ^- are boundary values of Φ on Γ from inside and outside Γ, respectively. This problem is also known as the linear conjugation problem, (see [97], [98]).

In Part I we seek the solution of (1) in the class of analytic functions represented by the Cauchy type integral with density in the spaces $L^{p(\cdot)}(\Gamma)$ with variable exponent assuming that g belongs to the same class. We consider the cases when the coefficient G is continuous or piecewise continuous as well as the case of oscillating coefficient. The solvability conditions are derived and in all the cases of solvability the explicit formulas are given. The related boundary singular integral equations in

$L^{p(\cdot)}(\Gamma)$ are treated. The solution of boundary value problem (1) allows us to obtain weight results for Cauchy singular integral operator in $L^{p(\cdot)}(\Gamma)$ spaces; among them we deal with some extension of the well known Helson-Szegö theorem.

In Sections 4.9–4.16 of Part II we study the Riemann boundary value problem

$$\Phi^+(\alpha(t)) = a(t)\Phi^-(t) + b(t), \quad t \in \Gamma \tag{2}$$

with shift, (the Haseman problem), when the coefficient $a(t)$ is a nonvanishing continuous function, while Section 4.17 deals with the case of piecewise continuous coefficient $a(t)$ for which $\inf |a(t)| > 0$.

A natural setting of the Haseman problem in the class of Cauchy type integrals with density from the weighted Lebesgue space with a variable exponent $p(t)$ is proposed. The influence of the discontinuity of the function $a(t)$ on a character of solvability of the problem is revealed; solvability conditions are established and the solutions are constructed.

Besides that the problem is studied in a more general statement than that considered earlier; we cover the case of non-smooth boundary satisfying the chord condition.

We make use of the following notations:

$$\mathcal{R}^{p(\cdot)} = \Big\{\Gamma : \ S_\Gamma \quad \text{is bounded in} \quad L^{p(\cdot)}(\Gamma)\Big\}. \tag{3}$$

Let

$$K^{p(\cdot)}(\Gamma,\rho) = \Big\{\Phi(z) : \Phi(z) = (K_\Gamma\varphi)(z) = \frac{1}{2\pi i}\int\limits_\Gamma \frac{\varphi(\tau)d\tau}{\tau - z}, z \notin \Gamma, \varphi \in L^{p(\cdot)}(\Gamma;\rho)\Big\} \tag{4}$$

$$\widetilde{K}_0^{p(\cdot)}(\Gamma,\rho) = \Big\{\Phi(z) : \Phi(z) = \Phi_0(z) + const, \quad \Phi_0 \in K^{p(\cdot)}(\Gamma,\rho)\Big\}. \tag{5}$$

We write $K^{p(\cdot)}(\Gamma) = K^{p(\cdot)}(\Gamma,1)$ and $\widetilde{K}_0^{p(\cdot)}(\Gamma,1) = \widetilde{K}_0^{p(\cdot)}(\Gamma)$.

Part I

The Riemann Boundary Problem with Data from the $L^{p(\cdot)}(\Gamma)$

4.1 Some properties of the Cauchy type integrals with densities in $L^{p(\cdot)}(\Gamma)$

In this section we present some auxiliary results which provide an extension of known properties of the Cauchy singular integrals in the Lebesgue spaces with constant p to the case of variable exponent.

Proposition 1. *Let Γ be a rectifiable Jordan curve. If $\frac{1}{\rho} \in L^{q(\cdot)}(\Gamma)$, $q(t) = p'(t) = \frac{p(t)}{p(t)-1}$, then the operator S_Γ is continuous in measure in $L^{p(\cdot)}(\Gamma; \rho)$, i.e., for any sequence f_n converging in $L^{p(\cdot)}(\Gamma, \rho)$ to function f_0 the sequence $S_\Gamma f_n$ converges in measure to $S_\Gamma f_0$.*

The validity of this statement may be obtained by word-for-word repetition of the proof of Theorem 2.1 from [40, p. 21], since $L^{p(\cdot)}(\Gamma, \rho) \subset L^1(\Gamma)$ according to our assumption.

Theorem 2. *Let Γ be a simple closed rectifiable curve bounding the domains D^+ and D^-. The following statements are valid:*
i) *Let S_Γ map $L^{p(\cdot)}(\Gamma, \rho)$ to $L^{\mu(\cdot)}(\Gamma, \omega)$ for some weight functions ρ and ω. Then $\frac{1}{\rho} \in L^{q(\cdot)}(\Gamma)$ and S_Γ is bounded from $L^{p(\cdot)}(\Gamma, \rho)$ into $L^{\mu(\cdot)}(\Gamma, \omega)$.*
ii) *Let S_Γ be bounded from $L^{p(\cdot)}(\Gamma, \rho)$ to $L^{\alpha}(\Gamma)$, $\alpha > 0$. Then for arbitrary $\varphi \in L^{p(\cdot)}(\Gamma, \rho)$ the Cauchy type integral $(K_\Gamma \varphi)(z)$ belongs to $E^{\alpha}(D^{\pm})$.*
iii) *Let S_Γ be bounded in $L^{p(\cdot)}(\Gamma)$. Then for arbitrary $\varphi \in L^{p(\cdot)}(\Gamma)$ we have*

$$(K_\Gamma \varphi)(z) \in E^{\underline{p}}.$$

iv) *For $\rho \in W^{p(\cdot)}(\Gamma)$ and $\varphi \in L^{p(\cdot)}(\Gamma)$ the function $K_\Gamma \left(\frac{\varphi}{\rho}\right)$ belongs to $E^1(D^{\pm})$.*

Proof. i). Since S_Γ is defined for any function in $L^{p(\cdot)}(\Gamma, \rho)$, we have the embedding $L^{p(\cdot)}(\Gamma, \rho) \subset L^1(\Gamma)$. Then for any $\varphi \in L^{p(\cdot)}(\Gamma)$ the function $\frac{\varphi}{\rho}$ is integrable on Γ. Therefore, $\frac{1}{\rho} \in L^{q(\cdot)}(\Gamma)$. According to the Proposition 1 we conclude that for the sequence of functions φ_n converging to φ in $L^{p(\cdot)}(\Gamma)$ the sequence $S_\Gamma \varphi_n$ converges to $S_\Gamma \varphi$ in measure. Thus, if S_Γ maps $L^{p(\cdot)}(\Gamma, \rho)$ into $L^{\mu(\cdot)}(\Gamma, \omega)$, then S_Γ is a closed operator and by the closed graph theorem we conclude that it is bounded.

ii). Let S_Γ be bounded from $L^{p(\cdot)}(\Gamma, \rho)$ into $L^{\alpha}(\Gamma)$, $\alpha > 0$, then $\frac{1}{\rho} \in L^{p'(\cdot)}(\Gamma)$ (see i)). Let $\varphi \in L^{p(\cdot)}(\Gamma, \rho)$ and let φ_n be a sequence of rational functions (with a unique pole in D^+) such that φ_n converges to φ in $L^{p(\cdot)}(\Gamma,\rho)$ (see Chapter 3 Proposition 4). Then for the functions $\Phi_n(z)=(K_\Gamma\varphi_n)(z)$ we have $\Phi_n(z)\in E^\alpha(D^\pm)$ and $\|\Phi_n^\pm\|_\alpha\le M\|\varphi_n\|_{p(\cdot),\rho}$ and by Proposition 1 $\Phi_n^\pm$ converges in measure to the function $\pm\frac{1}{2}\varphi+\frac{1}{2}S_\Gamma\varphi$. Applying G. Tumarkin's Theorem [108, p. 269], we conclude that $\Phi(z)=\lim\limits_{n\to\infty}\Phi_n(z)$ belongs to $E^\alpha(D^\pm)$. In our case $\Phi(z)=(K_\Gamma\varphi)(z)$.

iii). From the embedding $L^{p(\cdot)}(\Gamma)\subseteq L^{\underline{p}}(\Gamma)$ and the boundedness of S_Γ in $L^{p(\cdot)}(\Gamma)$ it follows that S_Γ maps $L^{p(\cdot)}(\Gamma)$ into $L^{\underline{p}}(\Gamma)$. Then by *i)* S_Γ is bounded from $L^{p(\cdot)}(\Gamma)$ into $L^{\underline{p}}(\Gamma)$. In view of *ii)* then $K_\Gamma\varphi\in E^{\underline{p}}(D^\pm)$ for arbitrary $\varphi\in L^{p(\cdot)}(\Gamma)$.

iv). Since $\rho\in W^{p(\cdot)}(\Gamma)$, we have $\frac{1}{\rho}\in L^{q(\cdot)}(\Gamma)$ and then $S_\Gamma\left(\frac{\varphi}{\rho}\right)\in L^1(\Gamma)$for any $\varphi\in L^{p(\cdot)}(\Gamma)$. The last follows from the equality $S_\Gamma\left(\frac{\varphi}{\rho}\right)=\frac{1}{\rho}\left(\rho S_\Gamma\frac{\varphi}{\rho}\right)$. Therefore, the operator $S_\Gamma\frac{1}{\rho}$ is defined on $L^{p(\cdot)}(\Gamma;\rho)$ and acts into $L^1(\Gamma)$. Then it is continuous in measure in $L^{p(\cdot)}(\Gamma;\rho)$ and consequently is a closed operator and therefore, it is bounded from $L^{p(\cdot)}(\Gamma;\rho)$ to $L^1(\Gamma)$. Applying *ii)* when $\alpha=1$ we conclude that $K_\Gamma\left(\frac{\varphi}{\rho}\right)(z)\in E^1(D^\pm)$. □

Corollary 3. *If $\Gamma\in\mathcal{R}^{p(\cdot)}$, then Γ is a Smirnov curve.*

Indeed, since $\Gamma\in\mathcal{R}^{p(\cdot)}$ and $L^{\overline{p}}(\Gamma)\subset L^{p(\cdot)}(\Gamma)\subset L^{\underline{p}}(\Gamma)$ follows that S_Γ maps $L^{\overline{p}}(\Gamma)$ into $L^{\underline{p}}(\Gamma)$. Then Γ is a Smirnov curve (see [31] and [40, p.22]).

Corollary 4. *Let $\Gamma\in\mathcal{R}^{p(\cdot)}$ (particular if Γ is Carleson curve and $p\in\mathcal{P}(\Gamma)$). Then for arbitrary bounded on Γ function φ we have $(K_\Gamma\varphi)(z)\in\bigcap\limits_{\beta>\underline{p}}E^\beta(D^\pm)$.*

Proof. Since $\varphi\in\bigcap\limits_{\alpha>1}L^{\alpha p(\cdot)}$ according to the statement *iii)* from Theorem 2 we obtain that $(K_\Gamma\varphi)(z)\in\bigcap\limits_{\alpha>1}E^{\alpha\underline{p}}(D^\pm)$. Therefore $(K_\Gamma\varphi)(z)\in\bigcap\limits_{\beta>\underline{p}}E^\beta(D^\pm)$. □

Theorem 5. *Let $\Gamma\in\mathcal{R}^{p(\cdot)}$, $\underline{p}>1$. Then S_Γ is also bounded in the space $L^{\alpha p(\cdot)}(\Gamma)$ for any $\alpha>1$ and the inequality*

$$\|S_\Gamma\|_{\alpha p(\cdot)}\le\operatorname{ctg}\frac{\pi}{4\alpha}\,\|S_\Gamma\|_{p(\cdot)}\tag{6}$$

holds.

Proof. We follow Cotlar's idea [8] and paper [101]. We base ourselves on the well known relation

$$(S_\Gamma\varphi)^2=-\varphi^2+2S_\Gamma(\varphi S_\Gamma\varphi),\tag{7}$$

(see for instance, [40, p.33]), which follows also as a particular case from the Poincaré-Bertrand formula (see, for example, [25, Section 7.2] or [43, p. 96])

$$\frac{1}{\pi i}\int\limits_\Gamma\frac{d\tau}{\tau-t}\frac{1}{\pi i}\int\limits_\Gamma\frac{a(\tau,\tau_1)}{\tau_1-\tau}d\tau_1=a(t,t)+\frac{1}{\pi i}\int\limits_\Gamma d\tau_1\frac{1}{\pi i}\int\limits_\Gamma\frac{a(\tau,\tau_1)}{(\tau-t)(\tau_1-\tau)}d\tau$$

under the choice $a(t,\tau) = \varphi(t)\varphi(\tau)$; we take φ a rational function.

We observe that

$$\|\varphi^2\|_{p(\cdot)} = \|\varphi\|^2_{2p(\cdot)}$$

and obtain from (7)

$$\|S_\Gamma\varphi\|^2_{2p(\cdot)} \le \|\varphi\|^2_{2p(\cdot)} + 2\|S_\Gamma\|_{p(\cdot)}\|\varphi S_\Gamma\varphi\|_{p(\cdot)}. \tag{8}$$

By the usual Hölder inequality we have $\|\varphi S_\Gamma\varphi\|_{p(\cdot)} \le \|\varphi\|_{2p(\cdot)} \cdot \|S_\Gamma\varphi\|_{2p(\cdot)}$ and then from (8)

$$\|S_\Gamma\varphi\|^2_{2p(\cdot)} - 2\|S_\Gamma\|_{p(\cdot)}\|S_\Gamma\varphi\|_{2p(\cdot)}\|\varphi\|_{2p(\cdot)} - \|\varphi\|^2_{2p(\cdot)} \le 0$$

whence the estimate

$$\|S_\Gamma\varphi\|_{2p(\cdot)} \le \left(\|S_\Gamma\|_{p(\cdot)} + \sqrt{\|S_\Gamma\|^2_{p(\cdot)} + 1}\right)\|\varphi\|_{2p(\cdot)} \tag{9}$$

follows for any rational function φ. By denseness of rational functions in $L^{2p(\cdot)}(\Gamma)$, this estimate is extended to the whole space $L^{2p(\cdot)}(\Gamma)$.

Further by induction we prove that

$$\|S_\Gamma\|_{2^{k-1}p(\cdot)} \le \operatorname{ctg}\frac{\pi}{2^{k+1}}\,\|S_\Gamma\|_{p(\cdot)}, \quad k \in \mathbb{N}. \tag{10}$$

Indeed, from (9) we obtain that

$$\|S_\Gamma\|_{2^k p(\cdot)} \le \|S_\Gamma\|_{p(\cdot)}\left(\operatorname{ctg}\frac{\pi}{2^{k+1}} + \sqrt{1 + \operatorname{ctg}^2\frac{\pi}{2^{k+1}}}\right)$$

$$\le \|S_\Gamma\|_{p(\cdot)}\left(\operatorname{ctg}\frac{\pi}{2^{k+1}} + \frac{1}{\sin\frac{\pi}{2^{k+1}}}\right) = \|S_\Gamma\|_{p(\cdot)}\operatorname{ctg}\frac{\pi}{2^{k+2}}. \tag{11}$$

Now we apply the Riesz type interpolation theorem known for the spaces $L^{p(\cdot)}$ (see [93, Theorem 14.16]) in the following form: *if a linear operator A is bounded in the spaces $L^{2^k p(\cdot)}(\Gamma)$ and $L^{2^{k+1}p(\cdot)}(\Gamma)$, then it is also bounded in the space $L^{\alpha p(\cdot)}(\Gamma)$ with $\alpha \in [2^k, 2^{k+1})$, $\frac{1}{\alpha} = \theta 2^{-k} + (1-\theta)2^{-k-1}$, and*

$$\|A\|_{\alpha p(\cdot)} \le \|A\|^\theta_{2^k p(\cdot)}\|A\|^{1-\theta}_{2^{k+1}p(\cdot)}$$

Then from (10) and (11) we get

$$\|S_\Gamma\|_{\alpha p(\cdot)} \le \|S_\Gamma\|_{p(\cdot)}\left\{\operatorname{ctg}\left(\frac{\pi}{2^{k+1}}\right)\right\}^\theta\left\{\operatorname{ctg}\left(\frac{\pi}{2^{k+2}}\right)\right\}^{1-\theta}.$$

Obviously,

$$\left\{\operatorname{ctg}\left(\frac{\pi}{2^{k+1}}\right)\right\}^\theta\left\{\operatorname{ctg}\left(\frac{\pi}{2^{k+2}}\right)\right\}^{1-\theta} \le \operatorname{ctg}\left(\frac{\pi}{2^{k+2}}\right) = \operatorname{ctg}\left(\frac{\pi}{4}\cdot\frac{1}{2^k}\right).$$

But $\alpha \geq 2^k$. Therefore,

$$\left\{\operatorname{ctg}\left(\frac{\pi}{2^{k+1}}\right)\right\}^{\theta}\left\{\operatorname{ctg}\left(\frac{\pi}{2^{k+2}}\right)\right\}^{1-\theta} \leq \operatorname{ctg}\left(\frac{\pi}{4}\cdot\frac{1}{\alpha}\right) = \operatorname{ctg}\left(\frac{\pi}{4\alpha}\right).$$

Consequently,

$$|S_\Gamma\|_{\alpha p(\cdot)} \leq \operatorname{ctg}\frac{\pi}{4\alpha}\,\|S_\Gamma\|_{p(\cdot)}\,. \tag{12}$$

The proof is complete. □

Theorem 6. *Let Γ be simple closed curve bounded the domains D^+, D^-, $(\infty \in D^-)$, and φ be a bounded measurable function on Γ. Assume that $z_0 \in D^+$ and $\Gamma \in \mathcal{R}^{p(\cdot)}$, $\underline{p} > 1$. Then*
i) *there exists an integer $k \geq 0$ such that*

$$\exp\{(K_\Gamma\varphi)(z)\} := X(z) \in E^\delta(D^+) \quad \textit{and} \quad \frac{X(z)-1}{(z-z_0)^k} \in E^\delta(D^-) \tag{13}$$

where

$$0 < \delta < \frac{\pi\underline{p}}{2(1+\ell)eM\|S_\Gamma\|_{p(\cdot)}}, \qquad M = \sup_{t\in\Gamma}|\varphi(t)|; \tag{14}$$

ii) *in case $\varphi \in C(\Gamma)$*

$$X(z) \in \bigcap_{\delta>1} E^\delta(D^+) \quad \textit{and} \quad X(z)-1 \in \bigcap_{\delta>1} E^\delta(D^-). \tag{15}$$

Proof. We use an idea developed in [53]. Let Γ_r be the image of $\gamma_r = \{z : |z| = r\}, r < 1$, under the conformal mapping of U onto D^+. We have

$$\int_{\Gamma_r} |X(z)|^\delta |dz| \leq \int_{\Gamma_r} \sum_{n=0}^\infty \frac{1}{n!}\,|\delta\Phi(z)|^n\,|dz| \quad \text{where} \quad \Phi(z) = \frac{1}{2\pi i}\int_\Gamma \frac{\varphi(\tau)\,d\tau}{\tau - z}. \tag{16}$$

According to Corollary 4 we have $\Phi(z) \in E^n(D^+)$ for any $n \geq 1$. Then by the known property of the class $E^p(D^+)$ (see [108, Chapter III]), we have

$$\int_{\Gamma_r} |\Phi(z)|^n |dz| \leq \int_\Gamma |\Phi^+(t)|^n |dt|$$

and then from (16) we obtain

$$\int_{\Gamma_r} |X(z)|^\delta |dz| \leq \sum_{n=0}^\infty \int_\Gamma |\delta\Phi^+(t)|^n |dt| \leq \sum_{n=0}^\infty \frac{1}{n!}\int_\Gamma \left|\frac{\delta\varphi(t)}{2} + \frac{\delta}{2}(S_\Gamma\varphi(t))\right|^n |dt|$$

$$\le \sum_{n=0}^{\infty} \frac{1}{n!} \int_{\Gamma} |\delta\varphi(t)|^n |dt| + \sum_{n=0}^{\infty} \frac{1}{n!} \int_{\Gamma} |\delta(S_\Gamma\varphi(t))|^n \, |dt|.$$

Hence

$$\int_{\Gamma_r} |X(z)|^\delta |dz| \le \ell e^{\delta M} + \left(\sum_{n=0}^{n_0-1} + \sum_{n=n_0}^{\infty} \right) \frac{1}{n!} \int_{\Gamma} |\delta(S_\Gamma\varphi)(t)|^n \, |dt|$$

where we take any $n_0 > \underline{p}$. It remains to show that the series $\sum_{n=n_0}^{\infty}$ converges. Let $\alpha_n = \frac{n}{\underline{p}} > 1$. Then $n = \alpha_n \underline{p} \le \alpha_n p(t)$. Since if $p(t) \le p_1(t)$ then

$$\|f\|_{p(\cdot)} \le (1+\ell)\|f\|_{p_1(\cdot)}, \quad \ell = \text{mes}\Gamma,$$

we have

$$\|S_\Gamma\varphi\|_n \le (1+\ell)\|S_\Gamma\varphi\|_{\alpha_n p(\cdot)}$$

Then by (6) we obtain

$$\|S_\Gamma\varphi\|_n \le (1+\ell)\,\text{ctg}\,\frac{\pi}{4\alpha_n}\,\|S_\Gamma\|_{p(\cdot)}\|\varphi\|_{\alpha_n p(\cdot)}$$

Take an inequality

$$\min\left(\ell^{\frac{1}{\underline{p}}}, \ell^{\frac{1}{\overline{p}}}\right) \le \|1\|_{p(\cdot)} \le \max\left(\ell^{\frac{1}{\underline{p}}}, \ell^{\frac{1}{\overline{p}}}\right),$$

we see that $\|\varphi\|_{\alpha_n p(\cdot)} \le M \max\left(1, \ell^{\frac{1}{n}}\right)$ and then

$$\|S_\Gamma\varphi\|_n \le c_0 n \max\left(1, \ell^{\frac{1}{n}}\right) \|S_\Gamma\|_{p(\cdot)}, \quad c_0 = \frac{4}{\pi \underline{p}}(1+\ell)M.$$

Therefore,

$$\sum_{n=n_0}^{\infty} \frac{1}{n!} \int_{\Gamma} |\delta(S_\Gamma\varphi(t))|^n \, |dt| \le \sum_{n=n_0}^{\infty} \frac{\delta^n}{n!}\|S_\Gamma\varphi\|_n^n \le \max(1,\ell) \sum_{n=n_0}^{\infty} \frac{(c_0\delta)^n n^n}{n!}\|S_\Gamma\|_{p(\cdot)}^n$$

where the series on the right-hand side converges if $c_0\delta\|S_\Gamma\|_{p(\cdot)}e < 1$.

Thus it was proved that $X(z) \in E^\delta(D^+)$ when

$$0 < \delta < \delta_0 = \frac{\pi \underline{p}}{4(1+\ell)eM\|S_\Gamma\|_{p(\cdot)}}, \tag{17}$$

In the case D^- and for $1 < \delta < \delta_0$ and arbitrary r we are enable to obtain the similar estimates by the same way as in case D^+. As to $\delta \le 1$ it is necessary to consider two cases: $0 < r < r_0$ and $r_0 < r < 1$ for some fixed r_0. In the last case the appropriate estimates can be proved as in the case $\delta > 1$. As to the case $0 < r < r_0$ the needed inequalities we obtain by means of choice of number $k > \left[\frac{1}{\delta}\right]$.

Now we are able to get more stronger result, namely, that $X(z) \in E^{\delta}(D^+)$ and $\frac{X(z)-1}{(z-z_0)^k} \in E^{\delta}(D^-)$ for $\delta < 2\delta_0$.

Indeed,

$$\int_{\Gamma} |X^{\pm}(t)|^{\delta} |dt| = \int_{\Gamma} \left| e^{\pm \frac{\delta \varphi(t)}{2}} \right| \left| e^{\frac{\delta}{2}(S_{\Gamma}\varphi)}(t) \right| |dt|$$

$$\leq e^{\frac{\delta M}{2}} \sum_{n=0}^{\infty} \frac{1}{n!} \int_{\Gamma} \left| \frac{\delta}{2} (S_{\Gamma}\varphi)(t) \right|^n |dt| \leq e^{\frac{\delta M}{2}} \left(\sum_{n=1}^{n_0-1} \frac{1}{n!} \int_{\Gamma} \left| \frac{\delta}{2} (S_{\Gamma}\varphi)(t) \right|^n |dt| \right.$$

$$\left. + \sum_{n=n_0}^{\infty} \frac{1}{n!} \int_{\Gamma} \left| \frac{\delta}{2} (S_{\Gamma}\varphi)(t) \right|^n |dt| \right),$$

where $n_0 > \underline{p}$.

From the previous proof it is clear that last series converges when $\delta < 2\delta_0$, therefore $X^{\pm} \in L^{\delta}(\Gamma)$.

Apply Smirnov's theorem we have $X(z) \in E^{\delta}(D^+)$ and $\frac{X(z)-1}{(z-z_0)^k} \in E^{\delta}(D^-)$ when $\delta < 2\delta_0$ with δ_0 from (17).

By this i) is proved.

Now prove ii). For arbitrary $\varepsilon > 0$ we can find a Hölder function ψ on Γ such that

$$\operatorname*{ess\,sup}_{t \in \Gamma} |\varphi(t) - \psi(t)| < \varepsilon.$$

On the other hand, for the Hölder function $\psi(t)$ there exist positive numbers a_1 and a_2 such that $0 < a_1 \leq |\exp(K_{\Gamma}\psi)(z)| \leq a_2 < \infty$.

Thus from i) and (17) we conclude ii). □

Remark 7. As it follows from the proof of the final part of previous theorem the number M in formula (17) can be replaced by $\nu(\varphi) = \inf \|\varphi - \psi\|_C$, where the infinum is taken over all rational functions ψ.

4.2 The problem of linear conjugation with continuous coefficients

In the present Section we proceed to the solution of problem (1) in the class $K^{p(\cdot)}(\Gamma, \rho)$ under various assumptions with respect to the data.

We begin with the case when $\Gamma \in \mathcal{R}^{p(\cdot)}(\Gamma)$ and G is a non-vanishing continuous function on Γ. The function g is assumed to be in $L^{p(\cdot)}(\Gamma)$. We look for a function $\Phi \in K^{p(\cdot)}(\Gamma)$ whose boundary values $\Phi^{\pm}$ satisfy relation (1) almost everywhere on Γ.

Let $\varkappa = \frac{1}{2\pi}[arg\ G(t)]_\Gamma$ be the index of G on Γ. Below we shall show that for the above formulated problem all the statements for its solvability known for constant p remain valid in the general case of variable exponent; namely, the following statement is valid.

Theorem 8. *Let $\Gamma \in \mathcal{R}^{p(\cdot)}$ and let $g \in L^{p(\cdot)}(\Gamma)$. Assume that $G \in C(\Gamma)$ and $G(t) \neq 0, t \in \Gamma$. Then for problem* (1) *the following statements hold:*
i) *for $\varkappa \geq 0$ problem* (1) *is unconditionally solvable in the class $K^{p(\cdot)}(\Gamma)$ and all its solutions are given by*

$$\Phi(z) = \frac{X(z)}{2\pi i}\int_\Gamma \frac{g(\tau)}{X^+(\tau)}\frac{d\tau}{\tau - z} + X(z)Q_{\varkappa-1}(z) \tag{18}$$

with

$$X(z) = \begin{cases} \exp h(z), & z \in D^+ \\ (z-z_0)^{-\varkappa}\exp h(z), & z \in D^-, \quad z_0 \in D^+, \end{cases}$$

where

$$h(z) = K_\Gamma\Big(ln\ G(t)(t-z_0)^{-\varkappa}\Big)(z)$$

and $Q_{\varkappa-1}(z)$ is an arbitrary polynomial of degree $\varkappa - 1$ ($Q_{-1}(z) \equiv 0$);
ii) *for $\varkappa < 0$ problem (1) is solvable in this class if and only if*

$$\int_\Gamma \frac{g(t)}{X^+(t)} t^k\, dt = 0, \qquad k = 0, 1, \dots, |\varkappa| - 1 \tag{19}$$

and under these conditions problem (1) *has the unique solution given by* (18) *with $Q_{\varkappa-1} = 0$.*

Proof. Consider first the case $\varkappa = 0$. We choose a rational function $\widetilde{G}(t)$ such that

$$\sup_{t\in\Gamma}\left|\frac{G(t)}{\widetilde{G}(t)} - 1\right| < \frac{1}{2}\left(1 + \|S_\Gamma\|_{p(\cdot)}\right)^{-1}. \tag{20}$$

Obviously $ind\ \widetilde{G} = 0$ and therefore the function $\widetilde{X}(z) = \exp\Big(K_\Gamma(\ln\ \widetilde{G})\Big)(z)$ is continuous in the domains $\overline{D^\pm}$. Now recall that if $\Phi \in K^{p(\cdot)}(\Gamma)$, then according to Theorem 2 we have $\Phi \in E^{\underline{p}}(D^\pm)$. Since $\underline{p} > 1$, the equality $\Phi(z) = (K_\Gamma(\Phi^+ - \Phi_-))(z)$ holds (see 0.21). Now we have

$$\left(\frac{\Phi}{\widetilde{X}}\right)^+ = \frac{G}{\widetilde{G}}\left(\frac{\Phi}{\widetilde{X}}\right)^- + \frac{g}{\widetilde{X}^+}, \tag{21}$$

where $\widetilde{X}(z) = \exp\left\{K_\Gamma(\ln\widetilde{G})(z)\right\}$. Let us show that $\frac{\Phi}{\widetilde{X}} \in K^{p(\cdot)}(\Gamma)$. To this end, we observe that $\Phi \in E^{\underline{p}}(D^\pm)$ and $\frac{1}{\widetilde{X}}$ is bounded so that $\frac{\Phi}{\widetilde{X}} \in E^{\underline{p}}(D^\pm)$ and therefore

$$\frac{\Phi}{\widetilde{X}} = K_\Gamma\left(\frac{\Phi^+}{\widetilde{X}^+} - \frac{\Phi^-}{\widetilde{X}^-}\right).$$

From the Plemelj–Sokhotskiĭ formula and from the condition $\Gamma \in \mathcal{R}^{p(\cdot)}$ it follows that $\Phi^{\pm} \in L^{p(\cdot)}(\Gamma)$ and hence $\frac{\Phi}{\widetilde{X}} \in K^{p(\cdot)}(\Gamma)$.

Let

$$\frac{\Phi(z)}{\widetilde{X}(z)} = (K_\Gamma \psi)(z), \quad \psi \in L^{p(\cdot)}(\Gamma).$$

Then equality (21) yields

$$\psi(t) = \left(\frac{G(t)}{\widetilde{G}(t)} - 1\right)\left(-\frac{1}{2}\psi(t) + \frac{1}{2}S_\Gamma \psi(t)\right) + \frac{g(t)}{\widetilde{X}(t)} \tag{22}$$

i.e. the function ψ is a solution of the equation of the type $\psi = M\psi$ in the space $L^{p(\cdot)}(\Gamma)$, where by inequality (20) M is a contractive operator. Therefore, equation (22) and consequently problem (1) has the unique solution in $K^{p(\cdot)}(\Gamma)$. Basing on Theorem 6 we construct the solution. Let

$$X(z) = \exp\left(K_\Gamma(ln\ G)\right)(z).$$

As far as $\varkappa = 0$, we find that $ln\ G(t) = ln\ |G(t)| + i \arg\ G(t)$ is a continuous function, and by Theorem 6

$$\left(\frac{1}{X(z)} - 1\right) \in \bigcap_{\delta > 1} E^{\delta}(D^{\pm}).$$

If Φ is a solution of problem (1), then $\Phi \in K^{\underline{p}}(\Gamma)$ and therefore, $\Phi \in E^{\underline{p}}(D^{\pm})$. Moreover, $\frac{\Phi}{X} \in E^{\underline{p}-\varepsilon}(D^{\pm})$ for arbitrary $\varepsilon \in \left(0, \underline{p}\right)$. Therefore $\frac{\Phi}{X} \in K^{\underline{p}-\varepsilon}(\Gamma)$. So that $\frac{\Phi}{X} \in K^1(\Gamma)$. At the same time

$$\left(\frac{\Phi}{X}\right)^{+} - \left(\frac{\Phi}{X}\right)^{-} = \frac{g}{X^{+}}.$$

Since this problem in $K^1(\Gamma)$ has only solution $K_\Gamma\left(\frac{g}{X^+}\right)$ then

$$\Phi(z) = X(z) K_\Gamma \left(\frac{g}{X^+}\right)(z)$$

is the solution of (1) in the class $K^{p(\cdot)}(\Gamma)$.

Let now $\varkappa > 0$. We choose a point $z_0 \in D^+$ and rewrite (1) in the form

$$\Phi^{+}(t) = G_1(t)(t - z_0)^{\varkappa}\Phi^{-}(t) + g(t)$$

where $G_1(t) = (t - z_0)^{-\varkappa}G(t)$ is a continuous function with zero index. We introduce a new unknown function

$$F(z) = \begin{cases} \Phi(z), & z \in D^{+} \\ (z - z_0)^{\varkappa}\Phi(z), & z \in D^{-} \end{cases}. \tag{23}$$

For $F(z)$ there exists a polynomial $Q_{\varkappa-1}(z)$ such that

$$\Psi(z) = F(z) - Q_{\varkappa-1}(z) \in E^1(D^-). \tag{24}$$

It is obvious that $\Psi(z) \in E^1(D^+)$. Then $\Psi(z) = K_\Gamma(\Psi^+ - \Psi^-)$. But

$$\Psi^+(t) - \Psi^-(t) = F^+(t) - F^-(t) = \Phi^+(t) - (t - z_0)^\varkappa \Phi^-(t) \in L^{p(\cdot)}(\Gamma)$$

so that $\Psi \in K^{p(\cdot)}(\Gamma)$. Moreover,

$$\Psi^+(t) = G_1(t)\Psi^-(t) + g_1(t),$$

where $g_1(t) = g(t) - Q_{\varkappa-1}(t) + G_1(t)Q_{\varkappa-1}(t)$. Since $ind\ G_1 = 0$, according to what was proved above,

$$\Psi(z) = X_1(z)K_\Gamma\left(\frac{g_1}{X_1^+}\right)(z) \qquad \text{where} \qquad X_1(z) = \exp\{(K_\Gamma ln\ G_1)(z)\}.$$

Talking into account that $G = \frac{X_1^+}{X_1^-}$ we have

$$K_\Gamma\left(\frac{g_1}{X_1^+}\right)(z) = K_\Gamma\left(\frac{g}{X_1^+}\right)(z) - \frac{1}{2\pi i}\int\limits_\Gamma \frac{Q_{\varkappa-1}(t)}{X_1^+(t)}\frac{dt}{t-z} + \frac{1}{2\pi i}\int\limits_\Gamma \frac{Q_{\varkappa-1}(t)}{X_1^-(t)}\frac{dt}{t-z}.$$

But

$$\frac{1}{2\pi i}\int\limits_\Gamma \frac{Q_{\varkappa-1}(t)}{X_1^+(t)}\frac{dt}{t-z} = \begin{cases} \frac{Q_{\varkappa-1}(z)}{X_1(z)}, & z \in D^+, \\ 0, & z \in D^-, \end{cases}$$

$$\frac{1}{2\pi i}\int\limits_\Gamma \frac{Q_{\varkappa-1}(t)}{X_1^-(t)}\frac{dt}{t-z} = \frac{1}{2\pi i}\int\limits_\Gamma \left[\frac{Q_{\varkappa-1}(t)}{X_1^-(t)} - Q_{\varkappa-1}(t)\right]\frac{dt}{t-z} + \frac{1}{2\pi i}\int\limits_\Gamma \frac{Q_{\varkappa-1}(t)}{t-z}dt$$

$$= \begin{cases} Q_{\varkappa-1}(z), & z \in D^+, \\ -\frac{Q_{\varkappa-1}(z)}{X_1(z)} + Q_{\varkappa-1}(z), & z \in D^-. \end{cases}$$

Therefore,

$$\Psi(z) = X_1(z)K_\Gamma\left(\frac{g_1}{X_1^+}\right)(z) = X_1(z)K_\Gamma\left(\frac{g}{X_1^+}\right)(z) + X_1(z)Q_{\varkappa-1}(z) - Q_{\varkappa-1}(z).$$

Then by (23) and (24) we arrive at formula (18).

It can be easily verified that the latter provides the solution of problem (1) for an arbitrary polynomial $Q_{\varkappa-1}(z)$ which does not depend on the choice of the point z_0.

Finally we consider the case $\varkappa < 0$. This time the function F given by (23) is in $K^{p(\cdot)}(\Gamma)$. Moreover, $F^+ = G_1 F^- + g$, whence

$$F(z) = X_1(z) K_\Gamma \left(\frac{g}{X_1^+}\right)(z)$$

and the condition $\Phi(z) = (z - z_0)^{-\varkappa} F(z) \in E^1(D^-)$ is fulfilled if and only if the conditions (19) are satisfied. □

Via the solution of (1) in $K^{p(\cdot)}(\Gamma)$ with non-vanishing $G \in C(\Gamma)$ and $g \in L^{p(\cdot)}(\Gamma)$ we are now able to derive the following weight result for Cauchy singular integrals.

Theorem 9. *Let φ be a real-valued function in $C(\Gamma)$ and $\Gamma \in \mathcal{R}^{p(\cdot)}$. Then the function*

$$\rho(t) = \left| \exp\left(\frac{1}{2\pi} \int_\Gamma \frac{\varphi(\tau) d\tau}{\tau - t}\right) \right|$$

belongs to the class $W^{p(\cdot)}(\Gamma)$.

Proof. Consider the problem (1) in the class $K^{p(\cdot)}(\Gamma)$ with $G(t) = \exp(i\varphi(t))$ and $g \in L^{p(\cdot)}(\Gamma)$. Obviously, $G \in C(\Gamma)$ and $ind\, G = 0$. Consequently, the function

$$\Phi(z) = X(z) K_\Gamma \left(\frac{g}{X^+}\right)(z), \quad X(z) = \exp(K_\Gamma \varphi)(z),$$

belongs to $K^{p(\cdot)}(\Gamma)$ for arbitrary $g \in L^{p(\cdot)}(\Gamma)$. This implies that the function

$$\Phi^+(t) = X^+(t) \left(S_\Gamma \frac{g}{X^+}\right)(t)$$

belongs to $L^{p(\cdot)}(\Gamma)$, i.e.

$$e^{iarg\, X^+(t)} \frac{\rho(t)}{\pi i} \int_\Gamma \frac{g(\tau) e^{-iarg\, X^+(\tau)}}{\rho(\tau)} \frac{d\tau}{\tau - t} \in L^{p(\cdot)}(\Gamma).$$

Since $\left| e^{\pm iarg\, X^+(\tau)} \right| = 1$, by statement i) of Theorem 2 we immediately conclude that $\rho \in W^{p(\cdot)}(\Gamma)$. □

4.3 The problem of linear conjugation with continuous coefficients in the weighted class $K^{p(\cdot)}(\Gamma, \rho)$

If we assume that $\rho \in W^{p(\cdot)}(\Gamma)$ and choose the function $\widetilde{G}$ in the proof of Theorem 8 such that instead of condition (20), the condition

$$\sup_{t \in \Gamma} \left| \frac{G(t)}{\widetilde{G}(t)} - 1 \right| < \frac{1}{2} \left(1 + \|S_\Gamma\|_{L^{p(\cdot)}(\Gamma, \rho)}\right)^{-1}$$

is fulfilled, then for $\varkappa = 0$ we conclude that problem (1) is uniquely solvable in $K^{p(\cdot)}(\Gamma,\rho)$ for arbitrary $g \in L^{p(\cdot)}(\Gamma,\rho)$. If, in addition, we assume that $\frac{1}{\rho} \in L^{q(\cdot)+\varepsilon}(\Gamma)$, then formula (18) remains valid because in this case $\frac{\Phi}{X} \in E^{1+\delta}(D^{\pm})$, for some $\delta > 0$. The last inclusion is valid since $\Phi \in E^{1+\varepsilon}(D^{\pm})$, $\frac{1}{X} \in \cap_{\delta>1} E^{\delta}(D^{\pm})$ and

$$\left(\frac{\Phi}{X}\right)^{+} - \left(\frac{\Phi}{X}\right)^{-} = \frac{g}{X^{+}}.$$

Consequently, we arrive at the following statement.

Theorem 10. *Let $\Gamma \in \mathcal{R}^{p(\cdot)}$, (particulary if Γ is Carleson curve and $p \in \mathcal{P}(\Gamma)$), $\rho \in W^{p(\cdot)}(\Gamma)$ and $\frac{1}{\rho} \in L^{q(\cdot)+\varepsilon}(\Gamma)$, for some $\varepsilon > 0$. Assume that $G \in C(\Gamma), G(t) \neq 0, t \in \Gamma, g \in L^{p(\cdot)}(\Gamma,\rho)$. Then for problem (1) in the class $K^{p(\cdot)}(\Gamma,\rho)$ all the statements of Theorem 8 remain valid.*

Corollary 11. *Let Γ is Carleson curve, $p \in \mathcal{P}(\Gamma)$, and $\rho \in W^{p(\cdot)}(\Gamma)$, $\frac{1}{\rho} \in L^{q(\cdot)+\varepsilon}(\Gamma)$ $(\varepsilon > 0)$. Then the function*

$$r(t) = \rho(t) \exp\left|\frac{1}{2\pi}\int_{\Gamma} \frac{\varphi(\tau)\, d\tau}{\tau - t}\right| \tag{25}$$

with real continuous φ, belongs to $W^{p(\cdot)}(\Gamma)$.

4.4 The problem of linear conjugation with a piecewise continuous coefficient

The results from Chapter 2 allows us to investigate problem (1) in the class $K^{p(\cdot)}(\Gamma)$ in the case of piecewise continuous coefficient G and $g \in L^{p(\cdot)}(\Gamma)$.

Thus let $G \in C(\Gamma, t_1, t_2, \ldots, t_m)$, i.e. G be continuous on the arcs $[t_k, t_{k+1}]$, $k = 1, \ldots, m+1$, $t_{m+1} = t_1$. Assume that $\inf\limits_{t\in\Gamma} |G(t)| > 0$. We will follow [24], [97], [42].

Let

$$\frac{G(t_k^-)}{G(t_k^+)} = e^{2\pi i \lambda_k}, \qquad k = 1, 2, \ldots, m, \tag{26}$$

where the real part of the complex numbers $\lambda_k = \alpha_k + i\beta_k$ is defined up to an arbitrary additive integer. We assume that

$$\alpha_k \neq \frac{1}{q_k} \mod (1), \quad q_k = q(t_k), \quad q(t) = \frac{p(t)}{p(t)-1} \tag{27}$$

and choose α_k in the interval

$$-\frac{1}{p_k} < \alpha_k < \frac{1}{q_k}, \qquad p_k = p(t_k), \qquad q_k = q(t_k). \tag{28}$$

In D^+ we choose an arbitrary point z_0. Let γ_k be a simple smooth curve (a cut) from the point z_0 to ∞ which crosses Γ only at the point t_k so that the function $(z-z_0)^{\lambda_k}$ is analytic on the complex plane cut along γ_k and

$$(t_k - z_0)_-^{\lambda_k} = (t_k - z_0)_+^{\lambda_k} \exp(2\pi i \lambda_k)$$

where $(t_k - z_0)_\pm^{\lambda_k} = \lim\limits_{\substack{t \to t_k \pm \\ t \in \Gamma}} (t - z_0)^{\lambda_k}$. Then the function

$$G_1(t) = \frac{G(t)}{\prod\limits_{k=1}^{m} (t-z_0)^{\lambda_k}} \tag{29}$$

is continuous ([25, p. 432]) and $G_1(t) \neq 0, t \in \Gamma$.

Consider now the function

$$\rho(z) = \prod_{k=1}^{m} \rho_{t_k}(z) \quad \text{with} \quad \rho_{t_k}(z) = \begin{cases} (z-t_k)^{\lambda_k}, & z \in D^+, \\ \left(\frac{z-t_k}{z-z_0}\right)^{\lambda_k} = \frac{(z-t_k)^{\lambda_k}}{(z-z_0)^{\lambda_k}}, & z \in D^-, \end{cases}$$

where the branch for the function $\left(\frac{z-t_k}{z-z_0}\right)^{\lambda_k}$ is chosen so that it tends to 1 as $z \to \infty, z \in D^-$, so that $\rho_{t_k}(z)$ is analytic in $D^\pm$.

Assuming that the curve Γ at the points t_k has at least one-sided tangents, and taking into account inequalities (28), we conclude from the equality

$$\rho_{t_k}(z) = e^{\alpha_k ln\,|z-t_k| - \beta_k\, arg(z-t_k)}\, e^{i(\beta_k ln\,|z-t_k| + \alpha_k\, arg\,(z-t_k))}$$

that there exists an $\varepsilon > 0$ such that

$$\rho_{t_k}(z) \in E^{p_k+\varepsilon}(D^\pm), \quad \left(\frac{1}{\rho_{t_k}(z)} - 1\right) \in E^{q_k+\varepsilon}(D^\pm).$$

Therefore, we have

$$\rho(z) \in E^{p_0}(D^\pm), \quad \left(\frac{1}{\rho(z)} - 1\right) \in E^{q_0}(D^\pm), \tag{30}$$

where $p_0 = \min\, p_k, q_0 = \min\, q_k$.

Let $X(z) = \rho(z) X_1(z)$ where

$$X_1(z) = \exp\{K_\Gamma(\ln\, G_1)(z)\},$$

and introduce a new unknown function

$$\Phi_1(z) = \frac{\Phi(z)}{\rho(z)}. \tag{31}$$

As far as $\Phi \in E^{\underline{p}}(D^{\pm})$ and (30) holds, we have that $\Phi_1(z) \in E^{\delta}(D^{\pm})$, for some $\delta > 0$. By results from Chapter 2 we have $\rho \in W^{p(\cdot)}(\Gamma)$ and by (28) $\frac{1}{\rho(t)} \in L^{q(\cdot)+\varepsilon}$, $\varepsilon > 0$. So that $\Phi_1^{\pm} \in L^{1+\eta}(\Gamma)$ for some $\eta > 0$. According to Smirnov's theorem we conclude that $\Phi_1 \in E^{1+\eta}(D^{\pm})$. Now the equality $\Phi_1 = K_{\Gamma}(\Phi_1^+ - \Phi_1^-)$ holds and therefore $\Phi_1 \in K^{1+\eta}(\Gamma)$. From (26) and (31) we derive that

$$\Phi_1^+(t) = G_1(t)\Phi_1^-(t) + \frac{g(t)}{\rho(t)}. \tag{32}$$

Having resolved (32), we find that all the possible solutions of problem (1) in the case $\varkappa = ind\, G_1(t) \geq 0$ are given by

$$\Phi(z) = \rho(z)X_1(z)K_{\Gamma}\left(\frac{g}{\rho^+ X_1^+}\right)(z) + \rho(z)X_1(z)Q_{\varkappa-1}(z) \tag{33}$$

where $Q_{\varkappa-1}(z)$ is an arbitrary polynomial of degree $\varkappa - 1$.

Since $\rho \in W^{p(\cdot)}(\Gamma)$ and $\frac{1}{\rho} \in L^{q(\cdot)+\varepsilon}(\Gamma)$. Therefore all the conditions of Corollary 11 are satisfied. Therefore, for the function Φ from (33) we get $\Phi^{\pm} \in L^{p(\cdot)}(\Gamma)$. Consequently, (33) with an arbitrary polynomial $Q_{\varkappa-1}$ provides us the solution of (1) in $K^{p(\cdot)}(\Gamma)$.

The case of negative index is considered in standard way.

As a result, we arrive at the following theorem.

Theorem 12. *Let Γ be a closed Carleson curve, $p \in \mathcal{P}(\Gamma)$,*

$$G \in C(\Gamma, t_1, t_2, \ldots, t_m), \qquad \inf_{t\in\Gamma} |G(t)| > 0$$

and conditions (26)–(28) are fulfilled. Suppose that the curve Γ has at least one-sided tangent lines at the points t_k, $k = 1, 2, \ldots, m$ and $\frac{1}{\rho} \in L^{q(\cdot)+\varepsilon}(\Gamma)$ when

$$\rho(t) = \prod_{k=1}^{n} (t - t_k)^{\lambda_k}.$$

Let $\varkappa = ind\, G_1(t)$, where G_1 is given by (29). Under assumptions (26)–(28), the statements of Theorem 8 holds for problem (1), if $X(z)$ is replaced by $X_1(z)$ and formula (18) is replaced by formula (33).

4.5 The problem of linear conjugation with bounded measurable coefficient

On the base of various approaches to investigate problem (1), we are able to study this problem in the classes $K^{p(\cdot)}(\Gamma)$ either when

I. Γ belongs to rather narrow class of curves (Lyapunov curves) and G is in a

sufficiently wide class of bounded measurable functions, or
II. Γ belongs to a wide class of curves but G is in a more narrow class than in the previous case.

I. In this situation we assume that $p \in \mathcal{P}(\Gamma)$, Γ is a Lyapunov curve. We assume that $G \in \mathcal{A}(\lambda)$, $\lambda > 1$, where $\mathcal{A}(\lambda)$ is the Simonenko class (see 0.23).

In [119] for every $G \in \mathcal{A}(\lambda)$ there was introduced an argument $\alpha(t)$ of $G(t)$ at every point $t \in \Gamma$ and its increment $2\pi\varkappa_\lambda(G)$ and it was shown that there exists a function $G_0(t) = \exp\{i\alpha_0(t)\}$ satisfying the Lipschitz condition such that

$$\operatorname*{ess\,sup}_{t\in\Gamma} |\alpha(t) - \alpha_0(t)| < \frac{a(\lambda)}{2} \quad \text{and} \quad \varkappa(G_0) = \varkappa_\lambda(G)$$

and thus

$$G(t) = |G(t)|e^{i\alpha_0(t)}e^{i[\alpha(t)-\alpha_0(t)]}. \tag{34}$$

In the sequel the index $\varkappa = \varkappa(G)$ of a function $G \in \mathcal{A}(\lambda)$ is interpreted, by definition given in (0.23),

$$\varkappa := \varkappa(G) = \varkappa_\lambda(G). \tag{35}$$

II. It is known that in the general case of $\Gamma \in \mathcal{R}^{p(\cdot)}$ even for constant $p, 1 < p < \infty$, problem (1) with real-valued function G under the condition $0 < m_1 \le |G(t)| \le m_2 < \infty$, the statements of Theorem 8 may become invalid (see [11]). In this general case of $\Gamma \in \mathcal{R}^{p(\cdot)}$ we restrict ourselves to oscillatory coefficients of the form

$$G(t) = e^{i\alpha(t)} \tag{36}$$

with a real-valued function $\alpha(t)$.

Theorem 13. *Let p belongs to $\mathcal{P}(\Gamma)$. Suppose that either*
I) *Γ is a Lyapunov curve and $G \in A(\lambda)$, where*

$$\lambda = \max\left(\frac{\pi}{2\operatorname{arcctg}\|S_{\Gamma_0}\|_{p(\cdot)}}, \overline{q}\right),$$

or
II) *$\Gamma \in \mathcal{R}^{p(\cdot)}$ and G is an oscillating function having form (36) with real-valued $\alpha(t), e^{i\alpha(t)} \in A(\lambda)$ where*

$$\lambda = \max\left(\frac{\pi}{2\operatorname{arcctg}\|S_\Gamma\|_{p(\cdot)}}, \frac{2(1+\ell)e\|S_\Gamma\|_{p(\cdot)}}{\underline{p}}\right).$$

Then for problem (1) in $K^{p(\cdot)}(\Gamma)$ all the conclusions of Theorem 8 hold, where we mean that

$$h(z) = K_\Gamma\left(\ln|G(t)| + i\alpha(t) - \varkappa\ln(t - z_0)\right)(z).$$

Proof. I. Let Φ be a solution of problem (1) in $K^{p(\cdot)}(\Gamma)$ and $Q_{\varkappa-1}(z)$ such a polynomial that the function $(z-z_0)^{\varkappa}\Phi(z)-Q_{\varkappa-1}(z)$ belongs to $K^{p(\cdot)}(\Gamma)$ $(Q_{\varkappa-1}(z)\equiv 0$ when $\varkappa \le 0)$. We rewrite problem (1) as

$$\Phi^+(t) = (t-z_0)^{-\varkappa}G(t)\left[\Phi^-(t)(t-z_0)^{\varkappa} - Q_{\varkappa-1}(t)\right]$$

$$+g(t) + (t-z_0)^{-\varkappa}G(t)Q_{\varkappa-1}(t).$$

By (34) we have

$$G(t) = e^{i[\alpha(t)-\alpha_0(t)]}\frac{X^+_{\alpha_0}(t)}{X^-_{\alpha_0}(t)} \tag{37}$$

where

$$X_{\alpha_0}(z) = \exp\left(\frac{1}{2\pi i}\int_\Gamma \frac{ln\left[(t-z_0)^{-\varkappa}\exp(i\alpha_0(t))\right]dt}{t-z}\right)\exp\left(\frac{1}{2\pi i}\int_\Gamma \frac{ln\,|G(t)|}{t-z}dt\right)$$

$$= X_1(z)X_2(z).$$

Since Γ is a Lyapunov curve, $X_2^\pm(z)$ are bounded functions (see e. g. [108, pp. 253, 260]). The function $G_1(t) = (t-z_0)^{-\varkappa}\exp(i\alpha_0(t))$ belongs to Lip 1 and *ind* $G_1 = 0$. Hence $X_1^\pm$ are continuous in $\overline{D^\pm}$ and $X_1^\pm \neq 0$. Therefore, $X_{\alpha_0}(z)$ and $\frac{1}{X_{\alpha_0}(z)}$ are bounded analytic functions in $D^\pm$. By (37) our boundary problem takes the form

$$\frac{\Phi^+(t)}{X^+_{\alpha_0}(t)} = \exp[i(\alpha(t)-\alpha_0(t))]\frac{\Phi^-(t)(t-z_0)^{\varkappa} - Q_{\varkappa-1}(t)}{X^-_{\alpha_0}(t)} + g_1(t),$$

where

$$g_1(t) = \frac{g(t) + (t-z_0)^{-\varkappa}G(t)Q_{\varkappa-1}(t)}{X^+_{\alpha_0}(t)}.$$

Put

$$\Psi(z) = \begin{cases} \frac{\Phi(z)}{X_{\alpha_0}(z)}, & z \in D^+, \\ \frac{\Phi(z)(z-z_0)^{-\varkappa}-Q_{\varkappa-1}(z)}{X_{\alpha_0}(z)}, & z \in D^-. \end{cases}$$

Then $\Psi(z) \in K^{p(\cdot)}(\Gamma)$ and

$$\Psi^+(t) = \exp(i(\alpha(t)-\alpha_0(t)))\Psi^-(t) + g_1(t). \tag{38}$$

Since $\Psi(z) \in K^{p(\cdot)}(\Gamma)$, we put $\Psi(z) = (K\psi)(z)$ with $\psi \in L^{p(\cdot)}(\Gamma)$ and rewrite (38) as

$$(G_1+1)\psi = (G_1-1)S_\Gamma\psi + 2g_1, \tag{39}$$

where

$$G_1(t) = \exp\{i(\alpha(t)-\alpha_0(t))\} =: \exp(i\beta(t)).$$

Equation (39) yields

$$\psi(t) = i\,tg\frac{\beta(t)}{2}(S_\Gamma\psi)(t) + \frac{2g_1(t)}{1+G_1(t)}.$$

Let $M\psi = i\,tg\frac{\beta(t)}{2}(S_\Gamma\psi)(t) + \frac{2g_1(t)}{1+G_1(t)}$ so that $\psi = M\psi$ where M will be a contraction operator in $L^{p(\cdot)}(\Gamma)$ when $|\beta(t)| = |\alpha(t) - \alpha_0(t)| < 2\,\text{arcctg}\,\|S_\Gamma\|_{p(\cdot)}$, i.e.

$$e^{i\alpha(t)} \in \mathcal{A}(\lambda) \quad \text{with} \quad \lambda = \frac{\pi}{2\ \text{arcctg}\ \|S_\Gamma\|_{p(\cdot)}}. \tag{40}$$

Thus under condition (40), the boundary value problem (1) has the unique solution in the case $\varkappa = 0$.

Now we need that the following condition

$$Y_\alpha(z) := \left\{ \exp\left(-\frac{1}{2\pi i} \int_\Gamma \frac{\alpha(t)\,dt}{t-z} \right) - 1 \right\} \in K^{q(\cdot)}(\Gamma) \tag{41}$$

should be satisfied. The last inclusion holds if, for example, (see [119]).

$$e^{i\alpha(t)} \in \mathcal{A}(\max(2,\overline{q})), \qquad \overline{q} = \operatorname*{ess\,sup}_{t\in\Gamma} q(t). \tag{42}$$

Assuming that conditions (40) and (41) (or (42)) are fulfilled and taking into account that in this case equation (38) is uniquely solvable, just in the same way as in Section 4.2 we establish that the function given by (18) provides us again the solution of BVP (1) in $K^{p(\cdot)}(\Gamma)$ when $\varkappa \geq 0$, while in the case $\varkappa < 0$ for the solvability it is necessary and sufficient that conditions (19) are fulfilled.

When the curve is such that the operator

$$(T\varphi)(t) = \frac{1}{\pi i} \int_0^\ell \psi(t(\sigma)) \left(\frac{t'(\sigma)}{t(\sigma) - t(s)} - \frac{i\eta e^{i\eta\sigma}}{e^{i\eta\sigma} - e^{i\eta s}} \right) d\sigma, \qquad \eta = \frac{2\pi}{\ell}$$

is compact in the space $L^{p(t(\cdot))}([0,\ell])$, then in the same manner as for constant p in [40, p. 101], we can conclude that for such curves statements *i)* and *ii)* of Theorem 8 are valid under the condition

$$|\beta(t)| < 2\,\text{arcctg}\,\|S_{\Gamma_0}\|_{p(\cdot)}. \tag{43}$$

From the compactness of the operators with a weak singularity in the spaces $L^{p(\cdot)}$ (see Theorem C in [70]) it follows that when Γ is a Lyapunov curve, then T is a compact operator in $L^{p(\cdot)}(\Gamma)$.

Therefore, when $p \in \mathcal{P}(\Gamma)$, then in the case of Lyapunov boundary Γ all the statements of Theorem 8 remain valid under conditions (41) and (43) (in particular, when (42) and (43) are fulfilled).

II). Condition (42) bring fulfilled, condition (41) is also satisfied, when

$$\operatorname*{ess\,sup}_{t\in\Gamma} |\alpha(t)| < \frac{\pi \underline{p}}{4(1+\ell)e\overline{q}\|S_\Gamma\|_{p(\cdot)}}$$

which can be easily verified by means of Theorem 6.

Then following the arguments in the proof of part **I)**, we again obtain an analogous statement. □

Remark 14. - In the case $p(t) = p = const$ one has $\|S_{\Gamma_0}\|_p = \operatorname{ctg} \frac{\pi}{2\max(p,q)}$, (see [26], Section 13.3), so that Theorem 13 is a generalization to the spaces $L^{p(\cdot)}(\Gamma)$ of the well-known Simonenko results [119], [120].

4.6 On the boundedness of the singular operator in weighted $L^{p(\cdot)}(\Gamma, \rho)$-spaces

As is well known, when investigating the problem of linear conjugation in $K^p(\Gamma)$ by the method of factorization, the most important is the fact that the singular operator S_Γ is bounded in Lebesgue weighted spaces, see for instance, [43, pp. 113–114].

In Section 4.4 it was shown that basing on the boundedness of S_Γ in $L^{p(\cdot)}(\Gamma;\rho)$ when ρ is a power weight, we can solve the problem of linear conjugation in the class $K^{p(\cdot)}(\Gamma)$, if the coefficient G is piecewise continuous.

Another approach is known when solving the problem of linear conjugation with a measurable bounded coefficient G in the explicit form, in this or other way, not making use of the boundedness results, one is able to conclude that the singular operator is bounded in the Lebesgue space with weight generated by the coefficient G. This approach was developed by I. Simonenko [119] in the case of constant p.

Basing on the solution of BVP with continuous coefficient in Section 4.2, we proved the weighted inequality for the singular integral operator, see Theorem 9. Now we base ourselves on the solution of problem (1) with oscillating coefficient in the class $K^{p(\cdot)}(\Gamma)$ given in Section 4.5 avoiding weighted boundedness results, and deduce the boundedness statements for the operator S_Γ in the weighted spaces $L^{p(\cdot)}(\Gamma,\rho)$ with weights more general than the power ones.

Theorem 15. *Let Γ be a close Carleson curve and $p \in \mathcal{P}(\Gamma)$. Assume that*

$$\operatorname*{ess\,sup}_{t\in\Gamma} |\alpha(t)| < \min\left(2\operatorname{arcctg}\|S_{\Gamma_0}\|_{p[t(\cdot)]}, \frac{\pi}{\overline{q}}\right). \tag{44}$$

Then the function

$$\rho(t) = \left|\exp\left(\frac{1}{2\pi}\int_\Gamma \frac{\alpha(\tau)}{\tau - t}\,d\tau\right)\right|$$

belongs to $W^{p(\cdot)}(\Gamma)$.

Proof. Taking into account that the function $G(t) = \exp(i\alpha(t))$ under condition (44) satisfies conditions (41) and (43), we can state that for arbitrary $g \in L^{p(\cdot)}(\Gamma)$ the function

$$\Phi(z) = \frac{X(z)}{2\pi i} \int_{\Gamma} \frac{g(\tau)}{X^+(\tau)} \frac{d\tau}{\tau - z}$$

with

$$X(z) = \exp\left(\frac{1}{2\pi i} \int_{\Gamma} \frac{lnG(\tau)}{\tau - z} d\tau\right) = \exp\left(\frac{1}{2\pi} \int_{\Gamma} \frac{\alpha(\tau)}{\tau - z} d\tau\right)$$

belongs to $K^{p(\cdot)}(\Gamma)$. Consequently, the function Φ^+ and thus

$$Y(t) = \frac{X^+(t)}{2\pi i} \int_{\Gamma} \frac{g(\tau)\, d\tau}{X^+(\tau)(\tau - t)}$$

belong to $L^{p(\cdot)}(\Gamma)$.

Then by virtue of statement *i)* of Theorem 2 we conclude that $\rho \in W^{p(\cdot)}(\Gamma)$ (see also the proof of Theorem 9). □

Remark 16. According to Remark 7, the number $\operatorname*{ess\,sup}_{t\in\Gamma} |\alpha(t)|$ in (44) can be replaced by $\nu(\alpha) = \inf \|\alpha - \psi\|_C$ in the case of bounded $\alpha(t)$, where the infinum is taken over all rational functions ψ.

The following example given on the basis of Theorem 15 is of interest, in our opinion. Let τ_j $(j = 1, \dots, m)$ and t_k $(k = 1, 2, \dots)$ be distinct points on Γ, $\tau_j \neq t_k$ such that the accumulation points of the sequence $\{t_k\}$ form a finite set $\{t^{(1)}, \dots, t^{(i)}\}$ and no point τ_j is in this set. For simplicity we assume that $i = 1$ and $t_k = t(s_k)$ where $\{s_k\}$ is monotone sequence. Then the function

$$\rho(t) = \prod_{j=1}^{m} |t - \tau_j|^{\alpha_j} \prod_{k=1}^{\infty} |t - t_k|^{\beta_k} \ =: \rho_1(t)\rho_2(t)$$

belongs to $W^{p(\cdot)}(\Gamma)$ under the conditions

1) $-\frac{1}{p(\tau_j)} < \alpha_j < \frac{1}{q(\tau_j)}$, $j = 1, 2, \dots, m$, $\quad -\frac{1}{p(t_k)} < \beta_k < \frac{1}{q(t_k)}$, $k = 1, 2, \dots$, and
2) there exists an N such that $|\beta_k| < \min\left(2 \operatorname{arcctg} \|S_{\Gamma_0}\|_{p(\cdot)}, \frac{\pi}{q}\right)$ for $k \geq N$.

Indeed, let $G(t) = \exp(i\alpha(t))$ where $\alpha(t)$ is the piecewise constant function equal to γ_k on arc $t_k t_{k+1}$, $\gamma_{k+1} - \gamma_k = \beta_k$, $k = N, N+1, \dots$. Then $\alpha(t)$ satisfies the condition (44) and by Theorem 15 we conclude that the function

$$\left| \exp\left\{\frac{1}{2\pi} \int_{\Gamma} \frac{\alpha(\tau)}{\tau - t} d\tau\right\} \right| \sim \prod_{k=N}^{\infty} |t - t_k|^{\beta_k}$$

belongs to $W^{p(\cdot)}(\Gamma)$. As regards the function

$$\rho_1(t)\prod_{k=1}^{N-1}|t-t_k|^{\beta_k}=\prod_{j=1}^{m}|t-\tau_j|^{\alpha_j}\prod_{k=1}^{N-1}|t-t_k|^{\beta_k},$$

it belongs to $W^{p(\cdot)}(\Gamma)$ so for us the points τ_j being isolated from the points t_k by assumption. Therefore $\rho=\rho_1\rho_2$ belongs to $W^{p(\cdot)}(\Gamma)$.

4.7 Reduction of the Riemann problem in weighted classes $K^{p(\cdot)}(\Gamma;\rho)$ to the Riemann problem in $K^{p(\cdot)}(\Gamma)$. Solution of the Riemann problem in weighted classes

Let Γ be a simple rectifiable closed curve bounding the domains $D^{\pm}$ and $\rho\in W^{p(\cdot)}(\Gamma)$. We consider the problem: find functions $\Phi\in K^{p(\cdot)}(\Gamma,\rho)$ whose boundary conditions satisfy almost everywhere on Γ the condition (1), where $g\in L^{p(\cdot)}(\Gamma,\rho)$ and G is a bounded measurable function on Γ such that $\operatorname*{ess\,inf}_{t\in\Gamma}|G(t)|>0$.

We show that under certain assumptions on the curve Γ, the exponent $p(t)$ and the weight function $\rho(t)$ in this general setting is reduced to the problem of conjugation without weight. We consider weight functions of the form

$$\rho(t)=\exp\left(\frac{i}{2}S_\Gamma\mu\right)\in W^{p(\cdot)}(\Gamma) \tag{45}$$

where μ is a real valued bounded function on Γ, and assume that

$$\Gamma\in\mathcal{R}^{p(\cdot)}. \tag{46}$$

We need the following auxiliary statements well known for constant p and easily proved for the variable exponent $p(\cdot)$ under assumptions (45)-(46).

Lemma 17. *If $\rho\in W^{p(\cdot)}(\Gamma)$, then $\frac{1}{\rho}\in W^{q(\cdot)}(\Gamma)$ and for all $\varphi\in L^{p(\cdot)}(\Gamma,\rho)$ and $\psi\in L^{q(\cdot)}(\Gamma,1/\rho)$ the equality*

$$\int_\Gamma\varphi(t)\,(S_\Gamma\psi)(t)\,dt=-\int_\Gamma\psi(t)\,(S_\Gamma\varphi)(t)\,dt \tag{47}$$

holds.

Proof. We observe that equality (47) is well known, for instance on, rational functions. Let Q be the set of rational functions on Γ. For functions $\psi\in Q$ according to equality

$$\|f\|_{L^{p(\cdot)}(\Gamma,\rho)}\sim\sup_{\|g\|_{L^{q(\cdot)}(\Gamma,\frac{1}{\rho})}\le1}\left|\int_\Gamma f(t)g(t)dt\right|$$

we have

$$\|S_\Gamma\psi\|_{L^{q(\cdot)}(\Gamma,1/\rho)} \sim \sup_{\|\varphi\|_{L^{p(\cdot)}(\Gamma,\rho)}\le 1} \left|\int_\Gamma \varphi(t)(S_\Gamma\psi)(t)\,dt\right|$$

$$= \sup_{\substack{\varphi\in Q\\ \|\varphi\|_{L^{p(\cdot)}(\Gamma,\rho)}\le 1}} \left|\int_\Gamma \varphi(t)(S_\Gamma\psi)(t)\,dt\right|$$

$$= \sup_{\substack{\varphi\in Q\\ \|\varphi\|_{L^{p(\cdot)}(\Gamma,\rho)}\le 1}} \left|\int_\Gamma \psi(t)(S_\Gamma\varphi)(t)\,dt\right| \le c\|\psi\|_{L^{q(\cdot)}(\Gamma,1/\rho)}\|S_\Gamma\|_{L^{p(\cdot)}(\Gamma,\rho)}.$$

The obtained estimate $\|S_\Gamma\psi\|_{L^{q(\cdot)}(\Gamma,1/\rho)} \le C\|\psi\|_{L^{q(\cdot)}(\Gamma,1/\rho)}$, $C = c\|S_\Gamma\|_{L^{p(\cdot)}(\Gamma,\rho)}$ is extended to all $\psi \in L^{q(\cdot)}(\Gamma, 1/\rho)$ by denseness of Q in weighted $L^{p(\cdot)}(\Gamma,\rho)$-spaces, (see Chapter 3 Proposition 4). Therefore, $\frac{1}{\rho} \in W^{q(\cdot)}(\Gamma)$. The validity of equality (47) on the whole range $L^{p(\cdot)}(\Gamma,\rho)\times L^{q(\cdot)}(\Gamma,1/\rho)$ follows in the same way since both the left-hand side and the right-hand side of (47) are bounded bilinear functionals in $L^{p(\cdot)}(\Gamma,\rho)\times L^{q(\cdot)}(\Gamma,1/\rho)$. □

Lemma 18. *If $\Phi \in \widetilde{K}_0^{p(\cdot)}(\Gamma,\rho)$ and $\Psi \in \widetilde{K}_0^{q(\cdot)}(\Gamma,1/\rho)$, then $\Phi\Psi \in \widetilde{K}_0^1(\Gamma)$.*

Lemma 17 having been proved, the proof of this lemma is obtained in the same way as in the case of constant p, (see [43, p. 98–99]).

Now we get back to problem (1). The following statement is valid.

Theorem 19. *Let $\Gamma \in \mathcal{R}^{p(\cdot)}$ and assume that condition (45) is fulfilled with bounded real function μ. If $\Phi(z)$ is a solution of problem (1) in the class $K^{p(\cdot)}(\Gamma,\rho)$ and*

$$Y(z) = \exp\left[-i\left(K_\Gamma\mu\right)(z)\right],$$

then the function

$$\Psi(z) = \frac{\Phi(z)}{Y(z)}$$

is a solution of the problem

$$\Psi^+(t) = G(t)e^{i\mu(t)}\Psi^-(t) + g_1(t) \tag{48}$$

in the class $K^{p(\cdot)}(\Gamma)$, where $g_1(t) = \frac{g(t)}{Y^+(t)}$.

Conversely, if $\Psi(z)$ is a solution of problem (48) in the class $K^{p(\cdot)}(\Gamma)$, then the function $\Phi(z) = \Psi(z)Y(z)$ is a solution of problem (1) in the class $K^{p(\cdot)}(\Gamma,\rho)$.

Proof. We follow the papers [53]. Since $\Gamma \in \mathcal{R}^{p(\cdot)}$ and μ is bounded, by Theorem 6 there exists a number $\delta > 0$ such that $1/Y \in E^{\delta}(D^+)$ and $(z-z_0)^k(1/Y-1) \in E^{\delta}(D^-)$ for some nonnegative integer k. Moreover, the functions

$$\left(\frac{1}{Y}\right)^{\pm} = \exp\left(\pm\frac{i\mu}{2} + \frac{i}{2}S_\Gamma\mu\right) = \rho\exp\left(\pm\frac{i\mu}{2}\right)$$

belongs to $L^{p(\cdot)}(\Gamma)$. Therefore, $\left(\frac{1}{Y}\right)^{\pm} \in L^{\underline{p}}(\Gamma)$. Since Γ is a Smirnov curve (see Corollary 3), we can take $\delta = \underline{p}$ and $k = 0$. But then $(1/Y-1) \in E^1(D^{\pm})$ and hence $(1/Y-1) \in K^1(\Gamma)$. In addition, $\left(\frac{1}{Y}\right)^{\pm} \in L^{q(\cdot)}(\Gamma, 1/\rho)$ and consequently, $1/Y \in K^{q(\cdot)}(\Gamma, 1/\rho)$. By virtue of Lemma 18, from the equality $\Psi = \Phi \cdot \frac{1}{Y}$ we conclude that $\Psi \in \widetilde{K}^1(\Gamma)$. But $\Psi(\infty) = 0$ so that $\Psi \in K^1(\Gamma)$. As far as

$$\Psi^{\pm}(t) = \frac{\Phi^{\pm}(t)}{Y^{\pm}(t)} = \frac{\Phi^{\pm}(t)}{\rho(t)}e^{\pm i\frac{\mu(t)}{2}},$$

we have $(\Psi^+ - \Psi^-) \in L^{p(\cdot)}(\Gamma)$. Then from the equality $\Psi = K_\Gamma(\Psi^+ - \Psi^-)$ the inclusion $\Psi \in K^{p(\cdot)}(\Gamma)$ follows.

The inverse statement can be analogously proved.

We have taken advantage of the fact that if $\Psi \in K^{p(\cdot)}(\Gamma)$, then $(\Psi^+ - \Psi^-) \in L^{p(\cdot)}(\Gamma)$ and $\Psi(z) = K_\Gamma(\Psi^+ - \Psi^-)(z)$. Indeed, since $\Psi \in K^{p(\cdot)}(\Gamma)$, we have $\Psi(z) = (K_\Gamma\psi)(z)$ with $\psi \in L^{p(\cdot)}(\Gamma)$. This implies that

$$\Psi^{\pm}(t) = \pm\frac{1}{2}\psi(t) + \frac{1}{2}(S_\Gamma\psi)(t)$$

and hence $\Psi^+ - \Psi^- = \psi$. □

Relying on the results of Sections 4.2, 4.4 and 4.5, we may use Theorem 19 to get a picture of solvability of the problem of linear conjugation in the weighted class $K^{p(\cdot)}(\Gamma, \rho)$ under certain assumptions on the coefficient G.

Theorem 20. *Let $\Gamma \in \mathcal{R}^{p(\cdot)}$ and let*

$$\rho \in W^{p(\cdot)}(\Gamma) \qquad \text{and has the form} \qquad \rho = \exp\left(\frac{i}{2}S_\Gamma\mu\right) \tag{49}$$

where μ is a bounded real measurable function. Suppose that for the function

$$G_\mu(t) = G(t)\exp(i\mu(t))$$

the appropriate conditions of one of Theorems 8, 12, 13 with $G(t)$ replaced by $G_\mu(t)$ are fulfilled and put $\varkappa = \operatorname{ind} G_\mu(t)$. Then for problem (1) in the weighted class $K^{p(\cdot)}(\Gamma, \rho)$ all the statements of Theorem 8 are valid.

The following remarks gives us a good reason to consider the assumptions $\rho \in W^{p(\cdot)}(\Gamma)$ and $\rho = \exp\left(\frac{i}{2}S_\Gamma\mu\right)$ of Theorem 20 as natural and rather general.

Remark 21. The latter condition in (49) is fulfilled automatically in the case when $p(t) = const$ and Γ is a Lyapunov curve: in this case all the weight functions of the class $W^p(\Gamma)$ have the form $\rho = \exp\left(\frac{i}{2}S_\Gamma\mu\right)$ with a real-valued bounded function μ (see [103], [40, p. 64]).

The first condition (49) is a necessary condition for the elementary BVP

$$\Phi^+(t) + \Phi^-(t) = g(t)$$

to have a solution in $K^{p(\cdot)}(\Gamma, \rho)$ for any $g \in L^{p(\cdot)}(\Gamma, \rho)$. If $\Phi = K_\Gamma\varphi$, $\varphi \in L^{p(\cdot)}(\Gamma; \rho)$, from the boundary condition it follows that $S_\Gamma\varphi = g$. Together with this, by the conditions $\underline{p} > 1$ and $\varphi \in L^{p(\cdot)}(\Gamma, \rho)$ we have $\varphi = \frac{\varphi_1}{\rho}$ with $\varphi_1 \in L^{p(\cdot)}(\Gamma)$. Consequently, by the statement *iv)* of Theorem 2 we derive that $K_\Gamma\left(\frac{\varphi_1}{\rho}\right) \in E^1(D^+)$. Therefore, $S_\Gamma^2\left(\frac{\varphi_1}{\rho}\right) = \frac{\varphi_1}{\rho}$ (see for instance [43, p. 103]), that is, $S_\Gamma g = \frac{\varphi_1}{\rho}$ so that $\rho S_\Gamma g \in L^{p(\cdot)}(\Gamma)$ and then $S_\Gamma g \in L^{p(\cdot)}(\Gamma, \rho)$ for any $g \in L^{p(\cdot)}(\Gamma, \rho)$. Thus, the equation $S_\Gamma\varphi = g$ being uniquely solvable in $L^{p(\cdot)}(\Gamma, \rho)$ for any $g \in L^{p(\cdot)}(\Gamma, \rho)$. Therefore the operator S_Γ maps the space $L^{p(\cdot)}(\Gamma, \rho)$ onto itself. Applying the statement *i)* of Theorem 2, we conclude that the operator S_Γ is bounded in $L^{p(\cdot)}(\Gamma, \rho)$, that is, $\rho \in W^{p(\cdot)}(\Gamma)$.

Remark 22. It should be noted that it is impossible to derive Theorem 9 from Theorem 20. By this reason it was separately presented.

4.8 On singular integral equations in $L^{p(\cdot)}(\Gamma, \rho)$

We apply now the above results to the singular integral equation

$$a(t)\varphi(t) + b(t)(S_\Gamma\varphi)(t) + V\varphi = f(t) \tag{50}$$

in the space $L^{p(\cdot)}(\Gamma, \rho)$, where V is any operator compact in the space $L^{p(\cdot)}(\Gamma, \rho)$.

We assume that the assumptions of Section 4.7 are fulfilled, and the coefficients $a(t)$ and $b(t)$ are bounded measurable functions on Γ.

Theorem 23. *Let $\Gamma \in \mathcal{R}^{p(\cdot)}$, the weight function ρ satisfy the assumptions in* (45) *and* $\operatorname*{ess\,inf}_{t\in\Gamma} |a(t) + b(t)| > 0$. *Assume that for the function*

$$G_\mu(t) = \frac{a(t) - b(t)}{a(t) + b(t)} e^{i\mu(t)}$$

the conditions of Theorem 13 *with G replaced by G_μ are fulfilled. Then for the equation* (50) *the Noether theorems are valid and its index in the space $L^{p(\cdot)}(\Gamma, \rho)$ is equal to $\varkappa = \varkappa(G_\mu)$, where $\varkappa$ is interpreted in accordance with* (35). *In the case $V = 0$ the solutions of equation* (50) *in the space $L^{p(\cdot)}(\Gamma, \rho)$ are given by the formula $\varphi = \Phi^+ - \Phi^-$, where $\Phi(z)$ is the solution of the following BVP*

$$\Phi^+(t) = G_\mu(t)\Phi^-(t) + g_1(t), \quad g_1(t) = \frac{f(t)}{a(t) + b(t)} \exp\left\{\frac{i}{2}\mu(t) - \frac{i}{2}(S_\Gamma\mu)(t)\right\}. \tag{51}$$

Proof. By Atkinson Theorem (see [4] and 0.22) it suffices to give the proof for the case $V = 0$. In this case equation (50) in the space $L^{p(\cdot)}(\Gamma, \rho)$ is equivalent to a boundary value problem of type (1) in the class $K^{p(\cdot)}(\Gamma, \rho)$, which is established in the usual way via the Cauchy integral

$$\Phi(z) = \frac{1}{2\pi i} \int_{\Gamma} \frac{\varphi(\tau)\, d\tau}{\tau - z} \in K^{p(\cdot)}(\Gamma, \rho) \tag{52}$$

so that equation (50) may be rewritten in the form

$$\Phi^+(t) = G(t)\Phi^-(t) + g(t) \tag{53}$$

with $G(t) = \frac{a(t)-b(t)}{a(t)+b(t)}$ and $g(t) = \frac{f(t)}{a(t)+b(t)}$ under the assumption that $\operatorname*{ess\,inf}_{t\in\Gamma} |a(t) + b(t)| \neq 0$. Thus any solution $\varphi \in L^{p(\cdot)}(\Gamma, \rho)$ of equation (50) generates a solution of (53) in $K^{p(\cdot)}(\Gamma, \rho)$ of form (52). Conversely, if $\Phi(z)$ is a solution of problem (53) in the class $K^{p(\cdot)}(\Gamma, \rho)$, then the function $\varphi(t) = \Phi^+(t) - \Phi^-(t)$ is a solution of equation (50).

Let

$$Y(z) = \exp\left\{-i \left(K_\Gamma \mu\right)(z)\right\}.$$

By Theorem 19 the function $\Psi(z) = \frac{\Phi(z)}{Y(z)} = \Phi(z) \exp\left\{i \left(K_\Gamma \mu\right)(z)\right\}$ is a solution in $K^{p(\cdot)}(\Gamma)$ of the BVP

$$\Psi^+(t) = G(t) \exp\{i\mu(t)\}\Psi^-(t) + g_1(t) \tag{54}$$

where $g_1(t) = \frac{g(t)}{Y^+(t)} = \frac{f(t)}{[a(t)+b(t)]Y^+(t)}$. Via $\Psi(z) = (K_\Gamma \psi)(z)$ this is equivalent in $L^{p(\cdot)}(\Gamma)$ to the equation

$$a_1(t)\psi(t) + b_1(t)(S_\Gamma \psi)(t) = g_1(t), \tag{55}$$

where

$$a_1(t) = \frac{1}{2}\left[a(t)(1 - e^{i\mu(t)}) + b(t)(1 + e^{i\mu(t)})\right],$$
$$b_1(t) = \frac{1}{2}\left[a(t)(1 + e^{i\mu(t)}) + b(t)(1 - e^{i\mu(t)})\right].$$

Thus any solution $\Psi \in K^{p(\cdot)}(\Gamma)$ of problem (54) generates the solution ψ of equation (55) via the equality

$$\psi = \Psi^+ - \Psi^- = \frac{\Phi^+}{Y^+} - \frac{\Phi^-}{Y^-} = \frac{\rho}{2}\left[\left(e^{i\frac{\mu}{2}} + e^{-i\frac{\mu}{2}}\right)\varphi + \left(e^{i\frac{\mu}{2}} + e^{-i\frac{\mu}{2}}\right) S_\Gamma \varphi\right]$$

where $\rho = \exp\left(\frac{i}{2} S_\Gamma \mu\right)$. Conversely, if $\operatorname*{ess\,inf}_{t\in\Gamma} |a_1(t) + b_1(t)| = \operatorname*{ess\,inf}_{t\in\Gamma} |a(t) + b(t)| > 0$, then for the solution $\psi \in L^{p(\cdot)}(\Gamma)$ of equation (55) the function

$$\varphi = \Phi^+ - \Phi^- = \Psi^+ Y^+ - \Psi^- Y^- = \frac{1}{2\rho}\left[\left(e^{i\frac{\mu}{2}} + e^{-i\frac{\mu}{2}}\right)\psi + \left(e^{i\frac{\mu}{2}} - e^{-i\frac{\mu}{2}}\right) S_\Gamma \psi\right]$$

is a solution of equation (50) in the space $L^{p(\cdot)}(\Gamma, \rho)$. □

Part II

The Haseman problem

In the Sections 4.9–4.16 we consider the case when in problem (2) coefficients $a(t)$ is continuous function, $a(t) \neq 0$ and in Section 4.17 we conclude case when $a(t)$ is piecewise continuous function, $\inf |a(t)| > 0$.

4.9 The curves with the chord–arc condition

We say that simple rectifiable a curve Γ is with the chord–arc condition or the Lavrentyev curve (see, e.g., [107, p. 163]), if there exists a constant $m > 0$ such that for any $t,\ \tau \in \Gamma$ we have

$$|t-\tau| \geq m\, s(t,\tau), \tag{56}$$

where $s(t,\tau)$ is the length of the least of the two arcs of Γ connecting the points t and τ. The set of such curves is denoted by Λ.

Easy proved that curve from Λ is regular.

Let $\alpha = \alpha(t)$ be a homeomorphism of Γ onto itself, and let at every point t there exist the derivative $\alpha'(t)$ satisfying $\alpha'(t) \neq 0$ and α' belongs to Hölder class $H(\Gamma)$, i.e., there exist constants M and μ, $0 < \mu \leq 1$, such that for any $t_1,\ t_2 \in \Gamma$ we have

$$\big|\alpha'(t_1) - \alpha'(t_2)\big| < M|t_1 - t_2|^{\mu}.$$

Lemma 24. *If $\Gamma \in \Lambda$ and*

$$K(\tau,t) = \frac{\alpha'(t)}{\alpha(\tau)-\alpha(t)} - \frac{1}{\tau - t}, \quad \tau,\ t \in \Gamma,$$

then there exists a constant c such that $|K(\tau,t)| < c[s(\tau,t)]^{\mu-1}$.

Proof. Let Γ be a simple rectifiable curve, $z = z(s)$, $0 \leq s \leq \ell$, be its equation with respect to the arc abscissa, $\tau = z(\sigma)$, $t = t(s)$ and $\gamma(\tau,t)$ be the least arc of Γ with the ends τ and t. Then

$$\begin{aligned} K(\tau,t) &= \frac{\alpha'(\tau)(\tau-t) - [\alpha(\tau)-\alpha(t)]}{(\tau-t)(\alpha(\tau)-\alpha(t))} = \frac{\alpha'(\tau) \int\limits_{\gamma(\tau,t)} du - \int\limits_{\gamma(\tau,t)} \alpha'(u)\, du}{(\tau-t)(\alpha(\tau)-\alpha(t))} \\ &= \frac{\int\limits_{\gamma(\tau,t)} [\alpha'(\tau)-\alpha'(u)]\, du}{(\tau-t)(\alpha(\tau)-\alpha(t))} = \frac{\varphi(\tau,t)}{(\tau-t)(\alpha(\tau)-\alpha(t))}. \end{aligned} \tag{57}$$

Next,

$$|\varphi(\tau,t)| = \bigg| \int\limits_{\gamma(\tau,t)} [\alpha'(z(\sigma)) - \alpha'(z(\zeta))] z'(\zeta)\, d\zeta \bigg|$$

$$\leq M \int\limits_{\gamma(\tau,t)} |z(\sigma) - z(\zeta)|^\mu \, d\zeta \leq M[s(\tau,t)]^{1+\mu}.$$

Now, by virtue of (56), from (57) we obtain

$$|K(\tau,t)| \leq \frac{M[s(\tau,t)]^{1+\mu}}{[ms(\tau,t)]^2 |\alpha(\tau) - \alpha(t)| \, |\tau - t|^{-1}} = \frac{M}{m^2} \frac{[s(\tau,t)]^{\mu-1}}{\psi(\tau,t)},$$

where $\psi(\tau,t) = |\alpha(\tau) - \alpha(t)| \, |\tau - t|^{-1}$.

Show that $\psi(\tau,t)$ tends uniformly to $\alpha'(t)$ as $\tau \to t$. This follows from the inequality

$$\left|\frac{\alpha(\tau) - \alpha(t)}{\tau - t} - \alpha'(t)\right| = \left|\frac{\int\limits_{\gamma(\tau,t)} [\alpha'(u) - \alpha'(t)] \, du}{\tau - t}\right| \leq \frac{M}{m} [s(\tau,t)]^\mu.$$

From the above-said and the assumption $\alpha'(t) \neq 0$ (i.e. $\min|\alpha'(t)| = k > 0$), we obtain

$$|K(\tau,t)| \leq \frac{2M}{km^2} [s(\tau,t)]^{\mu-1} = c[s(\tau,t)]^{\mu-1}, \quad c = \frac{2M}{km^2}. \qquad \square$$

4.10 The classes of functions $\widetilde{K}_\alpha^{p(\cdot)}(\Gamma,\rho)$ and $K_\alpha^{p(\cdot)}(\Gamma,\rho)$. The Statement of the Haseman Problem

Let $p_\alpha(t) = p(\alpha(t))$ and $\widetilde{p}_\alpha(t) = \max(p(t), p_\alpha(t))$.

Lemma 25. *If $p \in \mathcal{P}(\Gamma)$, $\alpha' \in H(\Gamma)$, $\alpha' \neq 0$, then the functions $p_\alpha(t)$ and $\widetilde{p}_\alpha(t)$ also belong to $\mathcal{P}(\Gamma)$.*

Proof. Let us first show that $p_\alpha \in \mathcal{P}(\Gamma)$. We have

$$|p(\alpha(t_1)) - p(\alpha(t_2))|$$
$$< \frac{A}{|\ln|\alpha(t_1) - \alpha(t_2)||} = \frac{A}{|\ln|\frac{\alpha(t_1)-\alpha(t_2)}{t_1-t_2}| + \ln|t_1 - t_2||}.$$

Since $0 < m_1 \leq |\frac{\alpha(t_1)-\alpha(t_2)}{t_1-t_2}| \leq m_2$, we have $\sup\limits_{t_1,t_2\in\Gamma} |\ln|\frac{\alpha(t_1)-\alpha(t_2)}{t_1-t_2}|| = M < \infty$, and hence for $|t_1 - t_2| < e^{-2M} = \delta$ we have

$$|p_\alpha(t_1) - p_\alpha(t_2)| = |p(\alpha(t_1)) - p(\alpha(t_2))| < 2A|\ln(t_1 - t_2)|^{-1}. \tag{58}$$

Moreover, $\min p_\alpha(t) = \min p(t) = \underline{p} > 1$, and thus $p_\alpha \in \mathcal{P}(\Gamma)$.

Consider now the function $\widetilde{p}_\alpha(t)$. Let t_1 be an arbitrary point on Γ and $\widetilde{p}_\alpha(t_1) = p(t_1) \neq p_\alpha(t_1)$. As far as the functions p and p_α are continuous, there exists $\delta_1 =$

$\delta_1(t_1)$ such that for $|t_1 - t_2| < \delta_1$ we have $\widetilde{p}_\alpha(t_2) = p(t_2)$. Then $\big|\widetilde{p}_\alpha(t_1) - \widetilde{p}_\alpha(t_2)\big| = |p(t_1) - p(t_2)|$, and the inequality (58) is fulfilled since $p \in \mathcal{P}(\Gamma)$. Analogously, if $\widetilde{p}_\alpha(t_1) = p_\alpha(t_1) \neq p(t_1)$, then there exists $\delta_2 = \delta_2(t_1)$ such that for $|t_1 - t_2| < \delta_2$ we have $\big|\widetilde{p}_\alpha(t_1) - \widetilde{p}_\alpha(t_2)\big| = |p_\alpha(t_1) - p_\alpha(t_2)|$, and the inequality (58) is fulfilled since $p_\alpha \in \mathcal{P}(\Gamma)$. When $\widetilde{p}_\alpha(t_1) = p(t_1) = p_\alpha(t_1)$, then if $\widetilde{p}_\alpha(t_2) = p(t_2)$ we write $\big|\widetilde{p}_\alpha(t_1) - \widetilde{p}_\alpha(t_2)\big| = |p(t_1) - p(t_2)|$, while if $\widetilde{p}_\alpha(t_2) = p_\alpha(t_2)$ we have $\big|\widetilde{p}_\alpha(t_1) - \widetilde{p}_\alpha(t_2)\big| = |p_\alpha(t_1) - p_\alpha(t_2)|$. In both cases the inequality (58) (for any t_2) is valid.

Thus for every point $t_1 \in \Gamma$ there is a neighborhood $u(t_1;\delta)$ $(\delta = \min(m\delta_1, m\delta_2))$ in which the inequality (58) is true. From the set of these intervals we select a finite covering in such a way that none of the intervals of that covering contains another one. Consider the set of all ends of those intervals. They divide Γ into a finite number of intervals. Let δ_0 be the length of the least interval. If $t_1, t_2 \in \Gamma$ and $|t_1 - t_2| < \delta < \frac{\delta_0}{4}$, then the points t_1 and t_2 get into some of the intervals of the initial covering, and hence the inequality (58) is fulfilled. Thus $\widetilde{p}_\alpha \in \mathcal{P}(\Gamma)$. □

We dwell especially on the case $\rho(t) = 1$. Then $L^{p_\alpha(\cdot)}(\Gamma;\rho_\alpha) = L^{p_\alpha(\cdot)}(\Gamma)$. By virtue of the inequalities $|f|^{\widetilde{p}_\alpha(t)} < |f|^{p(t)} + |f|^{p_\alpha(t)}$ and $|f|^{p(t)} < 1 + |f|^{\widetilde{p}_\alpha(t)}$, $|f|^{p_\alpha(t)} < 1 + |f|^{\widetilde{p}_\alpha(t)}$ we conclude that

$$L_{p(\cdot),\alpha}(\Gamma) = L^{p(\cdot)}(\Gamma) \cap L^{p_\alpha(\cdot)}(\Gamma) = L^{\widetilde{p}_\alpha(\cdot)}(\Gamma). \tag{59}$$

Consequently, the set $L_{p(\cdot),\alpha}(\Gamma)$ can be considered as the space $L^{\widetilde{p}_\alpha(\cdot)}(\Gamma)$.

Let $t_k \in \Gamma$ and

$$\rho(t) = \prod_{k=1}^{N} |t - t_k|^{\nu_k}, \nu_k \in \mathbb{R}, -\frac{1}{p(t_k)} < \nu_k < \frac{1}{p'(t_k)}, k = \overline{1,N}, p'(t) = \frac{p(t)}{p(t)-1}. \tag{60}$$

We denote by $\widetilde{K}^{p(\cdot)}(\Gamma;\rho)$ the set of the functions $\phi(z)$ analytic in the plane cut along Γ and representable in the form $\phi(z) = K_\Gamma(f) + q(z)$, $z \notin \Gamma$, $f \in L^{p(\cdot)}(\Gamma;\rho)$, where $q(z)$ is a polynomial. The subset of those functions from $\widetilde{K}^{p(\cdot)}(\Gamma;\rho)$ for which $q(z) = 0$ coincide with $K^{p(\cdot)}(\Gamma;\rho)$. The subset of those functions from $\widetilde{K}^{p(\cdot)}(\Gamma,\rho)$ for which $q = \text{const}$ we denote by $\widetilde{K}_0^{p(\cdot)}(\Gamma,\rho)$.

Assume $\widetilde{K}^{p(\cdot)}(\Gamma) = \widetilde{K}^{p(\cdot)}(\Gamma;1)$, $K^{p(\cdot)}(\Gamma) = K^{p(\cdot)}(\Gamma;1)$.

If Γ is a Carleson curve (in particular, if $\Gamma \in \Lambda$), $p \in \mathcal{P}(\Gamma)$, and for the weight ρ given by equality (60), then the function $\phi(z) = (K_\Gamma f)(z) \in K^{p(\cdot)}(\Gamma;\rho)$ for almost all $t \in \Gamma$ has angular boundary values $\phi^{\pm}(t) = \pm\frac{1}{2} f(t) + \frac{1}{2}(S_\Gamma f)(t)$, $t \in \Gamma$. Note that the functions $\phi^+(t)$ and $\phi^-(t)$ belong to $L^{p(\cdot)}(\Gamma;\rho)$ since the singular Cauchy operator is continuous in $L^{p(\cdot)}(\Gamma;\rho)$.

Here we present one more property of the solutions of problem (2).

Let $\Gamma \in \Lambda$ and $p \in \mathcal{P}(\Gamma)$. It follows from (60) that $\rho^{-1}(t) \in L^{p'(\cdot)+\varepsilon}(\Gamma)$ for some $\varepsilon > 0$. Moreover, for $f \in L^{p(\cdot)}(\Gamma;\rho)$, we have $f(t) = \varphi(t)\rho^{-1}(t)$, $\varphi \in L^{p(\cdot)}(\Gamma)$. To the above product we apply the generalized Hölder's inequality (see (10) from Chapter 2).

Taking as $p(t)$ the function $p(t)/(1+\eta)$, $\eta > 0$, we obtain

$$\int_\Gamma |f|^{1+\eta}\, d\zeta = \int_\Gamma |\varphi|^{1+\eta}\Big|\frac{1}{\rho}\Big|^{1+\eta} d\zeta \le c_0\|\varphi\|_{p(\cdot)}\Big\|\frac{1}{\rho}\Big\|_{p'(\cdot)+\widetilde{\eta}(\cdot)}, \quad \widetilde{\eta}(t) = \frac{\eta p(t)}{p(t)-1-\eta},$$

whence it follows that if $\max\widetilde{\eta}(t) < \varepsilon$, then $f \in L^{1+\eta}(\Gamma)$. For that it suffices to take $\eta \in \big(0; \varepsilon \min \frac{1}{p'(t)}\big)$. For such η, we have $K_\Gamma f \in E^{1+\eta}(D^\pm)$ [40, p. 29].

A solution of the problem (2) is sought in classes of functions representable by the Cauchy type integral with density integrable with a variable exponent. If $\phi \in K^{p(\cdot)}(\Gamma;\rho)$ satisfies the condition (2), then $\phi^+(t) = a(\beta(t))\phi^-(\beta(t)) + b(\beta(t))$, where $\beta = \beta(t)$ is the function inverse to $\alpha(t)$. Since $\phi^+ \in L^{p(\cdot)}(\Gamma;\rho)$, we must have $\big[a(\beta(t))\phi^-(\beta(t)) + b(\beta(t))\big] \in L^{p(\cdot)}(\Gamma;\rho)$. Therefore besides the assumptions $\phi \in K^{p(\cdot)}(\Gamma;\rho)$, $b \in L^{p(\cdot)}(\Gamma;\rho)$, it is natural to require the fulfilment of the conditions

$$\phi^-(\beta(t)) \in L^{p(\cdot)}(\Gamma;\rho), \quad b(\beta(t)) \in L^{p(\cdot)}(\Gamma;\rho).$$

This is equivalent to the requirement that

$$\phi^- \in L^{p(\alpha(t))}(\Gamma;\rho(\alpha(t))), \quad b \in L^{p(\alpha(t))}(\Gamma;\rho(\alpha(t))).$$

Introduce the notation

$$\rho_\alpha(t) = \rho(\alpha(t)), \quad L_{p(\cdot),\alpha}(\Gamma;\rho) = L^{p(\cdot)}(\Gamma;\rho) \cap L^{p_\alpha(\cdot)}(\Gamma;\rho_\alpha).$$

Since $p_\alpha \in \mathcal{P}(\Gamma)$, for the operator S_Γ to act continuously in $L^{p_\alpha(\cdot)}(\Gamma;\rho_\alpha)$, where

$$\rho_\alpha(t) = \prod_{k=1}^N |\alpha(t) - t_k|^{\nu_k} = \prod_{k=1}^N \big|\alpha(t) - \alpha(\beta(t_k))\big|^{\nu_k} \sim \prod_{k=1}^N |t - \beta(t_k)|^{\nu_k},$$

it is necessary and sufficient that the conditions

$$-\frac{1}{p(\alpha(\beta(t_k)))} < \nu_k < \frac{1}{p'(\alpha(\beta(t_k)))}, \quad k = 1,\dots,N, \tag{60_0}$$

be fulfilled. In view of the fact that $\alpha(\beta(t_k)) = t_k$, these conditions coincide with the conditions (60).

Thus if $p \in \mathcal{P}(\Gamma)$, $\alpha' \in H(\Gamma)$, $\alpha' \neq 0$ *and the conditions* (60) *are fulfilled, then the operator* S_Γ *is continuous in* $L^{p_\alpha(\cdot)}(\Gamma;\rho_\alpha)$.

Along with the classes $\widetilde{K}^{p(\cdot)}(\Gamma;\rho)$ and $K^{p(\cdot)}(\Gamma;\rho)$ we introduce the following classes of analytic functions

$$\widetilde{K}^{p(\cdot)}_\alpha(\Gamma;\rho) = \Big\{\phi:\ \phi \in \widetilde{K}^{p(\cdot)}(\Gamma;\rho),\ \phi^- \in L^{p_\alpha(\cdot)}(\Gamma;\rho_\alpha)\Big\},$$
$$p_\alpha(t) = p(\alpha(t)), \quad \rho_\alpha(t) = \rho(\alpha(t)),$$

$$K_\alpha^{p(\cdot)}(\Gamma;\rho)=\Big\{\phi:\ \phi\in K^{p(\cdot)}(\Gamma;\rho),\ \phi^-\in L^{p_\alpha(\cdot)}(\Gamma;\rho_\alpha)\Big\}.$$

If $\rho=1$, then according to (59) we have

$$K_\alpha^{p(\cdot)}(\Gamma;1)=K_\alpha^{p(\cdot)}(\Gamma)=\Big\{\phi:\ \phi\in K^{p(\cdot)}(\Gamma),\ \phi^-\in L^{\widetilde{p}_\alpha(\cdot)}(\Gamma)\Big\}.$$

We consider Haseman problem under the following assumptions.

Let $\Gamma\in\Lambda$, $p\in\mathcal{P}(\Gamma)$ and $\alpha=\alpha(t)$ be a preserving orientation homeomorphism of Γ onto itself, $\alpha'(t)\neq 0$, $\alpha'\in H(\Gamma)$, $a(t)$ be a continuous non-zero function on Γ, $b\in L_{p(\cdot),\alpha}(\Gamma;\rho)$. Find a function ϕ from $K_\alpha^{p(\cdot)}(\Gamma;\rho)$ (or $\widetilde{K}_\alpha^{p(\cdot)}(\Gamma;\rho)$), whose boundary values $\phi^+(t)$ and $\phi^-(t)$ almost everywhere on Γ satisfy the condition (2).

4.11 Conditions of solvability of the Haseman problem in the classes $K_\alpha^{p(\cdot)}(\Gamma;\rho)$ and $K_\alpha^{p(\cdot)}(\Gamma)$

Lemma 26. *For the problem (2) to be solvable in the class $K_\alpha^{p(\cdot)}(\Gamma;\rho)$, $\Gamma\in\Lambda$, with a non vanishing continuous coefficient $a(t)$ and $b(t)\in L_{p(\cdot),\alpha}(\Gamma;\rho)$, it is necessary that the condition*

$$\int\limits_\Gamma \widetilde{b}(t)v(t)\,dt=0 \tag{61}$$

be fulfilled, where

$$\widetilde{b}(t)=b(t)+\frac{1}{\pi i}\int\limits_\Gamma\frac{b(\tau)}{\tau-t}\,d\tau+\int\limits_\Gamma K(t,\tau)b(\tau)\,d\tau$$

and $v(t)$ is an arbitrary solution of the class $L^{p'(\cdot)}(\Gamma;\frac{1}{\rho})$ or $L^{p'_\alpha(\cdot)}(\Gamma;\frac{1}{\rho_\alpha})$ of the equation

$$K'(a)v=-\frac{1}{\pi i}\int\limits_\Gamma\frac{a(t)+a(t_0)}{t-t_0}\,v(t)\,dt-\frac{a(t_0)}{\pi i}\int\limits_\Gamma K(t,t_0)v(t)\,dt=0. \tag{62}$$

Proof. Since the functions from $E^\delta(D^+)$ and $E^\delta(D^-)$, $\delta\geq 1$, are representable by the Cauchy integrals in the domains D^+ and D^-, respectively, and the restrictions of the functions from $K^{p(\cdot)}(\Gamma;\rho)$ on D^+ and D^- belong to $E^{1+\eta}(D^+)$ and $E^{1+\eta}(D^-)$, $\eta>0$ (see Section 4.10), therefore every solution $\phi(z)$ of the problem (2) is representable in the form

$$\phi(z)=\begin{cases}-\dfrac{1}{2\pi i}\displaystyle\int\limits_\Gamma\frac{\phi^-(t)}{t-z}\,dt, & z\in D^-,\\[2ex] \dfrac{1}{2\pi i}\displaystyle\int\limits_\Gamma\frac{\phi^+(t)}{t-z}\,dt, & z\in D^+.\end{cases}$$

Denoting $\mu(t) = \phi^-(t)$, we have $\mu \in L_{p(\cdot),\alpha}(\Gamma;\rho)$. By the assumptions regarding $\phi^-(t)$ and $b(t)$, the function $a(\beta(t))\mu(\beta(t)) + b(\beta(t))$ belongs to $L^{p(\cdot)}(\Gamma;\rho)$, and taking into account the condition (2), we can write

$$\phi(z) = \begin{cases} -\dfrac{1}{2\pi i}\displaystyle\int_\Gamma \dfrac{\mu(t)}{t-z}\,dt, & z \in D^-, \\ \dfrac{1}{2\pi i}\displaystyle\int_\Gamma \dfrac{a(\beta(t)\mu(\beta(t)) + b(\beta(t))}{t-z}\,dt, & z \in D^+. \end{cases} \tag{63}$$

Using the Plemelj–Sokhotskiĭ formula from (63) and (2) we derive

$$K(a)\mu \equiv \frac{1}{\pi i}\int_\Gamma \frac{a(t)+a(t_0)}{t-t_0}\,\mu(t)\,dt + M(a)\mu = \widetilde{b}(t_0), \tag{64}$$

where

$$M(a)\mu = \frac{1}{\pi i}\int_\Gamma K(t_0,t)a(t)\mu(t)\,dt, \quad K(t_0,t) = \frac{\alpha'(t)}{\alpha(t)-\alpha(t_0)} - \frac{1}{t-t_0}.$$

Since the operator S_Γ is continuous in the spaces $L^{p(\cdot)}(\Gamma;\rho)$ and $L^{p_\alpha(\cdot)}(\Gamma;\rho_\alpha)$, we have $\widetilde{b} \in L_{p(\cdot),\alpha}(\Gamma;\rho)$, and moreover, by the assumption, $\mu \in L_{p(\cdot),\alpha}(\Gamma;\rho)$. Hence it is not difficult to show that every solution μ of the equation (64) belonging to $L_{p(\cdot),\alpha}(\Gamma;\rho)$ generates (by means of the formula (63)) a function $\phi \in K_\alpha^{p(\cdot)}(\Gamma;\rho)$ satisfying the condition (2).

We write the operator $K(a)\mu$ in the form

$$\begin{aligned} K(a)\mu &= \frac{1}{\pi i}\int_\Gamma \frac{a(t)-a(t_0)}{t-t_0}\mu(t)\,dt + \frac{2a(t_0)}{\pi i}\int_\Gamma \frac{\mu(t)}{t-t_0}\,dt + M(a)\mu \\ &= 2aS_\Gamma\mu + T(a)\mu + M(a)\mu, \end{aligned}$$

where

$$T(a)\mu = \frac{1}{\pi i}\int_\Gamma \frac{a(t)-a(t_0)}{t-t_0}\,\mu(t)\,dt.$$

According to Lemma 24, the kernel of the operator $M(a)\mu$ has a weak singularity. Therefore it is a compact operator in the spaces $L^{p(\cdot)}(\Gamma;\rho)$ and $L^{p_\alpha(\cdot)}(\Gamma;\rho_\alpha)$ (see [70]). Moreover, since $a(t)$ is continuous, $T(a)\mu$ is a compact operator in the above-mentioned spaces. This can be proved just in the same way as for the constant p (see, e.g., [43, p. 85]).

Thus

$$K(a)\mu = 2aS_\Gamma\mu + \big(T(a)\mu + M(a)\mu\big),$$

where $[T(a)+M(a)]\mu$ is compact operator in the spaces $L^{p(\cdot)}(\Gamma;\rho)$ and $L^{p_\alpha(\cdot)}(\Gamma;\rho_\alpha)$.

Since $a(t_0) \neq 0$ and Γ is a Carleson curve, the operator $K(a)\mu$ in those spaces is Noetherian, and its index in both cases is equal to zero. Hence $K(a)\mu$ is a Fredholm operator in $L^{p(\cdot)}(\Gamma;\rho)$ and $L^{p_\alpha(\cdot)}(\Gamma;\rho_\alpha)$ (see Section 4.6).

For the equation (64) to have a solution $\mu \in L^{p(\cdot)}(\Gamma;\rho)$, it is necessary and sufficient that the condition (61) be fulfilled in which $v(t)$ is any solution of the class $L^{p'(\cdot)}(\Gamma;\frac{1}{\rho})$ of the conjugate equation $K'(a)v = 0$. For the equation (64) to have a solution $\mu \in L^{p_\alpha(\cdot)}(\Gamma;\rho_\alpha)$, it is necessary and sufficient that the function $\widetilde{b} = \widetilde{b}(t_0)$ satisfy the condition (61), where v is any solution of the equation (62) of the class $L^{p'_\alpha(\cdot)}(\Gamma;\frac{1}{\rho_\alpha})$.

Consequently, for the equation (64) to have a solution $\mu \in L_{p(\cdot),\alpha}(\Gamma;\rho)$, it is necessary that the conditions formulated in the lemma be fulfilled. □

Suppose now that $\rho(t) = 1$. In this case we solve the equation (64) in the class $L^{\widetilde{p}_\alpha(\cdot)}(\Gamma)$, then using the formula (63) we obtain all solutions of the problem (2). Since the space $L^{\widetilde{p}'_\alpha(\cdot)}(\Gamma)$ is conjugate to the space $L^{\widetilde{p}_\alpha(\cdot)}(\Gamma)$, following the way of proving Lemma 26 we can prove the following

Theorem 27. *For the problem* (2) *to be solvable in the class* $K_\alpha^{p(\cdot)}(\Gamma)$, $\Gamma \in \Lambda$ *with a non vanishing continuous coefficient* $a(t)$ *and* $b(t) \in L^{\widetilde{p}_\alpha(\cdot)}(\Gamma)$, *it is necessary and sufficient that the conditions* (61) *be fulfilled, where* $\widetilde{b} = b + S_\Gamma b + \int\limits_\Gamma K(\cdot,\tau)b(\tau)\,d\tau$, *and* $v(t)$ *is an arbitrary solution of the class* $L^{\widetilde{p}'_\alpha(\cdot)}(\Gamma)$ *of the equation* (62).

4.12 The Haseman problem with the coefficient $[\alpha'(t)]^{-1}$

We will need the following assertion proved in [86] (here we join the results of two Lemmas 2.1 (p. 73) and 2.2 (p. 79)).

Let Γ be a simple, rectifiable curve bounding the domains D^+ and D^-. If $\omega^+(z)$ is analytic in D^+ and continuous in $\overline{D^+}$, $\omega^-(z) = Az + \omega_0^-(z)$, where $A = \text{const} \neq 0$, $\omega_0^-(z)$ is analytic in D^- and continuous in $\overline{D^-}$, the set $L:\ \tau = \omega^-(t)$, $t \in \Gamma$, has no inner points and $\omega^+(\alpha(t)) = \omega^-(t)$, $t \in \Gamma$, then the functions $\omega^+(z)$ and $\omega^-(z)$ are schlicht in the domains D^+ and D^-, respectively, and the curve L is simple. If, however, $A = 0$, then $\omega^+(z) = c$, $\omega^-(z) = c$, $c = \text{const}$.

Lemma 28. *In the class of functions* Ψ *analytic in the plane cut along* Γ, *with the restrictions* $\Psi^+(z)$ *and* $\Psi^-(z)$ *in the domains* D^+ *and* D^-, *respectively, belonging to* $E^1(D^+)$ *and* $E^1(D^-)$, *the boundary problem*

$$\alpha'(t)\Psi^+(\alpha(t)) = \Psi^-(t) \tag{65}$$

has only the trivial solution. In particular, the problem (65) *in any class* $K^{p(\cdot)}(\Gamma;\rho)$, $p \in \mathcal{P}(\Gamma)$ *and hence in the class* $K_\alpha^{p(\cdot)}(\Gamma;\rho)$, *has only the trivial solution.*

Proof. By $F^+(z)$ and $F^-(z)$ we denote the primitives of the functions $\Psi^+(z)$ and $\Psi^-(z)$, respectively. Since $\Psi^\pm \in E^1(D^\pm)$, the primitives are continuously extendable up to Γ and absolutely continuous on Γ with respect to the arc abscissa (see, e.g., [108, p. 208]), and by the condition (56), they are continuous with respect to the parameter t as well. Moreover, from the condition $\int\limits_\Gamma \Psi^-(t)\,dt = 0$ it follows that the function $F^-(z)$ is single-valued in D^-. The condition (65) yields $F^+(\alpha(t)) = F^-(t)$.

Thus for the function

$$F(z) = \begin{cases} F^+(z), & z \in D^+, \\ F^-(z), & z \in D^-, \end{cases}$$

the conditions given in the begin of present section are fulfilled with $A = 0$. Hence $F(z) \equiv 0$. Thus $\Psi(z) = F'(z) = 0$.

The final part of the assertion of the lemma follows from the fact that $\Psi(z) = K_\Gamma f \in E^1(D^\pm)$ when $f \in L^{p(\cdot)}(\Gamma;\rho)$ (see Section 4.10). □

4.13 The Haseman problem with the coefficient $a = 1$

Following the way of investigation of the problem (2) for $a(t) = 1$ described in [86, pp. 79–80], we, first of all, prove that the following lemma is valid.

Lemma 29. *If the equation* (62) *with* $a(t) = 1$ *has a solution* $v \in L_{p'(\cdot),\alpha}(\Gamma; 1/\rho)$, *then* $v = 0$.

Proof. Consider the function

$$N(z) = \begin{cases} -\dfrac{1}{2\pi i} \displaystyle\int\limits_\Gamma \frac{v(t)}{t-z}\,dt, & z \in D^-, \\ \dfrac{1}{2\pi i} \displaystyle\int\limits_\Gamma \frac{\beta'(t)v(\beta(t))}{t-z}\,dt, & z \in D^+. \end{cases} \tag{66}$$

By the assumption, $v \in L^{p'(\cdot)}(\Gamma; 1/\rho)$, $v(\beta(t)) \in L^{p'_\alpha(\cdot)}(\Gamma; 1/\rho_\alpha)$, and for ρ and ρ_α the conditions (60) and (60_0) are fulfilled. Hence it is not difficult to prove that $N \in E^1(D^\pm)$. Thus we have

$$\begin{aligned} N^+(t_0) &= \frac{1}{2}\beta'(t_0)v(\beta(t_0)) + \frac{1}{2\pi i}\int\limits_\Gamma \frac{\beta'(t)v(\beta(t))}{t-t_0}\,dt \\ &= \frac{v(\beta(t_0))}{2\alpha'(\beta(t_0))} + \frac{1}{2\pi i}\int\limits_\Gamma \frac{v(t)}{\alpha(t)-t_0}\,dt, \end{aligned}$$

i.e.,

$$N^+(\alpha(t_0)) = \frac{v(t_0)}{2\alpha'(t_0)} + \frac{1}{2\pi i}\int\limits_\Gamma \frac{v(t)}{\alpha(t)-\alpha(t_0)}\,dt,$$

whence

$$\alpha'(t_0)N^+(\alpha(t_0)) = \frac{1}{2}\,v(t_0) + \frac{\alpha'(t_0)}{2\pi i}\int\limits_\Gamma \frac{v(t)}{\alpha(t)-\alpha(t_0)}\,dt.$$

On the other hand, $N^-(t_0) = \frac{1}{2}\,v(t_0) - \frac{1}{2}(S_\Gamma v)(t_0)$ and

$$\begin{aligned} K'(1)v &= -\frac{1}{\pi i}\int\limits_\Gamma \frac{2v(t)}{t-t_0}\,dt + \frac{1}{\pi i}\int\limits_\Gamma \Big(\frac{\alpha'(t_0)}{\alpha(t_0)-\alpha(t)} - \frac{1}{t_0-t}\Big)v(t)\,dt \\ &= -\frac{1}{\pi i}\int\limits_\Gamma \frac{v(t)}{t-t_0}\,dt - \frac{\alpha'(t_0)}{\pi i}\int\limits_\Gamma \frac{v(t)}{\alpha(t)-\alpha(t_0)}\,dt = 0. \end{aligned}$$

Therefore

$$\frac{\alpha'(t_0)}{\pi i}\int\limits_\Gamma \frac{v(t)}{\alpha(t)-\alpha(t_0)}\,dt = -\frac{1}{\pi i}\int\limits_\Gamma \frac{v(t)}{t-t_0}\,dt.$$

Consequently,

$$N^-(t_0) = \frac{1}{2}\,v(t_0) + \frac{\alpha'(t_0)}{\pi i}\int\limits_\Gamma \frac{v(t)}{\alpha(t)-\alpha(t_0)}\,dt,$$

and from the above-said we obtain

$$\alpha'(t_0)N^+(\alpha(t_0)) = N^-(t_0).$$

By Lemma 28, we can conclude that $N(z) = 0$. Then from (66) it follows the existence of functions $\varphi^+ \in E^1(D^+)$, $\varphi^- \in E^1(D^-)$ such that

$$v(t) = \varphi^+(t), \quad \beta'(t)v(\beta(t)) = \varphi^-(t),$$

i.e., $\beta'(t)\varphi^+(\beta(t)) = \varphi^-(t)$. From the equality $\beta'(t) = [\alpha'(\beta(t))]^{-1}$ it follows that $\beta' \in H(\Gamma)$ and $\beta'(t) \neq 0$. On the basis of Lemma 28 with $\alpha(t)$ replaced by $\beta(t)$ we establish that $\varphi^\pm(z) = 0$, and hence $v = 0$. □

Theorem 30. *For $a(t) = 1$ and $b \in L_{p(\cdot),\alpha}(\Gamma;\rho)$, the problem* (2) *is uniquely solvable in the class $K_\alpha^{p(\cdot)}(\Gamma;\rho)$.*

Proof. First we consider the equation $K'(1)v=0$.

According to our assumptions regarding p and ρ, there exists a number $\eta > 0$ such that $L^{p'(\cdot)}(\Gamma;1/\rho) \subset L^{1+\eta}(\Gamma)$ and $L^{p'_\alpha(\cdot)}(\Gamma;1/\rho_\alpha) \subset L^{1+\eta}(\Gamma)$. Therefore all the solutions of the equation $K'(1)v = 0$ belonging to at least one of the above-mentioned classes lie in the set of solutions of this equation of the class $L^{1+\eta}(\Gamma)$.

Let us show that the equation $K'(1)v = 0$ has only the zero solution in this class. Towards this end, we consider the operator $K(1)\mu$ in the class $L^{\frac{1+\eta}{\eta}}(\Gamma)$. Then $p(t) = p_\alpha(t) = (1+\eta)/\eta$, $\rho(t) = 1$, $\rho_\alpha(t) = 1$, and hence $L^{p'_\alpha(\cdot)}(\Gamma;1/\rho) \subset L^{1+\eta}(\Gamma)$. By Lemma 29, $v(t) = 0$.

Since $K(1)\mu$ is a Fredholm operator in the spaces $L^{\delta}(\Gamma)$, $\delta > 1$, the equations

$$K(1)\mu = 0, \quad \mu \in L^{\frac{1+\eta}{\eta}}(\Gamma), \quad K'(1)v = 0, \quad v \in L^{1+\eta}(\Gamma)$$

and

$$K(1)\mu = 0, \quad \mu \in L^{1+\eta}(\Gamma), \quad K'(1)v = 0, \quad v \in L^{\frac{1+\eta}{\eta}}(\Gamma)$$

have the same number of linearly independent solutions.

By the above-proven, since the equation $K'(1)v = 0$ in $L^{1+\eta}(\Gamma)$ (and likewise in $L^{\frac{1+\eta}{\eta}}(\Gamma)$) has only the zero solution, we have that the equation $K(1)\mu = 0$ in $L^{1+\eta}(\Gamma)$ has the zero solution. All the more, it has only the zero solution in the classes $L^{p(\cdot)}(\Gamma;\rho)$ and $L^{p_\alpha(\cdot)}(\Gamma;\rho_\alpha)$, and thus in the class $L_{p,\alpha}(\Gamma;\rho)$ as well.

Consider now the problem (2) with $a(t) = 1$. Since the equation (64) is a Fredholm one in $L^{p(\cdot)}(\Gamma;\rho)$, it is solvable for those $b \in L^{p(\cdot)}(\Gamma;\rho)$ for which the condition (61) is fulfilled, where v is any solution of the equation $K'(1)v = 0$ of the class $L^{p'(\cdot)}(\Gamma;1/\rho)$. But, as we have shown, $v = 0$. Therefore (64) has a solution $\mu \in L^{p(\cdot)}(\Gamma;\rho)$.

Analogously, since $b \in L^{p_\alpha(\cdot)}(\Gamma;\rho_\alpha)$ and $K'(1)v = 0$ has only the zero solution in $L^{p'_\alpha(\cdot)}(\Gamma;1/\rho_\alpha)$, the equation (64) has a unique solution $\mu_\alpha \in L^{p_\alpha(\cdot)}(\Gamma;\rho_\alpha)$. The difference $\widetilde{\mu}(t) = \mu(t) - \mu_\alpha(t)$ is a solution of the equation $K(1)\mu = 0$ from $L^{1+\eta}(\Gamma)$, and consequently, $\widetilde{\mu} = 0$. Thus $\mu = \mu_\alpha$ is a unique solution of the equation (64) generating by means of the formula (63) a solution of the problem (2) of the class $K_\alpha^{p(\cdot)}(\Gamma;\rho)$. □

Corollary 31. *The operator*

$$R = \left(\Psi^+(\alpha(t)) - \Psi^-(t)\right) = K(1)\mu$$

in every space $L^{p(\cdot)}(\Gamma;\rho)$, $p \in \mathcal{P}(\Gamma)$, has the inverse operator $R_{p,\rho}^{-1}$ which is bounded in $L^{p(\cdot)}(\Gamma;\rho)$. In particular, the operators $R_{p',1/\rho}^{-1}$ and $R_{p'_\alpha,1/\rho_\alpha}^{-1}$ exist.

Remark 32. From the above-proven theorem it follows that the problem (2) for $a(t) = 1$, $b \in L_{p,\alpha}(\Gamma;\rho)$ in the class $K_\alpha^{p(\cdot)}(\Gamma;\rho)$ has only one solution, while in the class of the functions $\Psi \in \widetilde{K}_\alpha^{p(\cdot)}(\Gamma;\rho)$ with $\Psi(\infty) = 1$, it has only the solution $\phi(z) = 1$. In particular, this assertion is valid for the classes $K^{\delta}(\Gamma)$ and $\widetilde{K}^{\delta}(\Gamma)$, for any $\delta > 1$.

Remark 33. Following our reasoning when proving the above theorem and using Lemma 28, we can easily see that the problem

$$\Psi^+(\alpha(t)) = \frac{1}{\alpha'(t)}\,\Psi^-(t) + b(t)$$

with $b \in L_{p,\alpha}(\Gamma;\rho)$ is uniquely solvable in the class $K_\alpha^{p(\cdot)}(\Gamma;\rho)$. In particular, the problem is uniquely solvable in the classes $K^\lambda(\Gamma)$, $\lambda > 1$ (since $K_\alpha^\lambda(\Gamma;1) = K^\lambda(\Gamma)$). If the function $b(t)$ is bounded, then these solutions coincide. If ϕ is that solution, then $\phi \in \bigcap\limits_{\delta>1} K^\delta(\Gamma)$.

4.14 The Haseman problem with a coefficient close to unity

Lemma 34. *There exists a number $\varepsilon > 0$ such that if $\max|a(t)-1| < \varepsilon$, then the Haseman problem with coefficient $a(t)$ is uniquely solvable in the class $K_\alpha^{p(\cdot)}(\Gamma;\rho)$. At the same time this problem with the coefficient $\frac{1}{a(t)}$ is uniquely solvable in the class $K^{p'_\alpha}(\Gamma;\frac{1}{\rho_\alpha})$. Moreover homogeneous problem*

$$\Phi^+(\alpha(t)) = a(t)\Phi^-(t) \tag{67}$$

has a solution $X_0 \in \widetilde{K}_\alpha^{p(\cdot)}(\Gamma;\rho)$ possessing the property $[X_0]^{-1} \in \widetilde{K}_\alpha^{p'(\cdot)}(\Gamma;1/\rho)$, $X_0(\infty) = 1$.

Proof. Let $\Phi \in K_\alpha^{p(\cdot)}(\Gamma;\rho)$ be a solution of the problem (2). Then it is representable by the equality (63). Assume

$$R\mu = \Phi^+(\alpha(t)) - \Phi^-(t) = \frac{1}{2}\,(a(t)-1)\big(\mu(t) - (S_\Gamma\mu)(t)\big).$$

According to the corollary 31, there exists the bounded in $L^{p(\cdot)}(\Gamma;\rho)$ operator $R_{p,\rho}^{-1}$. The boundary condition (2) yields $\mu = R_{p,\rho}^{-1}\Big[\frac{1}{2}\,(a-1)(\mu + S_\Gamma\mu) + b\Big] = N\mu$.

We have

$$\big\|N(\mu_1-\mu_2)\big\|_{p(\cdot),\rho} \le \|R_{p(\cdot),\rho}^{-1}\|\frac{1}{2}\,\big(1+\|S_\Gamma\|_{p(\cdot),\rho}\big)\max|a(t)-1|\,\|\mu_1-\mu_2\|_{p(\cdot),\rho}.$$

Consequently, if

$$\max|a(t)-1| < \frac{2\nu}{\|R_{p,\rho}^{-1}\|(1+\|S_\Gamma\|_{p(\cdot),\rho})} = \varepsilon_1,\quad \nu \in (0,1),$$

then N is a contraction operator. This implies that the equation $K(a)\mu = \widetilde{b}$ is uniquely solvable in $L^{p(\cdot)}(\Gamma;\rho)$.

Analogously, if $b \in L^{p_\alpha(\cdot)}(\Gamma;\rho_\alpha)$ and

$$\max|a(t)-1| < \frac{2\nu}{\|R_{p_\alpha,\rho_\alpha}^{-1}\|(1+\|S_\Gamma\|_{p_\alpha,\rho_\alpha})} = \varepsilon_{1,\alpha},$$

then the equation $K(1)\mu = \widetilde{b}$ is uniquely solvable in $L^{p_\alpha(\cdot)}(\Gamma;\rho_\alpha)$. According to these two statements we find that the problem (2) with $\max|a(t)-1| < \min(\varepsilon_1;\varepsilon_{1,\alpha})$ is uniquely solvable in $K_\alpha^{p(\cdot)}(\Gamma;\rho)$.

Next, since $\max\left|\frac{1}{a(t)}-1\right| < \frac{\max|a(t)-1|}{\min|a(t)|}$, it is not difficult to see that if

$$\max|a(t)-1| < \frac{2\nu\min|a(t)|}{\|R^{-1}_{p',1/\rho}\|(1+\|S_\Gamma\|_{p'(\cdot),1/\rho})} = \varepsilon_2,$$

then the equation $K(1/a)\mu = c$, $c \in L^{p'(\cdot)}(\Gamma;1/\rho)$, is uniquely solvable. If, however,

$$\max|a(t)-1| < \frac{2\nu\min|a(t)|}{\|R_{p'_\alpha,1/\rho_\alpha}\|(1+\|S_\Gamma\|_{p(\cdot)'_\alpha,1/\rho_\alpha})} = \varepsilon_{2,\alpha},$$

then the equation $K(1/a)\mu = c$, $c \in L^{p'_\alpha(\cdot)}(\Gamma;1/\rho_\alpha)$, is uniquely solvable in the class $L^{p'_\alpha(\cdot)}(\Gamma;1/\rho_\alpha)$.

Consequently, if $\max|a(t)-1| < \min(\varepsilon_2,\varepsilon_{2,\alpha})$, then the problem (2) with the coefficient $[a(t)]^{-1}$ is uniquely solvable in $K_\alpha^{p'(\cdot)}(\Gamma;1/\rho)$. Thus if $\max|a(t)-1| < \epsilon$, where $\epsilon = \min(\epsilon_1,\epsilon_2;\epsilon_{1,\alpha},\epsilon_{2,\alpha})$ the statements from first part of lemma are proved.

Now we prove statement from of second part.

Bearing in mind that $a(t)$ and $[a(t)]^{-1}$ are bounded functions, we that find the boundary value problems

$$\Phi^+(\alpha(t)) = a(t)\Phi^-(t) + a(t), \quad \Phi \in K_\alpha^{p(\cdot)}(\Gamma;\rho),$$
$$\Psi^+(\alpha(t)) = \frac{1}{a(t)}\Psi^-(t) + \frac{1}{a(t)}, \quad \Psi \in K_\alpha^{p'(\cdot)}(\Gamma;1/\rho)$$

are uniquely solvable when $\max|a(t)-1| < \varepsilon$.

Note that we can choose $\eta > 0$ such that the condition (60) preserves if we replace in it $p(t)$ by $p(t)+\eta$. Replacing everywhere in the sequel $p(t)$ by $p(t)+\eta$ and denoting the obtained values for ε_k, $\varepsilon_{k,\alpha}$, $k=1,2$, respectively by $\widetilde{\varepsilon}_1$, $\widetilde{\varepsilon}_2$, $\widetilde{\varepsilon}_{1,\alpha}$, $\widetilde{\varepsilon}_{2,\alpha}$, we will find that if $\max|a(t)-1| < \varepsilon(\eta) = \min(\widetilde{\varepsilon}_1,\widetilde{\varepsilon}_2,\widetilde{\varepsilon}_{1,\alpha},\widetilde{\varepsilon}_{2,\alpha})$, then the solutions $\Phi(z)$ and $\Psi(z)$ satisfy the conditions

$$\Phi \in K_\alpha^{p(\cdot)+\eta}(\Gamma,;\rho), \quad \Psi \in K_\alpha^{p'(\cdot)+\eta}(\Gamma,;1/\rho).$$

Further, we have

$$\Phi^+(\alpha(t))\Psi^+(\alpha(t)) = \Phi^-(t)\Psi^-(t) + \Phi^-(t) + \Psi^-(t) + 1.$$

Assume

$$F(z) = \begin{cases} \Phi(z)\Psi(z), & z \in D^+, \\ [\Phi(z)+1]\,[\Psi(z)+1], & z \in D^-. \end{cases}$$

Then $F(z) \in E^{1+\lambda}(D^+)$, $[F(z)-1] \in E^{1+\lambda}(D^-)$, where λ is some positive number. Moreover, $F^+(\alpha(t)) = F^-(t)$, $F(\infty) = 1$. Owing to Remark 32, we have $F(z) \equiv 1$. Therefore

$$\Phi(z)\Psi(z) = 1, \ \ z \in D^+, \quad [\Phi(z)+1]\,[\Psi(z)+1] = 1, \ \ z \in D^-.$$

The function

$$X_0(z) = \begin{cases} \Phi(z), & z \in D^+, \\ \Phi(z)+1, & z \in D^-, \end{cases}$$

is a solution of the problem (67) of the class $\widetilde{K}_\alpha^{p(\cdot)}(\Gamma;\rho)$ and $X_0(\infty) = 1$. Owing to last equalities on $\Phi(z)$ can be seen that $X_0(z)$ possesses the required property. □

Remark 35. From the above-said we can conclude that

$$X_0 \in \widetilde{K}_\alpha^{p(\cdot)+\eta}(\Gamma;\rho), \quad [X_0]^{-1} \in K_\alpha^{p'(\cdot)+\eta}(\Gamma;1/\rho), \quad X_0(\infty) = 1.$$

Remark 36. If $\rho(t) = 1$, then $X_0 \in \widetilde{K}_\alpha^{p(\cdot)}(\Gamma)$ for any $p \in \mathcal{P}(\Gamma)$, and $[X_0]^{-1} \in \widetilde{K}_\alpha^{p(\cdot)}(\Gamma)$. In particular, $X_0 \in \bigcap\limits_{\delta>1} \widetilde{K}^\delta(\Gamma)$ and $[X_0]^{-1} \in \bigcap\limits_{\delta>1} \widetilde{K}^\delta(\Gamma)$ (this follows from the fact that $\widetilde{K}_\alpha^\delta(\Gamma) = \widetilde{K}^\delta(\Gamma)$).

4.15 Solution of homogeneous Haseman problem. Canonical solution

In solving the problem (67) we will need good properties of the solution of the problem

$$\omega^+(\alpha(t)) = \omega^-(t). \tag{68}$$

Denote the restrictions of the solution $\omega(z)$ to the domains D^+ and D^- by $\omega^+(z)$ and $\omega^-(z)$, respectively.

Lemma 37. *There exists a solution of the problem (68) possessing the properties: $\omega^+(z)$ $(\omega^-(z))$ is schlicht and continuous in $\overline{D^+}$ $(\overline{D^-})$, $\omega^-(z) = z + \omega_0^-(z)$, where $\omega_0^-(z)$ is analytic in D^-. Moreover,*

$$[\omega^+(z)]' \in \bigcap_{\delta>1} E^\delta(D^+), \quad [\omega^-(z)-z]' \in \bigcap_{\delta>1} E^\delta(D^-), \tag{69}$$

$L = \omega^-(\Gamma)$ is a simple rectifiable curve, and the domains G^+, G^- bounded by L are Smirnov domains.

Proof. Consider the boundary value problem

$$\alpha(t)\Phi^+(\alpha(t)) = \Phi^-(t) + 1.$$

Since the free term ($b(t) \equiv 1$) in this problem is a bounded function, then by Remark 33 this problem has a unique solution $\Phi \in \bigcap\limits_{\delta>1} K^{\delta}(\Gamma)$. By $\omega^{+}(z)$ ($\omega^{-}(z)$) we denote the primitive of the function $\Phi(z)$ ($\Phi(z)+1$) in the domain D^{+} (D^{-}). We choose the primitive functions in such a way that the equality (68) is fulfilled. Since $\Phi \in \bigcap\limits_{\delta>1} K^{\delta}(D^{\pm})$, we have $\Phi \in \bigcap\limits_{\delta>1} E^{\delta}(D^{\pm})$. Hence the functions $\omega^{+}(z)$ and $\omega^{-}(z)$ are continuous in the domains $\overline{D^{+}}$ and $\overline{D^{-}} \setminus \{\infty\}$ (see [108, p. 208]). By the result obtained in [86] and cited above in Section 4.12, they are schlicht in the corresponding domains. Moreover,

$$[\omega^{+}(z)]' = \Phi(z), \quad z \in D^{+}, \quad [\omega^{-}(z) - z]' = \Phi(z), \quad z \in D^{-}.$$

Hence owing to the above-said regarding $\Phi(z)$, the inclusions (69) are valid.

The rectifiability of $L = \omega^{-}(\Gamma)$ follows from the summability of the function $[\omega^{-}(z)]'$. Take into account (69) the statement on domains G^{+}, G^{-} follows from Lemma 4 of Chapter 1. □

Assume $\varkappa = \operatorname{ind} a(t)$.

Theorem 38. *If $a(t)$ is a non-zero continuous function, then all the solutions of the problem (67) of the class $\widetilde{K}^{p(\cdot)}_{\alpha}(\Gamma;\rho)$ are given by the equality*

$$\Phi(z) = X(z)P(\omega(z)),$$

where $\omega(z)$ is the function constructed in Lemma 37, $P(z)$ is an arbitrary polynomial, and $X(z)$ is a solution of the homogeneous problem (67) defined to within a constant multiplier and possessing the properties:

(1) $X \in \widetilde{K}^{p(\cdot)}_{\alpha}(\Gamma;\rho)$, $[X]^{-1} \in \widetilde{K}^{p'(\cdot)}_{\alpha}(\Gamma;1/\rho)$;

(2) the order of $X(z)$ at infinity is equal to $(-\varkappa) = -\operatorname{ind} a(t)$.

Proof. Choose a rational function $r(z)$

$$r(z) = \sum_{k=-n}^{m} A_k (z - z_0)^k, \quad z_0 \in D^{+}, \quad r(t) \neq 0, \quad t \in \Gamma. \tag{70}$$

such that $\max |a(t)r^{-1}(t) - 1| < \varepsilon$, where ε is the number mentioned in Lemma 34.

Since

$$\operatorname{ind} a(t) = \operatorname{ind} \left(\frac{a(t)}{r(t)}\, r(t)\right) = \operatorname{ind} \frac{a(t)}{r(t)} + \operatorname{ind} r(t),$$

and from the condition $|a(t)r^{-1}(t) - 1| < \varepsilon$ (assuming $\varepsilon < 1$) it follows $\operatorname{ind} \frac{a(t)}{r(t)} = 0$, we have $\operatorname{ind} r(t) = \operatorname{ind} a(t) = \varkappa$.

Assume $a_0(t) = a(t)r^{-1}(t)$ and write the condition (67) in the form

$$\Phi^{+}(\alpha(t)) = a_0(t) r(t) \Phi^{-}(t). \tag{71}$$

By Lemma 34, there exists a function $X_0(z)$ such that

$$X_0 \in \widetilde{K}_\alpha^{p(\cdot)}(\Gamma;\rho),\ \ [X_0]^{-1} \in \widetilde{K}_\alpha^{p'(\cdot)}(\Gamma;1/\rho),$$
$$X_0(\infty)=1,\ \ X_0^+(\alpha(t)) = a_0(t)X_0^-(t).$$

Therefore (71) can be written as follows:

$$[X_0^+(\alpha(t))]^{-1}\Phi^+(\alpha(t)) = r(t)[X_0^-]^{-1}\Phi^-(t). \tag{72}$$

Now the use will be made of the function $\omega(z)$ constructed in Lemma 37.

Denote the domains bounded by the curve $L=\omega^-(\Gamma)$ by G^+ and G^-.

Let ω_1^+ and ω_1^- be the inverse functions to ω^+ and ω^-, respectively, in the domains G^+ and G^-. Assuming $\tau=\omega^+(\alpha(t))=\omega^-(t)$, we obtain $\alpha(t)=\omega_1^+(\tau)$, $t=\omega_1^-(\tau)$, and (72) can be written in the form

$$\begin{aligned}&\big[X_0^+(\omega_1^+(\tau))\big]^{-1}\Phi^+(\omega_1^+(\tau))\\ &=\big[X_0^-(\omega_1^-(\tau))\big]^{-1}r(\omega_1^-(\tau))\Phi^-(\omega_1^-(\tau)),\ \ \tau\in L.\end{aligned} \tag{73}$$

From Remark 35 we can easily conclude that $\Phi(z)[X_0(z)]^{-1}$ belongs to $E^{1+\eta_1}(D^+)$, for some $\eta_1>0$. Using now the inclusion (69), we get

$$\varphi_1(\zeta) = \big[X_0^+(\omega_1^+(\zeta))\big]^{-1}\Phi(\omega_1^+(\zeta)) \in E^1(G^+). \tag{74}$$

Analogously,

$$\big[X_0^-(\omega_1^-(\zeta))\big]^{-1}\Phi(\omega_1^-(\zeta))r(\omega_1^-(\zeta)) = \varphi_1(\zeta)+q(\zeta),\ \ \zeta\in G^-, \tag{75}$$

where $q(\zeta)$ is a polynomial, and $\varphi_1\in E^1(G^-)$.

Here we have used the fact that if $\mu\in L^{\delta_0}(\Gamma)$, $\delta_0>1$, and $P(z)$ is a polynomial, then

$$P(z)\frac{1}{\pi i}\int\limits_\Gamma \frac{\mu(t)}{t-z}\,dt = \frac{1}{\pi i}\int\limits_\Gamma\frac{\mu_1(t)}{t-z}\,dt + P_1(z),$$

where $\mu_1\in L^{\delta_0}(\Gamma)$, and $P_1(z)$ is a polynomial whose order is less by unity than that of the polynomial $P(z)$ [43, p. 98] and $P_1(\omega_1(\zeta))=q(\zeta)+\varphi_2(\zeta)$, where $q(z)$ is a polynomial and $\varphi_2\in \bigcap\limits_{\delta>1}E^\delta(G^-)$. The latter follows from the equality $\omega^-(\zeta)=\zeta+\omega_{0,1}(\zeta)$, where $\omega_{0,1}^-(\zeta)$ is a function continuous in $\overline{G^-}$ and analytic in G^-.

Note here that if $r(z)$ given by (70), then the order of $r(z)$ at infinity is equal to m. If $\Phi\in K_\alpha^{p(\cdot)}(\Gamma;\rho)$, then $\Phi(\infty)=0$, and (75) implies that $q(z)$ is a polynomial of order $m-1$.

Getting back to the problem (71) which is reduced to the problem (73) and taking into account (74) and (75), we see that the equality (73) yields

$$\varphi_1^+(\tau)=\varphi_1^-(\tau)+q(\tau),\ \ \tau\in L,\ \ \varphi_1^\pm\in E^1(G^\pm),$$

whence $\varphi_1^+(z) = q(z)$, $z \in G^+$, $\varphi_1^-(z) = 0$, $z \in G^-$. But then

$$\begin{aligned} \Phi(z) &= X_0(z)q(\omega^+(z)), \quad z \in D^+, \\ \Phi(z) &= r^{-1}(z)X_0(z)q(\omega^-(z)), \quad z \in D^-. \end{aligned} \tag{76}$$

For $\Phi(z)$ to be analytic in D^-, the function $[r(z)]^{-1}q(\omega^-(z))$ must have no singularities in D^-.

We write $r(z)$ in the form

$$r(z) = (z - z_0)^n (z - z_1) \cdots (z - z_{m+n}),$$

where $z_0 \in D^+$, and $z_1, z_2, \dots, z_{m+n}$ are zeroes of $r(z)$ of which $z_1, \dots, z_j$ belong to D^+ while $z_{j+1}, \dots, z_{m+n}$ belong to D^-. Then $\operatorname{ind} r(t) = j - n$, and since $\operatorname{ind} r(t) = \varkappa$, therefore $j - n = \varkappa$. For the function Φ to be analytic in D^-, it is necessary that the function $q(\omega^-(z))$ have zeroes at the points $z_{j+1}, \dots, z_{m+n}$. Since $\omega^-(z)$ is schlicht, there exist points $\zeta_{j+1}, \dots, \zeta_{m+n}$ belonging to the domain G^- and such that $\omega^-(\zeta_k) = z_k$, $k = j+1, \dots, m+n$. Consequently, we must have $q(\zeta) = \widetilde{q}(\zeta)P(\zeta)$, where $\widetilde{q}(\zeta) = (\zeta - \zeta_{j+1}) \cdots (\zeta - \zeta_{m+n})$, and $P(\zeta)$ is an arbitrary polynomial.

Thus from (76) we conclude that all possible solutions of the problem (71) (and hence of the problem (67)) of the class $\widetilde{K}_\alpha^{p(\cdot)}(\Gamma; \rho)$ lie in the set of functions

$$\Phi(z) = \begin{cases} [X_0(z)\widetilde{q}(\omega^+(z))]P(\omega^+(z)), & z \in D^+, \\ [X_0(z)r^{-1}(z)\widetilde{q}(\omega^-(z))]P(\omega^-(z)), & z \in D^-, \end{cases} \tag{77}$$

where $X_0(z)$ is the solution constructed in Lemma 34, $\widetilde{q}(z)$ is the above-mentioned and $P(z)$ is an arbitrary polynomial. Since $X_0 \in \widetilde{K}_\alpha^{p(\cdot)}(\Gamma; \rho)$, it is easily seen that the equality (77) provides us with a general solution of the problem (67) of the class $\widetilde{K}_\alpha^{p(\cdot)}(\Gamma; \rho)$.

Assuming

$$X(z) = \begin{cases} X_0(z)\widetilde{q}(\omega^+(z)), & z \in D^+, \\ X_0(z)r^{-1}(z)\widetilde{q}(\omega^-(z)), & z \in D^-, \end{cases} \tag{78}$$

we can write the equality (77) in the form $\Phi(z) = X(z)P(\omega(z))$. □

The function $X(z)$ at infinity is of the order $m - (m - n - j) = n - j = -\varkappa$. It possesses the following properties:

(1) $X^+(\alpha(t)) = a(t)X^-(t)$;

(2) $X \in \widetilde{K}_\alpha^{p(\cdot)}(\Gamma; \rho)$, $1/X \in \widetilde{K}_\alpha^{p'(\cdot)}(\Gamma; 1/\rho)$;

(3) at infinity it has the order $(-\varkappa)$.

The function X possessing the properties (1)–(3) will be called a canonical solution of the problem (67) of the class $\widetilde{K}_\alpha^{p(\cdot)}(\Gamma;\rho)$, or a canonical function of the function $a(t)$.

Although our formal representation of $X(z)$ (see (78)) depends on the function $r(z)$, on the basis of Theorem 38 we can easily conclude that all canonical functions differ from each other by a non-zero constant multiplier.

Remark 39. From Remark 35 regarding the function $X_0(z)$ it follows that the canonical function $X(z)$ given by the equality (78) possesses the property: there exists $\eta > 0$ such that

$$X \in \widetilde{K}_\alpha^{p(\cdot)+\eta}(\Gamma;\rho), \quad 1/X \in \widetilde{K}_\alpha^{p'(\cdot)+\eta}(\Gamma;1/\rho).$$

Remark 40. If we consider the problem for $\rho(t) = 1$, i.e. in the class $\widetilde{K}_\alpha^{p(\cdot)}(\Gamma)$, then there is no need to coordinate the relations between p and ρ by means of the inequality (60), and the function X satisfies the conditions $X \in \bigcap\limits_{p\in\mathcal{P}(\Gamma)} \widetilde{K}_\alpha^{p(\cdot)}(\Gamma)$, $1/X \in \bigcap\limits_{p\in\mathcal{P}(\Gamma)} \widetilde{K}_\alpha^{p'(\cdot)}(\Gamma)$. In particular, the continuous non-zero function $a(t)$ has a canonical function X for which

$$X \in \bigcap_{\delta>1} \widetilde{K}^\delta(\Gamma), \quad 1/X \in \bigcap_{\delta>1} \widetilde{K}^\delta(\Gamma).$$

Remark 41. From the Theorem 38 follows that every solution of the problem (67) of the class $\widetilde{K}^{\delta_0}(\Gamma)$, $\delta_0 > 1$ belongs to the set $\bigcap\limits_{\delta>1} \widetilde{K}^\delta(\Gamma)$.

4.16 Solution of the Haseman problem in the classes $\widetilde{K}_\alpha^{p(\cdot)}(\Gamma;\rho)$ and $K_\alpha^{p(\cdot)}(\Gamma;\rho)$

Theorem 42. *All the solutions of the class $\widetilde{K}_\alpha^{p(\cdot)}(\Gamma;\rho)$ of Haseman problem (2) with the continuous coefficient $a(t)$ and the condition $b \in L_{p(\cdot),\alpha}(\Gamma;\rho)$ are given by the formula*

$$\Phi(z) = X(z)\big[\Phi_0(z) + P(\omega(z))\big], \tag{79}$$

where $X(z)$ is the canonical solution of the problem (67) given by the equality (78), $\omega(z)$ is a piecewise analytic solution of the problem (68) with the principal part equal at infinity to z, $P(z)$ is an arbitrary polynomial, and $\Phi_0(z)$ is a solution of the problem

$$\Phi_0^+(\alpha(t)) = \Phi_0^-(t) + b_0(t), \quad b_0(t) = \big[X^+(\alpha(t))\big]^{-1} b(t) \tag{80}$$

belonging to the set $\bigcup\limits_{\delta>1} \widetilde{K}^\delta(\Gamma)$.

Solutions vanishing at infinity, i.e., those of the class $K_\alpha^{p(\cdot)}(\Gamma;\rho)$ *exist if* $\varkappa = \operatorname{ind} a(t) \geq 0$, *and all such solutions are given by the formula* (79) *in which* $P(z)$ *is an arbitrary polynomial of order* $\varkappa - 1$ *for* $\varkappa > 0$, *and* $P(z) = 0$ *for* $\varkappa = 0$.

If $\varkappa < 0$, *then for the solvability of the problem* (2) *in* $K_\alpha^{p(\cdot)}(\Gamma;\rho)$ *it is necessary and sufficient that the conditions*

$$\int_\Gamma t^k \mu(t)\,dt = 0, \quad k = 0, \dots, |\varkappa| - 1, \tag{81}$$

be fulfilled, where μ *is a solution of the equation* $K(1)\mu = \widetilde{b}_0(t)$, $\widetilde{b}_0 = b_0 + S_\Gamma b_0 + \int_\Gamma K(\cdot,\tau) b_0(\tau)\,d\tau$.

If the conditions (81) *are fulfilled, then there exists a unique solution* Φ *which is given by the equality* (79) *with* $P(z) \equiv 0$.

Proof. Let us show that the inhomogeneous problem is solvable in $\widetilde{K}_\alpha^{p(\cdot)}(\Gamma;\rho)$ for any $b \in L_{p(\cdot),\alpha}(\Gamma;\rho)$. We choose a rational function $r(z)$ of the type (70) in such a way that $\max|a(t) - 1| < \varepsilon$ and consider the problem

$$\Psi^+(\alpha(t)) = a(t) r^{-1}(t) \Psi^-(t) + b(t).$$

This problem is solvable in $\widetilde{K}_\alpha^{p(\cdot)}(\Gamma;\rho)$ if ε is a sufficiently small number (see Lemma 34).

Having chosen the function $r(z)$ in the way mentioned above, we see that it has no poles in D^-. Therefore the function

$$\widetilde{\Psi}(z) = \begin{cases} \Psi(z), & z \in D^+, \\ r(z)\Psi(z), & z \in D^-, \end{cases}$$

is a solution of the problem (2) of the class $\widetilde{K}_\alpha^{p(\cdot)}(\Gamma;\rho)$.

Further, let $\Phi_*(z) \in \widetilde{K}_\alpha^{p(\cdot)}(\Gamma;\rho)$ be any solution of the problem (2). Assuming $\Phi_0(z) = [X(z)]^{-1}\Phi_*(z)$ and taking into account that $X^{-1} \in \widetilde{K}_\alpha^{p'(\cdot)}(\Gamma; 1/\rho)$, we find that $\Phi_0 \in \widetilde{K}^{\delta_0}(\Gamma)$ for some $\delta_0 > 1$, and Φ_0 is a solution of the problem (80). The difference $\varphi(z) = \Phi_*(z) - \widetilde{\Psi}(z) = \Phi_0(z)X(z) - \widetilde{\Psi}(z)$ is a solution of the problem (67) of the class $\widetilde{K}^{\delta_0}(\Gamma)$. Every such solution belongs, by Remark 41, to $\bigcap\limits_{\delta>1} \widetilde{K}^\delta(\Gamma)$, and hence $\varphi^\pm \in \bigcap\limits_{\delta>1} L^\delta(\Gamma)$.

Let us show that if $\varphi \in \bigcap\limits_{\delta>1} K_\alpha^\delta(\Gamma)$ and the weight $\rho(t)$ satisfies the condition (60), then $\varphi \in \widetilde{K}_\alpha^{p(\cdot)}(\Gamma;\rho)$. Towards this end, it suffices to show that $\varphi^\pm(t)$ belong to $L^{p(\cdot)}(\Gamma;\rho) \cap L^{p_\alpha(\cdot)}(\Gamma;\rho_\alpha)$.

Using the generalized Hölder inequality, we obtain

$$\int_\Gamma |\varphi^\pm(t)\rho(t)|^{p(t)}\,|dt| \leq c_0 \|\varphi^\pm\|_{\frac{p(\cdot)+\eta}{\eta}\,p(\cdot)} \|\rho\|_{p(\cdot)+\eta}\,.$$

Since for $p(t) \le p_1(t)$ we have $\|f\|_{p(t)} < (1+|\Gamma|)\|f\|_{p_1(\cdot)}$, we obtain

$$\int_{\Gamma} \left|\varphi^{\pm}(t)\rho(t)\right|^{p(t)} |dt| \le c_0(1+|\Gamma|)\|\varphi^{\pm}\|_{\frac{\overline{p}(\overline{p}+\eta)}{\eta}} \|\rho\|_{p(\cdot)+\eta}, \tag{82}$$

where $\overline{p} = \max p(t)$. By (60), there exists $\eta > 0$ such that $\rho \in L^{p(\cdot)+\eta}(\Gamma)$. Then from (82) it follows that $\varphi^{\pm} \in L^{p(\cdot)+\eta}(\Gamma;\rho)$, and hence $\varphi \in \widetilde{K}^{p(\cdot)+\eta}(\Gamma;\rho)$. Analogously, we can prove that $\varphi^{-}(t) \in L^{p_\alpha(\cdot)}(\Gamma;\rho_\alpha)$.

Thus $\varphi^{-} \in L_{p(\cdot),\alpha}(\Gamma;\rho)$ and hence $\varphi(z)$ is a solution of the problem (67) of the class $\widetilde{K}_{\alpha}^{p(\cdot)}(\Gamma;\rho)$. All the solutions of this class are given by the equality $\Phi(z) = X(z)P(\omega(z))$. Hence $\varphi(z) = X(z)P_{\varphi}(\omega(z))$ for some polynomial P_{φ}. Therefore the function $\Phi_0(z)X(z)(= \widetilde{\Psi(z)} + X(z)P_{\varphi}(\omega(z))$ is a particular solution of the problem (67) of the class $\widetilde{K}_{\alpha}^{p(\cdot)}(\Gamma;\rho)$. This implies that the general solution of the problem (2) is given by the equality (79).

Thus the proof the first part of the theorem is complete.

From the union of functions given by the equality (79), we select the solutions belonging to $K_{\alpha}^{p(\cdot)}(\Gamma;\rho)$.

First, let $\varkappa \ge 0$. In this case $X(z)$ has at infinity the order $(-\varkappa)$. Therefore any function given by the equality (79) belongs to $K_{\alpha}^{p(\cdot)}(\Gamma;\rho)$ if the order of the polynomial $P(z)$ does not exceed $\varkappa - 1$ $(P_{-1}(z) = 0)$. Thus all the solutions in this case are given by the equality

$$\Phi(z) = X(z)(\Phi_0(z) + P_{\varkappa-1}(\omega(z))).$$

If $\varkappa < 0$, then $X(z)$ has at infinity pole of order $(-\varkappa)$. Therefore in (79) we have to take $P(z) = 0$ and, moreover, it is necessary and sufficient that $\Phi_0(z)$ have a zero of order $(-\varkappa) + 1$ at infinity.

Let μ be a solution of the equation $K(1)\mu = \widetilde{b}_0(t)$, $b_0(t) = b(t)[X^{+}(\alpha(t))]^{-1}$. Then the solution Φ_0 of the problem (80) in the domain D^{-} is given by the equality $\Phi_0(z) = -\left(K_{\Gamma}\mu\right)(z)$. Obviously, Φ_0 has a zero of order $(-\varkappa + 1)$ if and only if the conditions (81) are fulfilled. □

4.17 The Haseman problem with a piecewise continuous coefficient

When $a(t)$ is a piecewise continuous function, despite the fact that one succeeds in investigating the problem just by the same way when $p(t)$ is constant (see [86]), we had, however, to bring to light some properties of solutions of the problem (2) from the class of functions under consideration, to solve a number of rather difficult questions arisen in the course of their investigation and, finally, to find those preconditions for the functions $a(t)$, $b(t)$ and $p(t)$ under fulfilment of which the problem (2) becomes solvable.

We consider Haseman problem when $a(t) \in C^1(\Gamma; b_1, b_2, \dots, b_m)$, $\inf |a(t)| > 0$.

Reduction of the Problem (2) with piecewise continuous Coefficients to the Problem with a Continuous Coefficient

Assume that in the neighborhood of the points b_k and $\alpha(b_k)$ the curve Γ is smooth. We put

$$\frac{a(b_k-)}{a(b_k+)} = \lambda_k, \quad \tau_k = \frac{1}{2\pi i} \ln \lambda_k, \quad -1 < \mu_k = \operatorname{Re} \tau_k \leq 0 \tag{83}$$

and assume

$$\mu_k \neq -\frac{1}{p(b_k)}. \tag{84}$$

Then if

$$\gamma_k = \begin{cases} \mu_k + 1 & \text{for} \quad -1 < \mu_k < -\dfrac{1}{p(b_k)}, \\ \mu_k & \text{for} \quad -\dfrac{1}{p(b_k)} < \mu_k \leq 0, \end{cases} \tag{85}$$

we have

$$-\frac{1}{p(b_k)} < \gamma_k < \frac{1}{p'(b_k)}. \tag{86}$$

In D^- we draw a cut, i.e. a continuous curve e_k connecting the points $\alpha(b_k)$ and $z = \infty$. Let $X_k^1(z) = [z - \alpha(b_k)]^{\tau_k}$ be a single-valued branch of the function, analytic on the plane, cut along e_k. Next, we choose the point z_0 from D^+ and connect it by means of the cut e_k' with the point b_k lying in D^+; let $X_k(z) = \left(\frac{z-b_k}{z-z_0}\right)^{\tau_k}$ be that single-valued branch of the function, analytic on the plane, cut along e_k', for which $X_k(\infty) = 1$.

First we show some properties of solution of the boundary problem (68).

Lemma 43. *Let $\Gamma \in \Lambda$ and $b \in \Gamma$. If $\omega(z)$ is solution of the problem* (68)*, pointed Lemma* 37*, then for arbitrary $\varepsilon \in (0,1)$ exist M_ε such that*

$$\begin{aligned} &\left|\omega^+(t) - \omega^+(\alpha(b))\right| < M_\varepsilon \left|t - \alpha(b)\right|^{1-\varepsilon}, \\ &\left|\omega^-(t) - \omega^-(b)\right| < M_\varepsilon \left|t - b\right|^{1-\varepsilon}. \end{aligned} \tag{87}$$

Proof. According to the Lemma 37, we have $[\omega^\pm(t)]' \in \bigcap_{\delta>1} L^\delta(\Gamma)$. Therefore

$$\left|\omega^-(t) - \omega^-(b)\right| = \left|\int_b^t [\omega^-(t)]' dt\right| \leq \left(\int_b^t [\omega^-(t))'|^{\frac{1}{\varepsilon}} |dt|\right)^\varepsilon |s(t,b)|^{1-\varepsilon} \leq M_\varepsilon |t - b|^{1-\varepsilon}.$$

Analogously

$$\left|\omega^+(t)-\omega^+(\alpha(b))\right| \le \left|\int\limits_{\alpha(b)}^{t} \left[\omega^+(t)\right]'|dt|\right|^{\varepsilon} \left|s(t,\alpha(b)\right|^{1-\varepsilon} \le M_\varepsilon |t-\alpha(b)|^{1-\varepsilon}. \qquad \square$$

Lemma 44. *Let $\Gamma \in \Lambda$, $b_k \in \Gamma$ and the curve Γ be smooth in the neighborhood of the points b_k and $\alpha(b_k)$. If*

$$\Omega_k^+(z) = \frac{\omega^+(z)-\omega^+(\alpha(b_k))}{z-\alpha(b_k)}, \quad z \in D^+, \qquad \Omega_k^-(z) = \frac{\omega^-(z)-\omega^-(b_k)}{z-b_k}, \quad z \in D^-,$$

then

$$\left[\Omega_k^+\right]^{\pm 1} \in \bigcap_{\delta>1} E^\delta(D^+), \quad \left(\left[\Omega_k^-\right]^{\pm 1} - 1\right) \in \bigcap_{\delta>1} E^\delta(D^-). \tag{88}$$

Proof. The equalities

$$\Omega_k^+(\alpha(t)) = \frac{\omega^+(\alpha(t))-\omega^+(\alpha(b_k))}{\alpha(t)-\alpha(b_k)}, \quad \Omega_k^-(t) = \frac{\omega^-(t)-\omega^-(b_k)}{t-b_k},$$

imply that

$$\frac{\Omega_k^+(\alpha(t))}{\Omega_k^-(t)} = \frac{\omega^+(\alpha(t))-\omega^+(\alpha(b_k))}{\omega^-(t)-\omega^-(b_k)} \cdot \frac{t-b_k}{\alpha(t)-\alpha(b_k)}.$$

but since $\omega^+(\alpha(t)) = \omega^-(t)$, $\omega^+(\alpha(b_k)) = \omega^-(b_k)$, we obtain

$$\Omega_k^+(\alpha(t)) = h(t)\,\Omega_k^-(t), \tag{89}$$

where $h(t) = (t-b_k)(\alpha(t)-\alpha(b_k))^{-1}$.

As far as the curve Γ is smooth in the neighborhood of the points b_k and $\alpha(b_k)$, the functions $\varphi_k(z) = \frac{1}{z-b_k}$ and $\psi_k(z) = \frac{1}{z-\alpha(b_k)}$ belong to certain Smirnov classes $E^{\delta_0}(D^\pm)$, for some $\delta_0 > 0$.

Indeed, let $z = z(w)$ be the conformal mapping of U onto D, and Γ_r be the image of the circumference $|w| = r < 1$ under the mapping.

Let

$$I_r = \int\limits_{\Gamma_r} \frac{|dz|}{|z-b_k|^{\delta_0}} = \int\limits_{|w|=r} \frac{|z'(w)||dw|}{|z(w)-z(B_k)|^{\delta_0}}, \quad z(B_k) = b_k.$$

We now take into account that if $\Gamma \in \Lambda$, then $z' \in H^{1+\varepsilon}$, for some $\varepsilon > 0$, ([107, p. 160]) and $z(w) - z(B) \sim (w-B)z_0(w)$, where $[z_0]^{\pm 1} \in \bigcap_{\delta>1} H^\delta$ ([40, p. 153]).

Thus we have

$$I_r \le \left(\int\limits_{|w|=r} |z'(w)|^{1+\varepsilon}|dw|\right)^{\frac{1}{1+\varepsilon}} \left(\int\limits_{|w|=r} \frac{|dw|}{|w-B_k|^{\frac{\delta_0(1+\varepsilon)}{\varepsilon}}|z_0(w)|^{\frac{\delta_0(1+\varepsilon)}{\varepsilon}}}\right)^{\frac{\varepsilon}{1+\varepsilon}}.$$

The above-said allows us to conclude that, if $\frac{\delta_0(1+\varepsilon)}{\varepsilon} \le \frac{1}{2}$, $\sup\limits_{r>0} I_r < \infty$, and hence we have $\varphi_k \in E^{\delta_0}(D^+)$. Analogously, we can show that $\varphi_k(z) \in E^{\delta_0}(D^-)$ and $\psi_k(z) \in E^{\delta_0}(D^\pm)$.

From the estimates (87) follows $\Omega_k^\pm(t) \in L^{1+\eta}(\Gamma)$ for some $\eta > 0$. But $(\Omega_k^\pm(z) - 1) \in E^\delta(D^\pm)$ (since $\omega^+(z)$ is the bounded function, $\varphi_k \in E^\delta(D^+)$, and $\Omega_k^\pm(z) - 1$ has the first order zero at infinity), hence by Smirnov's theorem we conclude that $(\Omega_k^\pm(z) - 1) \in E^{1+\eta}(D^\pm)$. Thus the function

$$\Omega_k(z) = \begin{cases} \Omega_k^+(z), & z \in D^+, \\ \Omega_k^-(z), & z \in D^-, \end{cases}$$

is a solution of the class $\widetilde{K}^{1+\eta}(\Gamma)$ of the problem (89). But such a solution possesses the properties (88) (see Remark 35). □

Corollary 45. *For the solution $\omega(z)$ of the problem (68) we have*

$$\begin{aligned} \omega^+(z) - \omega^+(\alpha(b_k)) &= (z - \alpha(b_k))\Omega_k^+(z), \quad k = \overline{1,m}, \\ \omega^-(z) - \omega^-(b_k) &= (z - b_k)\Omega_k^-(z), \quad k = \overline{1,m}, \end{aligned} \tag{90}$$

where for the functions $\Omega_k^\pm(z)$, the inclusions (88) are fulfilled.

Now we reduce the problem (2) to the problem with a continuous coefficient. Denote $T = \left\{k : -1 < \mu_k < -\frac{1}{p(b_k)}\right\}$ and introduce the function

$$\Psi(z) = \begin{cases} \Phi(z)\left[\prod\limits_{k=1}^m X_k^1(z) \prod\limits_{k\in T} \left(\omega^+(z) - \omega^+(\alpha(b_k))\right)\right]^{-1}, & z \in D^+, \\ \Phi(z)\left[\prod\limits_{k=1}^m X_k(z) \prod\limits_{k\in T} \left(\omega^-(z) - \omega^-(b_k)\right)\right]^{-1}, & z \in D^-. \end{cases} \tag{91}$$

Since $\omega^+(\alpha(t)) - \omega^+(\alpha(b_k)) = \omega^-(t) - \omega^-(b_k)$, it follows from equality (2) that

$$\Psi^+(\alpha(t)) = A(t)\,\Psi^-(t) + B(t), \tag{92}$$

where

$$A(t) = a(t)\left[\prod_{k=1}^m X_k^1(\alpha(t))\right]^{-1} \prod_{k=1}^m X_k(t),$$

$$B(t) = b(t)\left[\prod_{k=1}^m X_k^1(\alpha(t)) \prod_{k\in T} \left(\omega^+(\alpha(t)) - \omega^+(\alpha(b_k))\right)\right]^{-1}.$$

We write $A(t)$ in the form

$$A(t) = a(t)\prod_{k=1}^m (t - z_0)^{-\tau_k} \prod_{k=1}^m \left(\frac{t - b_k}{\alpha(t) - \alpha(b_k)}\right)^{\tau_k}.$$

Since the curve Γ is smooth in the neighborhood of the points b_k and $\alpha(b_k)$, we can conclude that $A(t)$ is continuous on Γ and $A(t)\neq 0$.

Some properties of solution of the problem (92). Let us find what class $\Psi(z)$ belongs to. Since

$$\begin{aligned}\big|X_k^1(z)\big| &= \big|(z-\alpha(b_k))^{\tau_k}\big| = \big|\exp(\tau_k\ln(z-\alpha(b_k))\big|\\ &= \exp\big[\mu_k\ln|z-\alpha(b_k))| - (\operatorname{Im}\tau_k|\arg(z-\alpha(b_k)|\big],\\ \big|X_k(z)\big| &= \Big|\Big(\frac{z-b_k}{z-z_0}\Big)^{\tau_k}\Big| \sim \exp\big[\mu_k\ln|z-b_k| - (\operatorname{Im}\tau_k|\arg(z-b_k)\big],\end{aligned}$$

therefore

$$\big|X_k^1(z)\big|\sim|z-\alpha(b_k)|^{\mu_k},\quad z\in D^+\cup D^-,\quad \big|X_k(z)\big|\sim|z-b_k|^{\mu_k},\ z\in D^+. \tag{93}$$

By virtue of the adopted assumptions for the curve Γ, by the condition (83), and using the fact stated when proving Lemma 44 that the functions $\varphi_k(z)=(z-b_k)^{-1}$, $\psi_k(z)=(z-\alpha(b_k))^{-1}$, belong to $E^{\delta_0}(D^\pm)$, $\delta_0>0$, respectively. The above estimates allow us to verify that there exist $\delta>0$, such that

$$\begin{gathered}\Big[\prod_{k=1}^m X_k^1(z)\Big]^{\pm1}\in E^\delta(D^+),\quad \Big(\prod_{k=1}^m X_k(z)\Big)\in E^\delta(D^+),\\ \Big\{\Big[\prod_{k=1}^m X_k(z)\Big]^{-1}-1\Big\}\in E^\delta(D^-).\end{gathered} \tag{94}$$

Since $\Phi\in\widetilde{K}^{p(\cdot)}(\Gamma)$, hence $\Phi\in E^{\underline{p}}(D^+)$, and there exists the polynomial $p_\Phi(z)$, such that $(\Phi(z)-p_\Phi(z))\in E^{\underline{p}}(D^-)$, therefore it follows from (91) that there exist $\delta_0>0$ and the polynomial $p_\Psi(z)$, such that

$$\Psi^+(z)\in E^{\delta_0}(D^+),\quad \big[\Psi^-(z)-p_\Psi(z)\big]\in E^{\delta_0}(D^-). \tag{95}$$

By virtue of (91) and (90), we have

$$\begin{aligned}\big|\Phi^-(t)\big| &= \Big|\Psi^-(t)\prod_{k=1}^m X_k(t)\prod_{k\in T}\big(\omega^-(t)-\omega^-(b_k)\big)\Big|\\ &\sim \Big|\Psi^-(t)\prod_{k=1}^m |t-b_k|^{\gamma_k}\prod_{k\in T}\Omega_k^-(t)\Big|,\quad T=\Big\{k:\ -1<\mu_k<-\frac{1}{p(b_k)}\Big\}.\end{aligned}$$

Analogously,

$$\big|b(t)\big| = \Big|B(t)\prod_{k=1}^m X_k^1(\alpha(t))\prod_{k\in T}\big(\omega^+(\alpha(t))-\omega^+(\alpha(b_k))\big)\Big|$$

$$\sim \left| B(t) \prod_{k=1}^{m} |\alpha(t) - \alpha(b_k)|^{\mu_k} \prod_{k \in T} \big(\alpha(t) - \alpha(b_k)\big) \Omega_k(\alpha(t)) \right|$$
$$\sim \left| B(t) \prod_{k=1}^{m} |t - b_k|^{\gamma_k} \prod_{k \in T} \Omega_k^{+}(\alpha(t)) \right|. \tag{96}$$

Since $\Phi^{-}(t)$ and $b(t)$ belong to $L^{p(\cdot)}(\Gamma)$, by using Lemma 44, we find that

$$\Psi^{-}(t) \in L^{p(\cdot)-\varepsilon}(\Gamma; \rho), \quad B(t) \in L^{p(\cdot)-\varepsilon}(\Gamma; \rho), \tag{97}$$

where ε is an arbitrarily small number, and

$$\rho(t) = \prod_{k=1}^{m} |t - b_k|^{\gamma_k}, \quad -\frac{1}{p(b_k)} < \gamma_k < \frac{1}{p'(b_k)}.$$

Solution of the Homogeneous Problem. The homogeneous problem with respect to the function $\Psi(z)$ has the form

$$\Psi^{+}(\alpha(t)) = A(t)\, \Psi^{-}(t), \tag{98}$$

where $A(t)$ is the other than zero continuous function.

We so far only know that for the function $\Psi(z)$ the conditions (95) and (97) are fulfilled. When $B(t) = 0$ from (92) follows

$$\Psi^{+}(t) = A(\beta(t))\, \Psi^{-}(\beta(t)).$$

But from (91) we have

$$\big|\Psi^{-}(\beta(t))\big| \sim \left| \Phi^{-}(\beta(t)) \prod_{k=1}^{m} |\beta(t) - b_k|^{\mu_k} \prod_{k \in T} \big(\omega^{-}(\beta(t)) - \omega^{-}(b_k)\big) \right|$$
$$\sim \left| \Phi^{-}(\beta(t)) \prod_{k=1}^{m} |\beta(t) - b_k|^{\gamma_k} \prod_{k \in T} \Omega_k^{-}(\beta(t)) \right|. \tag{99}$$

Here $\Omega_k^{-}(\beta(t)) \in \bigcap\limits_{\delta > 1} L^{\delta}(\Gamma)$ (this follows from (88), because the function $\beta'(t) = \frac{1}{\alpha'(t)}$ is bounded). Moreover,

$$\big|\beta(t) - b_k\big| = \big|\beta(t) - \beta(\alpha(b_k))\big| \sim \big|t - \alpha(b_k)\big|. \tag{100}$$

The function $\Phi^{-}(t)$ belongs to $L_{p(\cdot),\alpha}(\Gamma)$, i.e., $\Phi^{-} \in L^{p(\alpha(t))}(\Gamma)$. Therefore

$$\int_{\Gamma} \big|\Phi^{-}(\beta(t))\big|^{p(t)} |dt| = \int_{\Gamma} \big|\Phi^{-}(u)\big|^{p(\alpha(u))} |\alpha'(u)| |du| \le \sup_{u \in \Gamma} |\alpha'(u)| \int_{\Gamma} \big|\Phi^{-}(u)\big|^{p(\alpha(u))} |du|.$$

By virtue of (100),

$$\prod_{k=1}^{m} \left|\beta(t) - b_k\right|^{\gamma_k} \sim \prod_{k=1}^{m} \left|t - \alpha(b_k)\right|^{\gamma_k},$$

hence this function will belong to $L^{p(\cdot)+\varepsilon}(\Gamma)$ for some $\varepsilon > 0$, if and only if

$$-\frac{1}{p(\alpha(b_k))} < \gamma_k < \frac{1}{p'(\alpha(b_k))}. \tag{101}$$

Along with (86), we receive that the above conditions are fulfilled. Taking into account the inclusions (97), we can conclude from (99) that $\Psi^+(t) \in L^{1+\eta}(\Gamma)$, for some $\eta > 0$, and by Smirnov's theorem, $\Psi^+(z) \in E^{1+\eta}(D^+)$. (93) and (97) imply that $[\Psi^-(z) - p_\Psi(z)] \in E^{1+\eta}(D^-)$. Therefore $\Psi(z) \in \widetilde{K}^{1+\eta}(\Gamma)$.

Thus all solutions of the problem

$$\Phi^+(\alpha(t)) = a(t)\,\Phi^-(t) \tag{102}$$

of the class $\widetilde{K}_\alpha^{p(\cdot)}(\Gamma)$ lie in the set defined by means of equalities (91), in which $\Psi(z)$ is a solution of the problem (98) of the class $\widetilde{K}_\alpha^{1+\eta}(\Gamma)$.

Let $y(z)$ be a canonical function of $A(t)$. This means that

$$y^+(\alpha(t)) = A(t)\,y^-(t), \quad \left[y(z)\right]^{\pm 1} \in \bigcap_{\delta>1} \widetilde{K}^\delta(\Gamma) \ \text{ and } \ \operatorname{ord}\left[y(z), \infty\right] = \operatorname{ind} A(t).$$

Such function does exist (see Theorem 38). Put

$$Z(z) = \begin{cases} y(z) \prod_{k=1}^{m} X_k^1(z), & z \in D^+, \\ y(z) \prod_{k=1}^{m} X_k(z), & z \in D^-. \end{cases} \tag{103}$$

Then $Z^+(\alpha(t)) = a(t)Z^-(t)$, and since $-1 < \mu_k \le 0$ (see condition (83) above), we have $[Z(z)]^{-1} \in \bigcap_{\delta>1} \widetilde{K}^\delta(\Gamma)$. This implies that if $F(z) = \Phi(z)[Z(z)]^{-1}$, then $F \in \widetilde{K}^{1+\eta_1}(\Gamma)$, $\eta_1 > 0$ and $F^+(\alpha(t)) = F^-(t)$. All such functions lie in the set defined by the equality $F(z) = P(\omega(z))$, where $P(z)$ is an arbitrary polynomial (see Theorem 38). Consequently, all solutions of the problem (102) are in the set

$$\Phi(z) = Z(z)\,P(\omega(z)).$$

From that set we will now try to single out all solutions of the class $\widetilde{K}_\alpha^{p(\cdot)}(\Gamma)$. Towards this end, it is obvious that at the points $\omega^-(b_k)$, $k \in T$, $P(\omega(z))$ must necessarily vanish. Thus we put

$$P_T(z) = \prod_{k\in T} \left(z - \omega^-(b_k)\right) \tag{104}$$

and let $P(z) = P_T(z)\, Q(z)$, where $Q(z)$ is an arbitrary polynomial.

Show that the functions

$$\Phi(z) = Z(z)\, P_T(\omega(z))\, Q(\omega(z)) \tag{105}$$

belong to $\widetilde{K}^{p(\cdot)}_{\alpha}(\Gamma)$. To this end, it suffices to show that the function

$$Z_T(z) = Z(z)\, P_T(\omega(z))$$

belongs to $\widetilde{K}^{p(\cdot)}_{\alpha}(\Gamma)$.

Assume $\min\limits_{1\le k\le m} \mu_k = \mu_0$. By the condition $-1 < \mu_0 \le 0$, it is not difficult to see that

$$Z_T(z) \in \widetilde{K}^{1+\eta}_{\alpha}(\Gamma), \quad 0 < \eta < \frac{1}{|\mu_0|} - 1. \tag{106}$$

Moreover,

$$|Z^-_T(t)| \sim \Big|y^-(t)\prod_{k=1}^{m}|t-b_k|^{\mu_k}\prod_{k\in T}|t-b_k|\Omega^-_k(t)\Big| = \prod_{k=1}^{m}|t-b_k|^{\gamma_k}\Big|y^-(t)\prod_{k\in T}\Omega^-_k(t)\Big|,$$

$$|Z^+_T(t)| \sim \prod_{k=1}^{m}|t-\alpha(b_k)|^{\mu_k}\prod_{k\in T}|t-\alpha(b_k)|\Big|\prod_{k\in T}\Omega^+_k(t)y^+(t)\Big|$$

$$= \prod_{k=1}^{m}|t-\alpha(b_k)|^{\gamma_k}\Big|y^+(t)\prod_{k\in T}\Omega^+_k(t)\Big|.$$

By virtue of the assumptions (86) and (101), these functions belong to $L^{p(\cdot)}(\Gamma)$, but then according to (106) we obtain $Z_T \in \widetilde{K}^{p(\cdot)}_{\alpha}(\Gamma)$.

Let us show that $Z^-(t) \in L^{p_\alpha(\cdot)}(\Gamma)$. We have

$$|Z^-(z)| = \frac{|Z^+(\alpha(t))|}{|a(t)|} \sim |Z^+(\alpha(t))| \sim \prod_{k=1}^{m}|\alpha(t)-\alpha(b_k)|^{\gamma_k}|y^+(\alpha(t)|\prod_{k\in T}|\Omega^+_k(t)|$$

$$\sim \prod_{k=1}^{m}|t-b_k|^{\gamma_k}\Big|y^+(t)\prod_{k\in T}\Omega^+_k(t)\Big|,$$

and by (101) we can conclude that $Z^-(t) \in L^{p_\alpha(\cdot)}(\Gamma)$.

Hence $Z_T \in \widetilde{K}^{p(\cdot)}(\Gamma)$ and $Z^-_T \in L^{p_\alpha(\cdot)}(\Gamma)$. This implies that $Z_T(z)$ and, consequently, all the functions, given by equality (105), belong to $\widetilde{K}^{p_\alpha(\cdot)}(\Gamma)$. Among these solutions, to the class $K^{p(\cdot)}_{\alpha}(\Gamma)$ belong the functions for which $\Phi^-(\infty) = 0$.

Denote by $N(T)$ a number of points of the set T. If the number

$$\varkappa(a) = \operatorname{ind} A(t) - N(T) = \varkappa(A) - N(T)$$

is negative, then $Z_T(z)$ has a pole at infinity, therefore we take $Q(z) = 0$. When $\varkappa(a) \ge 0$, in this case in the capacity of $Q(z)$ we take an arbitrary polynomial of order $\varkappa(a) - 1$ ($Q_{-1}(z) \equiv 0$).

Finally, we sum up the obtained results.

Theorem 46. *Let $\Gamma \in \Lambda$; $p \in \mathcal{P}(\Gamma)$; $\alpha(t)$ be homeomorphism of Γ onto itself, $\alpha' \in H(\Gamma)$, $\alpha'(t) \neq 0$, $a(t) \in C^1(\Gamma; b_1, \dots, b_m)$, $\inf |a(t)| > 0$,*

$$\frac{a(b_k-)}{a(b_k+)} = \lambda_k, \ \tau_k = \frac{1}{2\pi i} \ln \lambda_k, \ \ -1 < \mu_k = \operatorname{Re} \tau_k \leq 0, \quad \mu_k \neq -\frac{1}{p(b_k)},$$

and

$$\gamma_k = \begin{cases} \mu_k + 1, & \text{if} \quad -1 < \mu_k < -\dfrac{1}{p(b_k)}, \\ \mu_k, & \text{if} \quad -\dfrac{1}{p(b_k)} < \mu_k \leq 0, \end{cases}$$

and the conditions

$$-\frac{1}{p(\alpha(b_k))} < \gamma_k < \frac{1}{p'(\alpha(b_k))}$$

be fulfilled. Then the problem (102) in the class $\widetilde{K}^p_\alpha(\Gamma)$ is solvable and all its solutions are given by the equality (105), i.e. by equality

$$\Phi(z) = Z(z)\, P_T(\omega(z))\, Q(\omega(z)),$$

in which $Z(z)$ is defined by equality (103) and $P_T(z)$ by (104), where $T = \{k : -1 < \mu_k < -\frac{1}{p(b_k)}\}$, and $\omega(z)$ is a solution of the problem (68) mentioned in Lemma 37, $Q(z)$ is an arbitrary polynomial.

The problem (102) is solvable in the class $K^{p(\cdot)}_\alpha(\Gamma)$, as well. A number of solutions depend on the value

$$\varkappa(a) = \operatorname{ind} A(t) - N(T),$$

where

$$A(t) = a(t) \left[\prod_{k=1}^{m} X^1_k(\alpha(t)) \right]^{-1} \prod_{k=1}^{m} X_k(t).$$

Solution of the Inhomogeneous Problem. Here we rely on the results of Section 4.16 in which it is stated that the Haseman problem (92) has for a continuous coefficient $A(t)$ a solution of the class $\widetilde{K}^p_\alpha(\Gamma; \rho)$, when $\rho(t)$ is the power function given by (60), and the function $B(t)$ belongs to the class $L^{p(\cdot)}(\Gamma; \rho) \cap L^{p_\alpha(\cdot)}(\Gamma; \rho_\alpha)$. As we already know, $B(t) \in L^{p(\cdot)-\varepsilon}(\Gamma; \rho)$ (see (97)).

Assuming

$$b(t) \in \bigcup_{\delta>0} L^{p(\cdot)+\delta}(\Gamma), \tag{107}$$

from the estimates (96) we find that $b(t) \in L^{p(\cdot)}(\Gamma; \rho)$. We require for $B(t)$ to belong to the class $L^{p_\alpha(\cdot)}(\Gamma; \rho_\alpha)$. But by virtue of (96), we have

$$\int_\Gamma |B(t)\, \rho(\alpha(t))|^{p(\alpha(t))} |dt| \sim \int_\Gamma \left| \frac{b(t)\, \rho(\alpha(t))}{\rho(t) \prod_{k=1}^m \Omega^+_k(\alpha(t))} \right|^{p(\alpha(t))} |dt|,$$

and if we assume that

$$b(t) \in \bigcup_{\delta>0} L^{p_\alpha(\cdot)+\delta}\Big(\Gamma; \frac{\rho}{\rho_\alpha}\Big), \tag{108}$$

then we will get $B(t) \in L^{p(\cdot)}(\Gamma; \rho_\alpha)$.

Thus if the conditions (107) and (108) are fulfilled, then $B(t)$ belongs to $L_{p(\cdot),\alpha}(\Gamma; \rho)$, and hence we can to the problem (92) apply Theorem 42 by which this problem has the solution $\Psi_0(z) \in \widetilde{K}_\alpha^{p(\cdot)}(\Gamma; \rho)$, and what is more, for $\varkappa(A) \geq 0$ we can select this solution from $K_\alpha^{p(\cdot)}(\Gamma; \rho)$. If, however, $\varkappa(A) < 0$, such a solution exists if the $|\varkappa(A)|$-conditions (81) are fulfilled.

Putting

$$\Phi_0(z) = \begin{cases} \Psi_0(z) \prod\limits_{k=1}^{m} X_k^1(z) \prod\limits_{k\in T} \big(\omega^+(z) - \omega^+(\alpha(b_k))\big), & z \in D^+, \\ \Psi_0(z) \prod\limits_{k=1}^{m} X_k(z) \prod\limits_{k\in T} \big(\omega^-(z) - \omega^-(b_k)\big), & z \in D^-, \end{cases} \tag{109}$$

and using the estimates (91), we can easily establish that $\Phi_0 \in \widetilde{K}_\alpha^{p(\cdot)}(\Gamma; \rho)$. Thus the functions

$$\Phi(z) = \Phi_0(z) + Z(z)\, P_T(\omega(z))\, Q(\omega(z)) \tag{110}$$

make up a general solution of the problem (2) of the class $\widetilde{K}_\alpha^{p(\cdot)}(\Gamma)$.

As a result of the above reasoning, we arrive at the following

Theorem 47. *Let all the assumptions of Theorem 46 be fulfilled, and the inclusions (107) and (108) hold. Then the problem (2) is solvable in the class $\widetilde{K}_\alpha^{p(\cdot)}(\Gamma)$, and its general solution is given by equality (110) in which $\Psi_0(z)$ is the solution of problem (92) and Φ_0, $Z(z)$ and $P_T(z)$ are given by equalities (109), (103) and (104), respectively, $Q(z)$ is an arbitrary polynomial, and $\omega(z)$ is a solution of the problem (68) pointed out in Lemma 37.*

For the solvability in $K_\alpha^{p(\cdot)}(\Gamma)$ we have

I. *When $\varkappa(A) < 0$ for solvability of the problem (2) in the class $K_\alpha^{p(\cdot)}(\Gamma)$ necessary and sufficient that conditions (81) are hold. If these conditions are hold then problem has a uniquely solution $\Phi_0(z)$.*

II. *If $\varkappa(A) \geq 0$ and $\varkappa(a) = [\varkappa(A) - N(T)] < 0$, where $N(T)$ is a number of points of the set $T = \big\{k : -1 < \mu_k < -\frac{1}{p(b_k)}\big\}$, the problem (2) is unconditionally solvable and has a unique solution.*

III. *In the case $\varkappa(a) \geq 0$, the problem is solvable and its general solution is given by equality (110), where $Q(z)$ is an arbitrary polynomial of order $\varkappa(a) - 1$, $(Q_{-1}(z) \equiv 0)$.*

Part III

The Riemann boundary problem with data from the grand Lebesgue spaces

In this Part, the boundary value problem is investigated in the class of analytic functions representable by Cauchy type integral with density in $L^{p),\theta}(\Gamma)$, $p > 1$, $\theta > 0$.

The space $L^{p),\theta}(\Gamma;\omega)$ is a Banach function space.

The following continuous embeddings are fulfilled:

$$L^{p}(\Gamma;\omega) \subset L^{p),\theta}(\Gamma;\omega) \subset L^{p-\varepsilon}(\Gamma;\omega)\,, \quad 0 < \varepsilon \leq p-1.$$

It is well known [50] that, for instance, the function $g_\lambda(t) = t^{-\frac{1}{p}} \ln^\lambda \frac{2}{t} \in L^{p),1}$ if and only if $\lambda \leq 0$. Along with this, $g_\lambda \notin L^p$.

In the previous studies, the discontinuous boundary value problems were studied under the assumption that the boundary function belongs to the class $L^{p),\theta}(\Gamma)$, $1 < p < \infty$, $\theta > 0$. Therefore functions of the form $g_\lambda(t)$ were inadmissible for the use as boundary functions.

4.18 Classes $W^{p),\theta}(\Gamma)$ and $K^{p),\theta}(\Gamma)$

Let ω be a measurable function different from zero almost everywhere on Γ. We will write $\omega \in W^{p),\theta}(\Gamma)$, $1 < p < \infty$, $\theta > 0$, if the Cauchy singular operator S_Γ is continuous in $L^{p),\theta}(\Gamma,\omega)$. This definition is equivalent to the following: if $\omega \in W^{p),\theta}(\Gamma)$, then the operator

$$\frac{\omega(t)}{\pi i} \int\limits_{\Gamma} \frac{f(\tau)}{\omega(\tau)} \frac{d\tau}{\tau - t}, \qquad t \in \Gamma,$$

is continuous in $L^{p),\theta}(\Gamma)$.

In Chapter 2 it is proved that: if Γ is a simple closed rectifiable Carleson curve, $t_k \in \Gamma$, $k = \overline{1,m}$, $\alpha_k \in \mathbb{R}$, $p > 1$, then the function

$$\omega(t) = \prod_{k=1}^{m} |t - t_k|^{\alpha_k}$$

belongs to $W^{p),\theta}(\Gamma)$ provided that the conditions $-\frac{1}{p} < \alpha_k < \frac{1}{p'}$ are fulfilled.

To investigate the Riemann problem, we need to have a more general function belonging to $W^{p),\theta}$, namely,

$$\rho(t) = \prod_{k=1}^{n} |t - t_k|^{\alpha_k} \exp \int\limits_{\Gamma} \frac{\psi(t)}{\tau - t}\, d\tau,$$

where $-\frac{1}{p} < \alpha_k < \frac{1}{p'}$ and $\psi(t)$ is a real continuous function given on Γ.

Below we give two proofs of this fact, of which one rests on the analysis of the boundary value problem and the other on the theory of weighted singular integrals in the Lebesgue spaces with weights from Muckenhoupt type classes. The investigation of each of the proofs of the above-mentioned facts is, in our opinion, of independent interest.

Assume

$$K^{p),\theta}(\Gamma;\omega) = \left\{\Phi:\ \Phi(z) = \frac{1}{2\pi i}\int\limits_{\Gamma} \frac{\varphi(\tau)}{\tau - z}\,d\tau,\ \ z \notin \Gamma,\ \ \varphi \in L^{p),\theta}(\Gamma;\omega)\right\},$$

$$K^{p),\theta}(\Gamma) \equiv K^{p),\theta}(\Gamma;1).$$

As is known (see 0.21), if Γ is a Carleson curve and $\varphi \in L^p(\Gamma)$, $p > 1$, then the function $(K_\Gamma\varphi)(z) = \frac{1}{2\pi i}\int_\Gamma \frac{\varphi(\tau)\,d\tau}{\tau - z}$ belongs to the Smirnov class $E^p(D^\pm)$. Hence it follows that if $\varphi \in L^{p),\theta}(\Gamma)$, then $(K_\Gamma\varphi) \in \bigcap\limits_{p>\varepsilon>0} E^{p-\varepsilon}(D^\pm)$.

One property of the operator S_Γ

Theorem 48. *Let Γ be a simple rectifiable curve, $p > 1$, $1/\omega \in L^{p'+\varepsilon}(\Gamma)$, $\varepsilon > 0$, and the operator S_Γ map the space $L^{p),\theta}(\Gamma;\omega)$ into $L^{p),\theta}(\Gamma;\omega)$, then the operator S_Γ is continuous in $L^{p),\theta}(\Gamma;\omega)$.*

Proof. First we will show that the operator S_Γ is measure-continuous in $L^{p),\theta}(\Gamma;\omega)$, i.e. if f_n tends to f_0 in $L^{p),\theta}(\Gamma;\omega)$, then $S_\Gamma f_n$ tends with respect to measure to $S_\Gamma f_0$. For this, we observe that $\forall f \in L^{p),\theta}(\Gamma;\omega)$ we have $f = \varphi/\omega$ where

$$\varphi \in L^{p),\theta}(\Gamma) \subset \bigcap_{0<\varepsilon<p-1} L^{p-\varepsilon}(\Gamma).$$

Since $\frac{1}{\omega} \in L^{p'+\varepsilon}(\Gamma)$, hence it follows that $f \in L^1(\Gamma)$, i.e. $L^{p),\theta}(\Gamma;\omega) \subset L^1(\Gamma)$. Since the sequence $\{f_n\}_{n=1}^{\infty}$ converges to f_0 in $L^{p),\theta}(\Gamma;\omega)$, the more so $\{f_n\}$ converges to f_0 in $L^1(\Gamma)$. But the operator S_Γ is continuous from $L^1(\Gamma)$ into the space of convergence in measure ([40, Theorem 2.1, p. 21]). This implies the continuity of the operator S_Γ from $L^{p),\theta}(\Gamma;\omega)$ into the space of convergence in measure.

Let now $S_\Gamma(L^{p),\theta}(\Gamma;\omega)) \subset L^{p),\theta}(\Gamma;\omega)$. We will show that S_Γ is a closed operator in $L^{p),\theta}(\Gamma;\omega)$. Indeed, if $f_n \overset{L^{p),\theta}(\Gamma;\omega)}{\longrightarrow} f_0$ and $S_\Gamma f_n \overset{L^{p),\theta}(\Gamma;\omega)}{\longrightarrow} \psi$ and we know that $S_\Gamma f_n$ converges in measure to $S_\Gamma f_0$, then we conclude that $\psi = S_\Gamma f_0$. This means the closure of the operator S_Γ in $L^{p),\theta}(\Gamma;\omega)$. Since $L^{p),\theta}(\Gamma;\omega)$ is a Banach space, by virtue of the well known theorem on a closed graph we conclude that S_Γ is a continuous operator in $L^{p),\theta}(\Gamma;\omega)$. □

Problem (1) in the Class $K^{p),\theta}(\Gamma)$ for Continuous G

Solution of problem (1) with continuous G is given by the following

Theorem 49. *Let Γ be a simple closed rectifiable Carleson curve, $G(t) \in C(\Gamma)$, $G(t) \neq 0$, $t \in \Gamma$, $p > 1$, and $g(t) \in L^{p),\theta}(\Gamma)$. If $\varkappa = \operatorname{ind} G(t) = \frac{1}{2\pi}[\arg G(t)]_\Gamma$, then*

i) *for $\varkappa \geq 0$, problem* (1) *is unconditionally solved in $K^{p),\theta}(\Gamma)$ and its general solution is given by the equality*

$$\Phi(z) = \frac{X(z)}{2\pi i} \int\limits_\Gamma \frac{g(t)}{X^+(t)(t-z)} + X(z) Q_{\varkappa-1}(z), \tag{111}$$

where

$$X(z) = \begin{cases} \exp h(z), & z \in D^+, \\ (t-z_0)^{-\varkappa} \exp h(z), & z \in D^-, \quad z_0 \in D^+, \end{cases}$$
$$h(z) = K_\Gamma \left(\ln G(t)(t-z_0)^{-\varkappa} \right)(z),$$

and $Q_{\varkappa-1}(z)$ is an arbitrary polynomial of order $\varkappa - 1$ ($Q_{-1}(z) \equiv 0$).

ii) *for $\varkappa < 0$, problem* (1) *is solvable in $K^{p),\theta}(\Gamma)$ if and only if*

$$\int\limits_\Gamma \frac{g(t)}{X^+(t)} t^k dt = 0, \quad k = \overline{0, \varkappa - 1}.$$

If the latter conditions are fulfilled, then problem (1) *has a unique solution given by equality* (111) *for $Q_{\varkappa-1}(z) \equiv 0$.*

Proof. We will follow to the proof of Theorem 8. Choose a rational function $\widetilde{G}$ so that

$$\sup_{t \in \Gamma} \left| \frac{G(t)}{\widetilde{G}(t)} - 1 \right| < \frac{1}{2} \left(1 + \|S_\Gamma\|_{L^{p),\theta}(\Gamma)} \right)^{-1}.$$

Put

$$\widetilde{X} = \exp\left(K_\Gamma(\ln \widetilde{G}) \right)(z).$$

Observe that the function $1/\widetilde{X}$ is bounded, therefore $\Phi/\widetilde{X} \in K^{p),\theta}(\Gamma)$.

Let

$$\frac{\Phi(z)}{\widetilde{X}(z)} = K_\Gamma \psi(z)$$

with $\psi \in L^{p),\theta}(\Gamma)$. In view of the selection of $\widetilde{G}$ we claim that equation (22) is uniquely solvable in $L^{p),\theta}(\Gamma)$.

Let $\varkappa = 0$ and

$$X(z) = \exp\left(K_\Gamma(\ln G) \right)(z).$$

Taking into account the inclusion $L^{p),\theta}(\Gamma) \subset L^{p-\varepsilon}(\Gamma)$ we conclude that

$$\Phi/X \in K^1(\Gamma).$$

Hence, a solution of (1) in $K^{p),\theta}(\Gamma)$ belongs also to $K^1(\Gamma)$. But problem (1) in $K^1(\Gamma)$ is uniquely solvable and this solution is

$$\Phi(z) = X(z) K_\Gamma \left(\frac{g}{X^+}\right)(z).$$

Therefore the latter function is a unique solution of (1) belonging also to $K^{p),\theta}(\Gamma)$.

If we take into account that the function $\Phi^+(t) - (t - z_0)^{\varkappa}\Phi^-(t)$ belongs to $L^{p),\theta}(\Gamma)$, then the remaining part of the proof of Theorem 8 for the case $\varkappa \neq 0$ is valid. □

On a weight function from $W^{p),\theta}(\Gamma)$

Theorem 50. *If Γ is a simple closed rectifiable Carleson curve, $\omega \in W^{p),\theta}(\Gamma)$, $p > 1$, $\frac{1}{\omega(t)} \in L^{p'+\varepsilon}(\Gamma)$, $\varepsilon > 0$, and ψ is a real continuous function on Γ, then*

$$\rho(t) = \omega(t) \exp \int_\Gamma \frac{\psi(\tau)\, d\tau}{\tau - t} \in W^{p),\theta}(\Gamma). \tag{112}$$

Proof. Consider problem (1) in the class $K^{p),\theta}(\Gamma;\omega)$ for $G(t) = \exp 2\pi i \psi(t)$. Observe that the function G is continuous on Γ and $\varkappa(G) = 0$. Choose a rational function $\widetilde{G}(z)$ such that

$$\sup_{t\in\Gamma} \left| \frac{G(t)}{\widetilde{G}(t)} - 1 \right| < \frac{1}{2}\left(1 + \|S_\Gamma\|_{L^{p),\theta}(\Gamma;\omega)}\right)^{-1}. \tag{113}$$

It is obvious that $\operatorname{ind} \widetilde{G}(t) = 0$. Consider the function

$$\widetilde{X}(z) = \exp\left\{\left(K_\Gamma(\ln \widetilde{G}(t))\right)(z)\right\}, \quad z \notin \Gamma.$$

Notice that this function is continuous in the domains $\overline{D^\pm}$-bounded by Γ and that $\widetilde{X}(z) \neq 0$. Since $\widetilde{X}^+(t)[\widetilde{X}^-(t)]^{-1} = \widetilde{G}(t)$, condition (1) can be written in the form

$$\left(\frac{\Phi}{\widetilde{X}}\right)^+ = \frac{G}{\widetilde{G}}\left(\frac{\Phi}{\widetilde{X}}\right)^- + \frac{g}{\widetilde{X}^+}, \tag{114}$$

i.e.

$$F^+ = \frac{G}{\widetilde{G}} F^- + \frac{g}{\widetilde{X}^+},$$

where

$$F(z) = \frac{\Phi(z)}{\widetilde{X}(z)}.$$

Since $\Phi \in K^{p),\theta}(\Gamma;\omega)$, we have

$$\Phi(z) = \frac{1}{2\pi i}\int\limits_{\Gamma}\frac{\varphi(t)\,dt}{t-z}, \quad \varphi \in L^{p),\theta}(\Gamma;\omega),$$

i.e. $\varphi = \frac{\mu}{\omega}$, where $\mu \in L^{p),\theta}(\Gamma)$ and, by assumption, $\frac{1}{\omega} \in L^{p'+\varepsilon}(\Gamma)$.

Since $L^{p),\theta}(\Gamma) \subset \bigcap\limits_{0<\delta<p-1} L^{p-\delta}(\Gamma)$, $p>1$, it is easy to establish that $\varphi \in L^{\alpha}(\Gamma)$, for some $\alpha > 1$. Since Γ is a Carleson curve, we have $F \in E^{\alpha}(D^{\pm})$. From the condition $\alpha > 1$ it follows that $F(z) = K_{\Gamma}(F^{+}-F^{-})(z) = \frac{1}{2\pi i}\int\limits_{\Gamma}\frac{f(t)\,dt}{t-z}$, $f = F^{+}-F^{-}$. The functions $F^{\pm}$ belong to $L^{p),\theta}(\Gamma)$. Indeed, by the Plemelj–Sokhotskiĭ formulas we obtain

$$\Phi^{\pm} = \pm\frac{1}{2}\varphi + \frac{1}{2}S_{\Gamma}\varphi, \quad \varphi \in L^{p),\theta}(\Gamma;\omega)$$

and since $\omega \in W^{p),\theta}(\Gamma)$, we conclude that $\Phi^{\pm}L^{p),\theta}(\Gamma;\omega)$, whereas by virtue of the boundedness of the function $\frac{1}{\widetilde{X}(z)}$ the more so $F^{\pm} \in L^{p),\theta}(\Gamma;\omega)$. Thus F is a solution of problem (1) from the class $K^{p),\theta}(\Gamma;\omega)$. Let us write (114) in the form

$$F^{+} - F^{-} = \left(\frac{G}{\widetilde{\widetilde{G}}} - 1\right)F^{-} + g_1, \quad g_1 = \frac{g}{\widetilde{\widetilde{X}}^{+}},$$

i.e.

$$f = \left(\frac{G}{\widetilde{\widetilde{G}}} - 1\right)\left(-\frac{1}{2}f + \frac{1}{2}S_{\Gamma}f\right) + g_1. \tag{115}$$

By virtue of condition (113), the right-hand part of (115) contains the contracting operator in $L^{p),\theta}(\Gamma;\omega)$ and therefore equation (115) is uniquely solvable in $L^{p),\theta}(\Gamma;\omega)$. Therefore problem (114) and thereby problem (1) have a unique solution in $K^{p),\theta}(\Gamma;\omega)$. According to Theorem 49, it defined given by equality (111) where $Q_{\varkappa-1} = 0$. This means that the function

$$\Phi(z) = \frac{X(z)}{2\pi i}\int\limits_{\Gamma}\frac{g(\tau)}{X^{+}(\tau)}\frac{d\tau}{\tau - z},$$

where $X(z)$ is given in Theorem 49, belongs to $K^{p),\theta}(\Gamma;\omega)$ for any $g \in L^{p),\theta}(\Gamma;\omega)$. Hence it follows that the operator $X^{+}S_{\Gamma}\frac{g}{X^{+}}$ acts from $L^{p),\theta}(\Gamma;\omega)$ into $L^{p),\theta}(\Gamma;\omega)$. Therefore

$$Tf = \frac{\omega(t)X^{+}(t)}{\pi i}\int\limits_{\Gamma}\frac{f(\tau)}{\omega(\tau)X^{+}(\tau)}\frac{d\tau}{\tau - t} \in L^{p),\theta}(\Gamma)$$

for any $f \in L^{p),\theta}(\Gamma)$.

By Theorem 48 we conclude that T is continuous in $L^{p),\theta}(\Gamma)$, i. e. $\omega X^{+} \in W^{p),\theta}(\Gamma)$. Since $G(t) = \exp 2\pi i\psi(t)$, we have $X^{+}(t) \sim \exp\int_{\Gamma}\frac{\psi(\tau)\,d\tau}{\tau - t}$, $t \in \Gamma$. Therefore $\rho(t) = \omega(t)X^{+}(t) \in W^{p),\theta}(\Gamma)$. □

As has been mentioned above, we are going to give another proof of the fact that a certain function belongs to the class $W^{p),\theta}(\Gamma)$. We begin with

Definition 51. Let Γ be a rectifiable curve. We say that $\omega \in A_p(\Gamma)$ if the condition

$$\sup_{\substack{z\in\Gamma\\ r>0}} \frac{1}{r}\int\limits_{B(z,r)} \omega(t)\,ds\left(\frac{1}{r}\int\limits_{D(z,r)} \omega^{1-p'}(t)ds\right)^{p-1} < \infty \quad (A_p(\Gamma)\text{ condition})$$

is fulfilled, where $D(z,r) = B(z,r)\cap\Gamma$ and $B(z,r)$ is a circle with center at the point z and radius r.

The following theorem is the known one.

Theorem A [5], [40]. *Let $1<p<\infty$. For a function ρ to belong to the class $W^p(\Gamma)$ it is necessary and sufficient that Γ be Carleson curve and the condition $\rho^p \in A_p(\Gamma)$ be fulfilled.*

In particular the function $\omega(t) = \prod_{k=1}^n |t-t_k|^{\alpha_k}$ belongs to the class $W^p(\Gamma)$ if and only if Γ is Carleson curve and the conditions $-\frac{1}{p} < \alpha_k < \frac{1}{p'}$ are fulfilled.

Remark 52. Theorem A can be reformulated as follows: a Cauchy singular operator is bounded in the space $L^p_\omega(\Gamma)$ for $1<p<\infty$ if and only if Γ is a Carleson curve and $\omega \in A_p(\Gamma)$. Here the norm in the space $L^p_\omega(\Gamma)$ is understood in the sense

$$\|f\|_{L^p_\omega(\Gamma)} = \left(\int\limits_\Gamma |f(t)|^p\omega(t)\,ds\right)^{\frac{1}{p}}.$$

A generalization of Theorem A in the latter formulation for the spaces $L^{p),\theta}_\omega(\Gamma)$, where the norm is understood in the sense

$$\|f\|_{L^{p),\theta}_\omega(\Gamma)} = \sup_{0<\varepsilon<p-1}\left(\frac{\varepsilon^\theta}{|\Gamma|}\int\limits_\Gamma |f(t)|^{p-\varepsilon}\omega(t)\,ds\right)^{\frac{1}{p-\varepsilon}},$$

is obtained in [50].

Let now return to the classes $W^{p),\theta}$. Note that the equivalency

$$g \in L^{p),\theta}_\omega \Longleftrightarrow g\omega^{\frac{1}{p}} \in L^{p),\theta}_\omega$$

does not hold for the Grand Lebesgue Spaces.

Theorem 53. *Let for some semisublinear operator T the weighted operator*

$$T_\rho : f \to \rho T\left(\frac{f}{\rho}\right)$$

be bounded in $L^p(\Gamma)$ $(1<p<\infty)$ for $\rho^p \in A_p(\Gamma)$ and at the same time T_ρ be bounded in $L^{p-\sigma}(\Gamma)$ for $\rho^{p-\sigma} \in A_{p-\sigma}(\Gamma)$ for some small σ. Then T_ρ is bounded in $L^{p),\theta}(\Gamma)$, i.e. $\rho \in W^{p),\theta}(\Gamma)$.

Proof. First we will prove that there exists a number $M > 0$ such that

$$\|T_\rho\|_{L^{p-\varepsilon}(\Gamma)\to L^{p-\varepsilon}(\Gamma)} \le M \tag{116}$$

for any $\varepsilon \in (0,\sigma)$.

As is known, if $\rho^p \in A_p(\Gamma)$, then there exists a number $\sigma \in (0, p-1)$ such that $\rho^p \in A_{p-\sigma}(\Gamma)$, and also $\rho^{p\alpha} \in A_{p-\sigma}(\Gamma)$ for arbitrary α, $0 < \alpha < 1$. Assume now $\alpha = \frac{p-\sigma}{p}$. Then $0<\alpha<1$ and $\rho^{p-\sigma} \in A_{p-\sigma}(\Gamma)$.

By virtue of the condition of the theorem we have

$$\|T_\rho f\|_{L^p(\Gamma)} \le M_1 \|f\|_{L^p(\Gamma)}$$

and

$$\|T_\rho f\|_{L^{p-\sigma}(\Gamma)} \le M_2 \|f\|_{L^{p-\varepsilon}(\Gamma)}\,.$$

For $p-\varepsilon$ with the condition $\varepsilon \in (0,\sigma)$ there exists $t_\varepsilon \in (0,1)$ such that

$$\frac{1}{p-\varepsilon} = \frac{t_\varepsilon}{p} + \frac{1-t_\varepsilon}{p-\sigma}\,.$$

By the Riesz–Thorin interpolation theorem (see [136]), the inequality

$$\|T_\rho f\|_{L^{p-\varepsilon}(\Gamma)} \le M_1^{t_\varepsilon} M_2^{1-t_\varepsilon} \|f\|_{L^{p-\varepsilon}(\Gamma)}$$

is fulfilled for arbitrary $\varepsilon \in (0,\sigma)$. From the latter inequality follows (116).

We have

$$\|T_\rho f\|_{L^{p),\theta}(\Gamma)} \le \max\{A,B\} \|f\|_{L^{p),\theta}(\Gamma)}, \tag{117}$$

where

$$A = \sup_{0<\varepsilon\le\sigma} \varepsilon^{\frac{\theta}{p-\varepsilon}} \|T_\rho f\|_{L^{p-\varepsilon}(\Gamma)}$$

and

$$B = \sup_{\sigma<\varepsilon\le p-1} \varepsilon^{\frac{\theta}{p-\varepsilon}} \|T_\rho f\|_{L^{p-\varepsilon}(\Gamma)}.$$

Fix $\varepsilon \in (\sigma, p-1)$. Then $\frac{p-\sigma}{p-\varepsilon} > 1$.

Using the Hölder inequality with the exponent $\frac{p-\sigma}{p-\varepsilon}$ and observing that

$$\left(\frac{p-\sigma}{p-\varepsilon}\right)' = \frac{p-\sigma}{\varepsilon-\sigma},$$

We find that

$$\|T_\rho f\|_{L^{p-\varepsilon}(\Gamma)} \le \|T_\rho f\|_{L^{p-\sigma}(\Gamma)} |\Gamma|^{\frac{\varepsilon-\sigma}{(p-\varepsilon)(p-\sigma)}} \le c\|T_\rho f\|_{L^{p-\sigma}(\Gamma)}, \tag{118}$$

where c is a constant not depending on f and ε.

Further, since $\varepsilon \in (\sigma, p-1)$, we have

$$0 < \frac{\varepsilon - \sigma}{(p-\sigma)(p-\varepsilon)} < \frac{p-1-\sigma}{p-\sigma}, \quad (p-1)\sigma^{-\frac{1}{p-\sigma}} > 1.$$

Then

$$\|T_\rho f\|_{L^{p),\theta}(\Gamma)} \le A \max\left\{1, \varepsilon^{\frac{\theta}{p-\varepsilon}} \sigma^{-\frac{\theta}{p-\sigma}}\right\}$$
$$\le \sup_{0<\varepsilon\le\sigma} \left(\varepsilon^{\frac{\theta}{p-\varepsilon}} \|T_\rho f\|_{L^{p-\varepsilon}(\Gamma)} \max\left\{1, \varepsilon^{\frac{\theta}{p-\varepsilon}} \sigma^{-\frac{\theta}{p-\sigma}}\right\}\right).$$

Taking advantage of (116), from (117) and (118) we finally get

$$\|T_\rho f\|_{L^{p),\theta}} \le c \max\left\{1, (p-1)\sigma^{-\frac{\theta}{p-\sigma}}\right\} \|f\|_{L^{p),\theta}}. \qquad \square$$

Corollary 54. *Let $1 < p < \infty$, $\theta > 0$. Let, further, ψ be a real continuous function given on Γ.*

If Γ is a Carleson curve, then

$$\rho(t) = \prod_{k=1}^{n} |t - t_k|^{\alpha_k} \exp \int_\Gamma \frac{\psi(\tau)\, d\tau}{\tau - t},$$

where $-\frac{1}{p} < \alpha_k < \frac{1}{p'}$, belongs to the class $W^{p),\theta}(\Gamma)$.

Proof. We know (see Corollary 11) that $\rho \in W^p(\Gamma)$, i.e. $\rho^p \in A_p(\Gamma)$. Then Theorem 53 implies that $\rho \in W^{p),\theta}(\Gamma)$. $\square$

4.19 Problem (1) in the class $K^{p),\theta}(\Gamma)$ with a piecewise continuous coefficient $G(t)$ and $g \in L^{p),\theta}(\Gamma)$

First we give the condition with respect to the curve Γ which is assumed to be fulfilled when considering problem (1) in $K^{p),\theta}(\Gamma)$. Let Γ be a Carleson curve such that if $\arg(z - t_0)$, $z \in D$, changes continuously for any point $t_0 \in \Gamma$, then we obtain a uniformly bounded function with respect to t_0, i.e.

$$|\arg(z - t_0)| < M, \quad z \in D^+, \quad t_0 \in \Gamma. \tag{119}$$

Below we give one more requirement for Γ.

Now we introduce conditions for G. Let $G \in C(\Gamma; t_1, \dots, t_m)$. It is assumed that the points t_k are numbered according to the growth of the arc abscissa. Suppose that

$$\inf_{t\in\Gamma} |G(t)| > 0. \tag{120}$$

We will show that the method adopted in [24], [97], [42] for factorization of piece-wise continuous functions can be successfully applied in the case under consideration if the theorems proved above are used.

Let

$$\frac{G(t_k-)}{G(t_k+)} = e^{2\pi i\lambda_k}, \quad k = \overline{1,m}, \tag{121}$$

where $\lambda_k = \alpha_k + i\beta_k$ is a complex number whose real part is defined to an accuracy of an integer summand.

Assume that

$$\alpha_k \neq \frac{1}{p'} \pmod 1. \tag{122}$$

Then α_k can be chosen so that

$$-\frac{1}{p} < \alpha_k < \frac{1}{p'}\,. \tag{123}$$

Choose in D^+ a point z_0 and draw from it a simple smooth curve γ_k connecting it with $z = \infty$ and having only one common point t_k with Γ. The function $(z - z_0)^{\lambda_k}$ is analytic in the plane cut along γ_k and

$$(t_k - z_0)_+^{\lambda_k} = (t_k - z_0)_-^{\lambda_k} \exp(2\pi i\lambda_k),$$

where $(t_k - z_0)_\pm^{\lambda_k} = \lim\limits_{t\to t_k\pm,\, t\in\Gamma} (t - z_0)^{\lambda_k}$. This and (121) imply that the function

$$G_1(t) = \frac{G(t)}{\prod\limits_{k=1}^{m} (t - z_0)^{\lambda_k}} \tag{124}$$

is continuous on Γ ([25, p. 432]) and $G_1(t) \neq 0$, $t \in \Gamma$.

Let

$$\rho_k(z) = \begin{cases} (z - z_0)^{\lambda_k}, & z \in D^+, \\ \left(\frac{z-t_k}{z-z_0}\right)^{\lambda_k} = \frac{(z-t_k)^{\lambda_k}}{(z-z_0)^{\lambda_k}}, & z \in D^-, \end{cases} \quad k = \overline{1,m}, \tag{125}$$

where the branch of functions $\left(\frac{z-t_k}{z-z_0}\right)^{\lambda_k}$ is chosen so that it tends to 1 as $z \to \infty$. These functions are analytic in the domains $D^\pm$.

We have

$$\rho_k(z) = e^{\alpha_k \ln|z-t_k| - \beta_k \arg(z-t_k)} e^{i(\beta_k \ln|z-t_k| + \alpha_k \arg(z-t_k))}.$$

By virtue of condition (119) we conclude that

$$|\rho_k(z)| \sim |z - t_k|^{\alpha_k}.$$

Let us now require that

$$(\rho_k(z) - 1) \in \bigcup_{\delta>0} E^{\delta}(D^{\pm}). \tag{126}$$

Since Γ has the Carleson property and using (123) the function $r(t) = |t - t_k|^{\alpha_k}$ belongs to $W^p(\Gamma)$. Hence it follows that for some $\eta > 0$

$$\rho_k^{\pm} \in L^{p+\eta}(\Gamma), \quad \frac{1}{\rho_k^{\pm}} \in L^{p'+\eta}(\Gamma). \tag{127}$$

Since Γ is a Carleson curve, the domains $D^{\pm}$ are Smirnov domains (see 0.21) and therefore to the functions $\rho_k(z)$ we can apply the Smirnov's theorem by virtue of (126) and (127) we obtain

$$(\rho_k(z) - 1) \in E^{p+\eta}(D^{\pm}), \quad \left(\frac{1}{\rho_k(z)} - 1\right) \in E^{p'+\eta}(D^{\pm}).$$

Now assuming

$$\rho(z) = \prod_{k=1}^{m} \rho_k(z), \tag{128}$$

by virtue of (126) we obtain

$$(\rho(z) - 1) \in E^{p+\eta}(D^{+}), \quad \left(\frac{1}{\rho(t)} - 1\right) \in E^{p'+\eta}(D^{-}). \tag{129}$$

Consider the function

$$X_1(z) = \exp\{(K_\Gamma \ln G_1(t))(z)\} \tag{130}$$

and let

$$X(z) = \rho(z) X_1(z). \tag{131}$$

Since G_1 is continuous, we have $(X_1 - 1) \in \bigcap_{\delta>1} E^{\delta}(D^{\pm})$ and therefore from (126) and (131) it follows that for some $\delta_0 > 0$,

$$(X - 1) \in E^{p+\delta_0}(D^{\pm}), \quad \left(\frac{1}{X} - 1\right) \in E^{p'+\delta_0}(D^{\pm}).$$

Let us introduce a new unknown function

$$\Phi_1(z) = \frac{\Phi(z)}{\rho(z)}. \tag{132}$$

Since $\Phi \in K^{p),\theta}(\Gamma) \subset \bigcap_{0<\delta<p-1} L^{p-\delta}$, $p > 1$, and Γ is a Carleson curve, we have $\Phi \in \bigcap_{0<\delta<p-1} E^{p-\delta}(D^{\pm})$. This and (129) imply that $\Phi_1 \in E^{\delta_1}(D^{\pm})$, for some $\delta_1 > 0$.

Moreover, $\Phi_1^{\pm}(t) \in L^{1+\eta_1}(\Gamma)$, for some $\eta_1 > 0$. Thus, by virtue of Smirnov's theorem, we obtain $\Phi_1 \in E^{1+\eta_1}(D^{\pm})$ and in that case the representation $\Phi_1(z) = K_\Gamma(\Phi_1^+ - \Phi_1^-)$ implies $\Phi_1 \in K^{1+\eta_1}(\Gamma)$. Therefore Φ_1 is a solution of the class $K^{1+\eta_1}(\Gamma)$ for the following boundary value problem

$$\Phi_1^+ = G_1\Phi_1^- + g/\rho^+ .$$

If $\varkappa = \operatorname{ind} G_1 \geq 0$, then all solutions of this problem are given by the equality

$$\Phi_1(z) = X_1(z)K_\Gamma\left(\frac{g}{\rho^+X_1^+}\right)(z) + X_1(z)Q_{\varkappa-1}(z)$$

(see e.g. [43, p. 113]).

Hence, taking the equality $\Phi(z) = \Phi_1(z)\rho(z)$ into account, we obtain that all solutions of problem (1) of the class $K^{p),\theta}(\Gamma)$ lie in the set of functions

$$\Phi(z) = \rho(z)X_1(z)K_\Gamma\left(\frac{g}{\rho^+X_1^+}\right)(z) + \rho(z)X_1(z)Q_{\varkappa-1}(z). \tag{133}$$

Let us show that all functions given by equality (133) belong to the class $K^{p),\theta}(\Gamma)$.

In the first place we note that $\rho(z)X_1(z) = X(z)$ and from (131) it follows that $X(z)Q_{\varkappa-1}(z) \in E^p(D^{\pm})$ and $[X^+(t) - X^-(t)]Q_{\varkappa-1}(t) \in L^p(\Gamma)$. So, the second summands from (133) are contained in $K^p(\Gamma)$ and thereby in $K^{p),\theta}(\Gamma)$, too.

For the first summand $F(z) = \rho(z)X_1(z)K_\Gamma\left(\frac{g}{\rho^+X_1^+}\right)(z)$ from (133) we have

$$F^{\pm}(t) = \frac{1}{2}\,g(t) + \frac{1}{2}\,\rho^+X_1^+\left(\int_\Gamma \frac{g}{\rho^+X_1^+}\right)(t_0), \quad g \in L^{p),\theta}(\Gamma). \tag{134}$$

Here $\rho^+ \in W^{p),\theta}(\Gamma)$ since $\rho \sim \prod\limits_{k=1}^{m} |t - t_k|^{\alpha_k}$ and, for the numbers α_k, conditions (123) are fulfilled. Hence, $\frac{1}{\rho^+} \in L^{p'+\varepsilon}(\Gamma)$, $\varepsilon > 0$. Moreover,

$$X_1^+(t) = \sqrt{|G(t)|}\,\exp\frac{1}{2\pi}\int_\Gamma \frac{\arg G_1(\tau)(\tau - z_0)^{-\varkappa}}{\tau - r}\,d\tau = \sqrt{|G(t)|}\,\exp\int_\Gamma \frac{\psi(\tau)\,d\tau}{\tau - t},$$

where $\psi(t) = 2\pi \arg G_1(\tau)(\tau - z_0)^{-\varkappa}$ is a real continuous function. Thus $X_1^+ \sim \exp\int_\Gamma \frac{\psi(\tau)\,d\tau}{\tau - t}$. Therefore due to Theorem 50 we conclude that $\rho^+X_1^+ \in W^{p),\theta}(\Gamma)$. By virtue of this fact, (134) implies that $F(z) \in K^{p),\theta}(\Gamma)$.

As a result we obtain that the functions Φ given by equality (133) belong to $K^{p),\theta}(\Gamma)$ for $\varkappa \geq 0$.

For $\varkappa < 0$, a function Φ given by equality (133) belongs to $K^{1+\eta_1}(\Gamma_1)$ if and only if $Q_{\varkappa-1}(z) \equiv 0$ and the conditions

$$\int_\Gamma \frac{g(\tau)}{\rho^+(\tau)X_1^+(\tau)}\,\tau^k d\tau = 0, \quad k = \overline{0, |\varkappa| - 1}, \tag{135}$$

are fulfilled.

Assuming that these conditions are fulfilled, in the same manner as above we will show that $\Phi \in K^{p),\theta}(\Gamma)$.

So, if conditions (119) and (126) (for Γ) and conditions (120) and (122) (for G) are fulfilled, then we have the situation in which problem (1) is solvable.

We observe that any piecewise smooth curve is a Carleson one and for such curves condition (119) is fulfilled. As to condition (126), it is fulfilled at least for those curves which do not have internal cusp.

Lemma 55. *If a domain D is bounded by a simple piecewise smooth closed curve Γ with one angular point t_0 where an angle value with respect to D is equal to $\nu\pi$, $0 < \nu \le 2$, then the function $\frac{1}{(z-t_0)^\lambda}$, $\lambda = \alpha + i\beta$, $\alpha \in \left(-\frac{1}{p}, \frac{1}{p'}\right)$, belongs to some class $E^\delta(D)$, $\delta > 0$.*

Proof. Let Γ_r be the images of circumferences $|w| = r < 1$ for the conformal mapping of $z = z(w)$ of the unit circle U onto D. To prove the lemma, it suffices to show that for some $\delta > 0$

$$I = \sup_r \int_{\Gamma_r} \frac{|dz|}{|z-t_0|^{\alpha\delta}} < \infty, \quad \alpha \in \left(-\frac{1}{p}, \frac{1}{p'}\right).$$

Assuming $z(c) = t_0$, we obtain

$$I = \sup_r \int_0^{2\pi} \frac{|z'(w)|\,|dw|}{|z(w)-z(c)|^{\alpha\delta}}\,.$$

If $\alpha < 0$, then $I < \infty$ since $z' \in H^1$ (see e.g. [27, Ch. X]). So we should consider the case $\alpha \in \left(0, \frac{1}{p'}\right)$. Using the relations (see 0.26).

$$z'(w) \sim (w-c)^{\nu-1} z_0(w), \quad z(w) - z(c) \sim (w-c)^\nu z_1(w), \quad z_k^{\pm 1} \in \bigcap_{\delta>0} H^\delta, \;\; k = 0,1,$$

we obtain

$$I = \sup_r \int_0^{2\pi} |w-c|^{\nu-1-\nu\alpha\delta} |z_0(w)|\,|z_1(w)|^{-\alpha\delta} |dw|\,.$$

Hence we see that $I < \infty$ if $\nu(1-\alpha\delta) - 1 > -1$, i.e. if $1 - \alpha\delta > 0$. Now, for δ we obtain the inequality $\delta < \frac{1}{\alpha}$ and since $\frac{1}{\alpha} > p'$, when $\alpha > 0$, we finally have that $I < \infty$ for $\delta < p'$. □

Thus the following theorem is valid.

Theorem 56. *Let Γ be a piecewise smooth simple closed curve without a zero internal angle and $G(t)$ be a function of the class $C(\Gamma; t_1, \ldots, t_m)$, $t_k \in \Gamma$, given on Γ and being such that conditions* (121)–(122) *are fulfilled for it. Let further the function $\rho(z)$ be given by equalities* (125) *and* (128), *the function G_1 by equality* (124), *$X_1(z)$ by formula* (130) *and $\varkappa = \operatorname{ind} G_1$.*

Then for $\varkappa \geq 0$, problem (1) *is solvable in the class $K^{p),\theta}(\Gamma)$, $p > 1$, and its general solution is given by equality* (133) *where $Q_{\varkappa-1}(z)$ is an arbitrary polynomial of order $\varkappa - 1$ $(Q_{-1}(z) \equiv 0)$. If however $\varkappa < 0$, then the problem is solvable if and only if conditions* (135) *are fulfilled and in that case it is solvable uniquely. Then a solution is given by equality* (133) *where $Q_{\varkappa-1}(z) \equiv 0$.*

4.20 Notes and comments on Chapter 4

The problem of linear conjugation for analytic functions was for the first time encountered by B. Riemann [111], [112]. Important results on which the posterior solution of problem (1) was based, were obtained by Yu. Sokhotskiĭ, D. Hilbert, I.Plemelj and T.Carleman. The solution of the Riemann problem in a continuous setting was given by F. D. Gakhov [24], [25]. In the case when the continuity is violated only at a finite number of boundary points a complete solution of the problem was given by N. Muskhelishvili [95], [96], [97].

Discontinuous BVPs when the right side of boundary condition belongs to the Lebesgue space L^p $(1 < p < \infty)$ and the sought functions are from the set of Cauchy type integrals in the domains with sufficiently smooth boundaries has been studied by B. V. Khvedelidze [42, 43], I. B. Simonenko [119, 120, 121], I. I. Danilyuk [12], I. Ts. Gohberg and N. Ya. Krupnik [26], A. Böttcher and Yu. I. Karlovich [5], G. F. Manjavidze and B. V. Khvedelidze [87], etc.

The Riemann boundary value problem for analytic functions within the frame of variable exponent Lebesgue spaces for the first time was explored in [65].

C. Haseman [30] was the first who investigated the boundary value problem with shift

$$\phi^+(\alpha(t)) = a(t)\phi^-(t).$$

Later T. Carleman [7] studied this problem in more general setting. D. A. Kveselava [81], [82] gave a complete solution of the problem when boundary is a Lyapunov curve, coefficients belong to the Hölder class, and a shift is a preserving orientation homeomorphism of boundary curve into itself with nonvanishing derivative and belonging to the Hölder class.

The Haseman problem in the vector case was considered by N. P. Vekua. His results, as well as those obtained by other authors in this direction can be found in [131].

The boundary value problem with shifts in the class of Cauchy type integrals with density from $L^p(\Gamma)$, $p > 1$ has been studied by B. V. Khvedelidze [42], B. V.

Khvedelidze and G. F. Manjavidze [87], I. B. Simonenko [118], G. F. Manjavidze [86], etc.

In the sequel, the boundary value problems with shifts became an investigation topic for a great deal of authors. For their results we refer the reader to [83] and [80]. The related Riemann problem for matrix function is studied in the monograph of G. S. Litvinchuk and I. M. Spitkovski [84].

The boundary value problems with shifts in the class of Cauchy type integrals with densities from variable exponent Lebesgue spaces have been investigated for the first time in the papers [104], [105].

The investigation of boundary value problems for analytic and harmonic functions with data from grand Lebesgue spaces was studied in [64].

The Chapter 4 is based on the papers [65], [104], [105].

The Sections 4.18 and 4.19 are included in unpublished paper [64].

Chapter 5

The Riemann-Hilbert Problem in Weighted Classes of Cauchy Type Integrals

Let D be a simply connected domain not containing $z = \infty$ and bounded by a simple closed rectifiable curve Γ. Let $A(D)$ be a subset of the set of analytic functions ϕ possessing boundary values $\phi^+(t)$, $t \in \Gamma$. The Riemann–Hilbert problem is formulated as follows [98, p. 144]: Find a function ϕ from the given class $A(D)$, which satisfies the condition

$$\operatorname{Re}\left[(a(t) + ib(t))\,\phi^+(t)\right] = c(t), \quad t \in \Gamma, \tag{1}$$

where $a(t)$ and $b(t)$ are real functions given on Γ.

In this chapter, the Riemann–Hilbert problem is considered in the class of functions representable in the form $\phi(z) = \omega^{-1}(z)(K_\Gamma\varphi)(z)$, where $(K_\Gamma\varphi)(z)$ is a Cauchy type integral with density from the class $L^{p(\cdot)}(\Gamma)$ and $\omega(z)$ is an arbitrary function of the form

$$\omega(z) = \prod_{k=1}^{\nu}(z - t_k)^{\alpha_k}, \quad t_k \in \Gamma, \quad \alpha_k \in \mathbb{R}. \tag{2}$$

The set of all such functions ϕ is called by the weighted class of Cauchy type integrals with density from $L^{p(\cdot)}(\Gamma;\omega)$ and is denoted by $K^{p(\cdot)}(D;\omega)$ as different from the set of Cauchy type integrals with density from $L^{p(\cdot)}(\Gamma;\omega)$ denoted by $K^{p(\cdot)}(\Gamma;\omega)$. When a weight function $\omega \in W^{p(\cdot)}(\Gamma)$, we show that $K^{p(\cdot)}(D;\omega)$ and $K^{p(\cdot)}(\Gamma;\omega)$ coincide for the wide class of curves Γ and functions $p(t)$ (see Section 5.2).

This Chapter contains two Parts.

In Part I problem (1) is considered in the classes $K^{p(\cdot)}(D;\omega)$, where D is a simply connected domain bounded by a simple piecewise Lyapunov curve with angular points A_k, at which the angle sizes with respect to D are equal to $\pi\nu_k$,

$$0 < \nu_k \leq 2. \tag{3}$$

The weight ω is assumed to be an arbitrary power function of form (2), while the coefficients a, b are piecewise Hölder with the condition $\inf(a^2(t) + b^2(t)) > 0$ and $c \in L^{p(\cdot)}(\Gamma; \omega)$. For the function $p(t)$ it is required that it belongs to $\mathcal{P}(\Gamma)$. Under these assumptions, we obtain a complete picture of the solvability – the conditions for the problem to be solvable are derived and solutions are constructed. These conditions and solutions largely depend both on the values of $p(t)$ at the angular points of Γ and on the angle sizes at these points. An important contribution to the picture of solvability is made by the exponents α_k of the weight function and the jump values at the discontinuity points of $a(t)$ and $b(t)$.

In Part II the problem (1) is studied when a boundary Γ is piecewise smooth curve, the coefficients a and b are piecewise continuous functions however the function $p(t)$ belongs to the class $\widetilde{\mathcal{P}}(\Gamma)$ which is rather narrow than $\mathcal{P}(\Gamma)$.

The presented results are new even for the constant exponents $p > 1$.

Part I

The case of domains with piecewise Lyapunov boundaries

5.1 Some definitions and auxiliary statements

Let now Γ be a simple closed piecewise Lyapunov curve with angular points t_k, $k = \overline{1, \nu}$ bounding the domain D and let Γ_k be a simple smooth curve connecting the points t_k and ∞ and having with $\overline{D}$ only one common point t_k. Then on the set $\mathbb{C} \setminus \bigcup\limits_{k=1}^{\nu} \{\Gamma_k\}$ we define an analytic function giving by (2). We fix one of its branches. In particular, $\omega(z)$ is analytic in D, and it is also analytic at the points $t \in \Gamma$ different from t_k. Its restriction on Γ is denoted by $\omega(t)$.

Recall the

$$\mathcal{R}^{p(\cdot)} = \Big\{\Gamma : \; S_\Gamma \text{ is bounded in } L^{p(\cdot)}(\Gamma)\Big\}.$$

If $p \in \mathcal{P}(\Gamma)$, then Γ belongs to $\mathcal{R}^{p(\cdot)}$ if and only if Γ is a regular curve.

Since piecewise smooth curve Γ is regular that if $p \in \mathcal{P}(\Gamma)$, then $\Gamma \in \mathcal{R}^{p(\cdot)}$.

Definition 1. We denote by $K^{p(\cdot)}(D; \omega)$ the set of functions ϕ representable in the form

$$\phi(z) = \frac{1}{\omega(z)} \frac{1}{2\pi i} \int\limits_\Gamma \frac{\varphi(t)\, dt}{t - z}, \quad z \in D, \quad \varphi \in L^{p(\cdot)}(\Gamma). \tag{4}$$

Suppose $K^{p(\cdot)}(D) = K^{p(\cdot)}(D; 1)$.

Definition 2. By $K^{p(\cdot)}(\Gamma;\omega)$ we denote the set of functions ϕ of the form

$$\phi(z) = (K_\Gamma f)(z) = \frac{1}{2\pi i}\int_\Gamma \frac{f(t)\,dt}{t-z}, \quad z \in D, \quad f \in L^{p(\cdot)}(\Gamma;\omega). \tag{5}$$

Lemma 3. *If the curve belongs to $\mathcal{R}^{p(\cdot)}$, then for almost all t from Γ each function $\phi \in K^{p(\cdot)}(D;\omega)$ given by equality (4) has an angular boundary value $\phi^+(t)$; moreover, $\phi^+(t)\omega(t) = \frac{1}{2}(\varphi(t) + (S_\Gamma\varphi)(t))$ and $\phi^+\omega \in L^{p(\cdot)}(\Gamma)$, i.e. $\phi^+ \in L^{p(\cdot)}(\Gamma;\omega)$.*

Proof. Since $K_\Gamma(\varphi)$, $\varphi \in L(\Gamma)$, admits almost everywhere an angular boundary value for arbitrary rectifiable curve Γ (see e. g. [5, p. 137] and well-known Privalov's lemma, [108], Ch. III, §2) we are able to use the Plemelj–Sokhotskiĭ formulas. Then the proof follows from the definition of $\mathcal{R}^{p(\cdot)}$. □

Lemma 4. *If D is a Smirnov domain, $\phi \in E^\delta(D)$, $\delta > 0$, and $\phi^+ \in L^{p(\cdot)}(\Gamma;\omega)$, where $\underline{p} > 1$ and $\omega^{-1} \in L^{p'(\cdot)}(\Gamma)$, then $\phi \in K^{p(\cdot)}(\Gamma;\omega)$.*

Proof. Indeed, from the conditions $\phi^+ \in L^{p(\cdot)}(\Gamma,\omega)$ and $\omega^{-1} \in L^{p'(\cdot)}(\Gamma)$ by the Hölder inequality it follows that $\phi^+ \in L(\Gamma)$. Then, by virtue of the Smirnov theorem, $\phi \in E^1(D)$. Therefore $\phi(z) = (K_\Gamma\phi^+)(z)$. But $\phi^+ \in L^{p(\cdot)}(\Gamma;\omega)$, which means that $\phi \in K^{p(\cdot)}(\Gamma;\omega)$. □

Definition 5. Let Γ be a simple oriented rectifiable curve, $t_k \in \Gamma$, $k = \overline{1,n}$, where t_k are numbered with respect to orientation. A function f is said to be piecewise Hölder if it is Hölder continuous on the arcs (t_k, t_{k+1}), $k = \overline{1,n}$, $t_{n+1} = t_1$ and one-sided limits at the points t_k exist.

5.2 Conditions of the coincidence of the classes $K^{p(\cdot)}(D;\omega)$ and $K^{p(\cdot)}(\Gamma;\omega)$

In this section, we establish the conditions for Γ, ω and p, which are sufficient for the equality $K^{p(\cdot)}(D;\omega) = K^{p(\cdot)}(\Gamma;\omega)$.

Theorem 6. *Let a Γ belongs to $\mathcal{R}^{p(\cdot)}$, $\underline{p} > 1$, $1/\omega \in E^\delta(D)$ for some $\delta > 0$ and $\omega(t) = \omega^+(t) \in W^{p(\cdot)}(\Gamma)$. Then the equality*

$$K^{p(\cdot)}(D;\omega) = K^{p(\cdot)}(\Gamma;\omega) \tag{6}$$

is fulfilled.

Proof. Let $\phi \in K^{p(\cdot)}(D;\omega)$. Since $\omega \in W^{p(\cdot)}(\Gamma)$, we have $1/\omega \in L^{p'(\cdot)}(\Gamma)$ (See Chapter 4, Theorem 2). Hence from the equality $\phi^+ = \frac{1}{\omega^+}(\omega\phi^+)$ it follows that $\phi^+ \in L(\Gamma)$. Since $\underline{p} > 1$, for $\varphi \in L^{p(\cdot)}(\Gamma)$ we have $K_\Gamma\varphi \in E^{\underline{p}}(D)$. But $\phi(z) = \omega^{-1}(z)(K_\Gamma\varphi)(z)$ and the condition $1/\omega \in E^\delta(D)$ implies that $\phi \in E^\eta(D)$, for some

$\eta > 0$. Along with this, $\phi^+ \in L(\Gamma)$. Moreover, since $\Gamma \in \mathcal{R}^{p(\cdot)}$, the Cauchy singular operator is continuous in $L^{p(\cdot)}(\Gamma)$. Thus D is a Smirnov domain (see Chapter 4 , Corollary 3). This gives us the right to apply the Smirnov theorem. Thus $\phi \in E^1(D)$ and therefore ϕ is representable by a Cauchy integral, i.e. $\phi(z) = (K_\Gamma \phi^+)(z)$. But, according to Lemma 3, $\phi^+ \in L^{p(\cdot)}(\Gamma;\omega)$ and the latter equality means that $\phi \in K^{p(\cdot)}(\Gamma;\omega)$. We have thereby proved that $K^{p(\cdot)}(D;\omega) \subset K^{p(\cdot)}(\Gamma;\omega)$.

Let us prove the validity of inverse inclusion.

For $\phi \in K^{p(\cdot)}(\Gamma;\omega)$ we have $\phi = K_\Gamma(\frac{\varphi}{\omega})$, $\varphi \in L^{p(\cdot)}(\Gamma)$. Then $\phi(z) = \frac{1}{\omega(z)}\phi_0(z)$ where $\phi_0(z) = \omega(z)(K_\Gamma\frac{\varphi}{\omega})(z)$. Since $\omega \in W^{p(\cdot)}(\Gamma)$ and $\varphi \in L^{p(\cdot)}(\Gamma)$, we have $K_\Gamma(\frac{\varphi}{\omega}) \in E^1(D)$ (see Chapter 4, Theorem 2) and therefore $\phi_0 \in E^\eta(D)$ for some $\eta > 0$. But

$$\phi_0^+(t) = \frac{1}{2}(\varphi(t) + \omega(t)(S_\Gamma\frac{\varphi}{\omega})(t))$$

belongs to $L^{p(\cdot)}(\Gamma)$. Therefore, again by virtue of the Smirnov theorem, we have $\phi_0 \in E^{\underline{p}}(D)$, $\underline{p} > 1$. Thus $\phi_0(z) = (K_\Gamma\phi_0^+)(z)$, where $\phi_0^+ \in L^{p(\cdot)}(\Gamma)$. Now from the representation $\phi(z) = \omega^{-1}(z)\phi_0(z)$ it follows that $\phi \in K^{p(\cdot)}(D;\omega)$. □

Corollary 7. *Let Γ be a regular curve, $\omega(z)$ be given by equality (2), and at the points t_k the curve Γ have the one-sided tangents forming a nonzero angle. If $p \in \mathcal{P}(\Gamma)$ and $\omega^+ \in W^{p(\cdot)}(\Gamma)$, then equality (6) is fulfilled.*

Proof. Under the assumptions we have made for the points t_k the function $\frac{1}{\omega(z)}$ belongs to some Smirnov class (see Chapter 4, Lemma 55). From the fact that Γ is regular and the condition $p \in \mathcal{P}(\Gamma)$, we conclude, that curve Γ belongs to $\mathcal{R}^{p(\cdot)}$. All the assumptions of the theorem are thereby fulfilled and equality (6) is valid. □

5.3 Reduction of the Riemann-Hilbert problem to the Riemann problem

Let $\Gamma \in C^{1,L}(A_1,\dots,A_i;\nu_1,\dots,\nu_i)$, $0 < \nu_k \le 2$, $k = \overline{1,i}$, be the boundary of given domain D; $a(t)$, $b(t)$ be piecewise Hölder real functions with the condition $\inf(a^2(t) + b^2(t)) > 0$. Furthermore, let ω be a weight function of form (2), $p \in \mathcal{P}(\Gamma)$ and let $c \in L^{p(\cdot)}(\Gamma;\omega)$, and $z = z(w)$ is conformal map of U on D and $w = w(z)$ is it inverse function.

Let $\phi \in K^{p(\cdot)}(D;\omega)$ be a solution of problem (1) and

$$\Psi(w) = \phi(z(w)) = \frac{1}{\omega(z(w))}\int_\Gamma \frac{\varphi(t)\,dt}{t - z(w)}, \quad \varphi \in L^{p(\cdot)}(\Gamma). \tag{7}$$

Then $\Psi(w)$ satisfies the boundary condition

$$\mathrm{Re}\left[(A(\tau) + iB(\tau))\,\Psi^+(\tau)\right] = C(\tau), \quad \tau \in \gamma,$$

where $A(\tau) = a(z(\tau))$, $B(\tau) = b(z(\tau))$, $C(\tau) = c(z(\tau))$. Hence it follows that

$$[A(\tau) + iB(\tau)]\,\Psi^+(\tau) + [\,\overline{A(\tau) + iB(\tau)}\,]\,\overline{\Psi^+(\tau)} = 2C(\tau),$$

i.e.

$$\Psi^+(\tau) = -[A(\tau) - iB(\tau)]\,[A(\tau) + iB(\tau)]^{-1}\,\overline{\Psi^+(\tau)} + 2C(\tau)\,[A(\tau) + iB(\tau)]^{-1}.$$

Assuming that

$$G(\tau) = -\,[A(\tau) - iB(\tau)]\,[A(\tau) + iB(\tau)]^{-1}, \quad c_1(\tau) = 2C(\tau)\,[A(\tau) + iB(\tau)]^{-1},$$

we rewrite the latter boundary condition in the form

$$\Psi^+(\tau) = G(\tau)\overline{\Psi^+(\tau)} + c_1(\tau), \quad \tau \in \gamma, \tag{8}$$

where the coefficient $G(\tau)$ is a piecewise Hölder function. Let $b_1, b_2, \ldots, b_\lambda$ be its discontinuity points, then $|G(\tau)| = 1$ for $\tau \neq b_k$, and $|G(b_k\pm)| = 1$.

In the case when Ψ is a continuous function in $\overline{U}$, N. Muskhelishvili constructed a analytic function Ω on the plane cut along γ such that it is continuously left and right extendable; coincides inside the disk with $\Psi(w)$, and $\Omega^-(\tau) = \overline{\Psi^+(\tau)}$. Thus (8) reduces to a linear conjugation problem with respect to $\Omega(w)$. Solutions of problem (1) are found by restriction of the solutions of the obtained problem onto U. However it is not all solutions that are suitable, since they must satisfy a certain additional condition. One may succeed in obtaining such solutions.

We follow the above-mentioned way. However it is not quite clear in what class solutions of the obtained linear conjugation problem should be sought for. Let $X(w)$ be a canonical function of the function G (see 0.24). Multiplying the function ΩX^{-1} by a specially chosen rational function Q, we try to make it belong to the class $K^{\ell(\cdot)}(\gamma;\rho)$, where $\ell(\tau) = p(z(\tau))$ and ρ is a power weight function from $W^{\ell(\cdot)}(\gamma)$. Solving the obtained problem for the function $Q(w)\Omega(w)[X(w)]^{-1}$, from the set of solutions we choose the ones by means of which, upon the substitution of $w = w(z)$, all solutions of problem (1) are found.

In the case of factorization of piecewise continuous functions in various classes of analytic functions, we preliminarily find a piecewise analytic function $r_k(w)$ in $\mathbb{C}\setminus\gamma$ such that $r_k(w) \neq 0$ for $|w| \neq 1$ and $R_k(\tau) = [r_k^+(\tau)]\,[r_k^-(\tau)]^{-1}$ has only one discontinuity point b_k on γ and also

$$[R_k(b_k-)][R_k(b_k+)]^{-1} = G(b_k+)[G(b_k-)]^{-1}.$$

Then the function $G_1(\tau) = G(\tau)\prod_{k=1}^{\lambda} R_k(\tau)$ belongs to the Hölder class on γ. If $X_1(w)$ is its canonical function, then $G(\tau) = X^+(\tau)[X^-(\tau)]^{-1}$, where $X(w) = X_1(w)\prod_{k=1}^{\lambda} r_k(w)$.

Let $G(b_k-)[G(b_k+)]^{-1} = \exp(2\pi i u_k)$. Since $|G(b_k\pm)| = 1$, u_k are real numbers. Using this fact and the above reasoning, we slightly modify the traditional way of choosing functions $r_k(w)$ (see [25], [98], [42]), taking those which are more suitable for our further consideration.

Let Γ_k be a simple smooth curve connecting the points b_k and ∞ and having with $\overline{U}$ only one common point b_k; $w \to (w-b_k)^{u_k}$ be an arbitrary branch of this analytic function in $\mathbb{C} \setminus \Gamma_k$. Its restriction on U is also denoted by $(w-b_k)^{u_k}$. It is assumed that

$$r_k(w) = \begin{cases} (w-b_k)^{u_k}, & |w|<1, \\ \overline{\left(\frac{1}{\overline{w}}-b_k\right)^{u_k}}, & |w|>1, \end{cases} \qquad r(w) = \prod_{k=1}^{\lambda} r_k(w). \tag{9}$$

We have

$$R_k(\tau) = \frac{r_k^+(\tau)}{r_k^-(\tau)} = \exp 2i \arg(\tau - b_k).$$

Hence $R_k(b_k-)[R_k(b_k+)]^{-1} = \exp(-2\pi i u_k)$ and therefore $G_1(\tau) = G(\tau)\prod\limits_{k=1}^{\lambda} R_k(\tau)$ is Hölder-continuous on γ and different from zero.

Let $X_1(w)$ be a canonical function for $G_1(\tau)$, i.e.

$$X_1(w) = \begin{cases} C\exp\left(\frac{1}{2\pi i}\int_\gamma \frac{\ln \widetilde{G}_1(\tau)\,d\tau}{\tau-w}\right), & |w|<1, \\ C(w-w_0)^{-\varkappa_1}\exp\left(\frac{1}{2\pi i}\int_\gamma \frac{\ln \widetilde{G}_1(\tau)\,d\tau}{\tau-w}\right), & |w|>1, \ |w_0|<1, \end{cases} \tag{10}$$

where C is for the time being an arbitrary constant (further it will be defined) and $\widetilde{G}_1(\tau) = G_1(\tau)(\tau-w_0)^{-\varkappa_1}$ and $\varkappa_1 = \operatorname{ind} G_1(\tau) = (2\pi)^{-1}[\arg G_1(\tau)]_\gamma$. We rewrite (8) in the form

$$\Psi^+(\tau)[X^+(\tau)]^{-1} = \overline{\Psi^+(\tau)}\,[X^-(\tau)]^{-1} + c_1(\tau)[X^+(\tau)]^{-1},$$

where

$$X(w) = X_1(w)r(w) = X_1(w)\prod_{k=1}^{\lambda} r_k(w). \tag{11}$$

Let

$$\Omega(w) = \begin{cases} \Psi(w), & |w|<1, \\ \overline{\Psi\left(\frac{1}{\overline{w}}\right)}, & |w|>1, \end{cases} \tag{12}$$

where $\Psi(w) = \phi(z(w))$ is defined by (7).

Then $\Omega^-(\tau) = \overline{\Psi^+(\tau)}$ and the latter boundary condition takes the form

$$\Omega^+(\tau)[X^+(\tau)]^{-1} = \Omega^-(\tau)[X^-(\tau)]^{-1} + c_2(\tau), \quad c_2(\tau) = c_1(\tau)[X^+(\tau)]^{-1}. \tag{13}$$

It should be also noted that if for an analytic function f in $\mathbb{C}\setminus\gamma$ we set

$$f_*(w) = \overline{f\left(\frac{1}{\overline{w}}\right)}, \quad |w| \neq 1,$$

then f_* is analytic in $\mathbb{C} \setminus \gamma$ and $(f_*)_* = f$.

From (12) we see that $\Omega_*(w) = \Omega(w)$. Therefore we should look for those solutions Ω of problem (13) for which the condition

$$\Omega_*(w) = \Omega(w), \quad |w| \neq 1, \tag{14}$$

is fulfilled.

5.4 The properties of the function Ω when $\phi \in K^{p(\cdot)}(D;\omega)$

In order to study of function $\Omega(w) = \omega^{-1}(z(w))(K_\Gamma\varphi)(z(w))$, $|w| < 1$, the lemmas 8, 10, 11 are useful to us.

Lemma 8. *If* $\Gamma \in C^{1,L}(A_1, \ldots, A_i; \nu_1, \ldots, \nu_i)$, $0 < \nu_k \leq 2$, $k = \overline{1,i}$, $\varphi \in L^{p(\cdot)}(\Gamma)$, $p \in \mathcal{P}(\Gamma)$, *then the function*

$$M(w) = (K_\Gamma\varphi)(z(w)),$$

belongs to H^δ *for some* $\delta > 0$.

Proof. Since $\varphi \in L^{p(\cdot)}(\Gamma)$, we have $\varphi \in L^{\underline{p}}(\Gamma)$ and therefore $K_\Gamma\varphi \in E^{\underline{p}}(D)$, i.e.

$$\sup_{\rho\in(0,1)} \int\limits_{|w|=\rho} |M(w)|^{\underline{p}} |z'(w)|\,|dw| < \infty. \tag{15_1}$$

For $0 < \delta < \underline{p}$ we obtain

$$\int\limits_{|w|=\rho} |M(w)|^\delta\,|dw| = \int\limits_{|w|=\rho} \left|M(w)|z'(w)|^{\frac{1}{\underline{p}}}\right|^\delta |z'(w)|^{-\frac{\delta}{\underline{p}}}\,|dw|$$

$$\leq \Bigg(\int\limits_{|w|=\rho} |M(w)|^{\underline{p}} |z'(w)|\,|dw|\Bigg)^{\frac{\delta}{\underline{p}}} \Bigg(\int\limits_{|w|=\rho} |z'(w)|^{-\frac{\delta}{\underline{p}-\delta}}\,|dw|\Bigg)^{\frac{\underline{p}-\delta}{\underline{p}}}. \tag{15}$$

By virtue of the assumption for ν_k (see (3)), we have

$$z(w) - z(a_k) \sim (w - a_k)^{\nu_k} z_{0,k}(w), \quad z'(w) \sim (w - a_k)^{\nu_k - 1} z_{1,k}(w), \tag{16}$$

where $z(a_k) = A_k$, $z_{0,k}$, $z_{1,k}$ are nonzero continuous functions belonging to the Hölder class (see 0.25). From here conclude that $[z']^{-1} \in \bigcap\limits_{\varepsilon\in(0,1)} H^\varepsilon$. If $\delta > 0$ is taken so small that $\delta(\underline{p} - \delta)^{-1} < 1$, then $[z']^{-1} \in H^{\frac{\delta}{\underline{p}-\delta}}$. By virtue of this fact and inequality (15_1), it follows from (15) that $M \in H^\delta$. □

Corollary 9. *If $\phi \in K^{p(\cdot)}(D;\omega)$ and $\Psi(w) = \phi(z(w)) = \frac{1}{\omega}(z(w))M(w)$, then*

$$\Psi \in \bigcup_{\eta>0} H^\eta.$$

This result easily follows by using equalities (16).

Lemma 10. *The function $M^+(\tau)$ belongs to $L^{\ell(\cdot)}(\gamma; |z'(\tau)|^{\frac{1}{\ell(\tau)}})$, $\ell(\tau) = p(z(\tau))$.*

Proof. Indeed, the fact that the function $K_\Gamma\varphi \in E^{\underline{p}}(D)$ implies $(K_\Gamma\varphi)(z) = (K_\Gamma(K_\Gamma\varphi)^+)(z)$. Since $\Gamma \in \mathcal{R}^{p(\cdot)}$, we have $(K_\Gamma\varphi)^+ \in L^{p(\cdot)}(\Gamma)$ and therefore

$$\int_\gamma \left(|M^+(\tau)|\,|z'(\tau)|^{\frac{1}{\ell(\tau)}}\right)^{\ell(\tau)}|d\tau| = \int_\Gamma |(K_\Gamma\varphi)^+(t)|^{p(t)}|dt| < \infty. \qquad \square$$

We set

$$T = \{\tau_k : \tau_k = w(t_k)\}, \quad A = \{a_k : a_k = w(A_k)\}, \quad B = \{b_k\}.$$

Among the points τ_k, a_k, b_k some may coincide. Let us renumber the points from $T \cup A \cup B$ as follows

$$\begin{cases} w_1 = \tau_1 = a_1 = b_1, \dots, \quad w_\mu = \tau_\mu = a_\mu = b_\mu, \\ w_{\mu+1} = \tau_{\mu+1} = a_{\mu+1}, \dots, \quad w_{\mu+r} = \tau_{\mu+r} = a_{\mu+r}, \\ w_{\mu+r+1} = \tau_{\mu+r+1} = b_{\mu+1}, \dots, \quad w_{\mu+r+q} = \tau_{\mu+r+q} = b_{\mu+q}, \\ w_{\mu+r+q+1} = a_{\mu+r+1} = b_{\mu+q+1}, \dots, \quad w_{\mu+r+q+p} = a_{\mu+r+p} = b_{\mu+q+p}, \\ w_{\mu+r+q+p+1} = \tau_{\mu+r+q+1}, \dots, \quad w_{\mu+r+q+p+m} = \tau_{\mu+r+q+m}, \\ w_{\mu+r+q+p+m+1} = a_{\mu+r+p+1}, \dots, \quad w_{\mu+r+q+p+m+n} = a_{\mu+r+p+n}, \\ w_{\mu+r+q+p+m+n+1} = b_{\mu+q+p+1}, \dots, \quad w_{\mu+r+q+p+m+n+s} = b_{\mu+q+p+s}. \end{cases} \tag{17}$$

Now let us try to construct a rational function $Q(w)$ such that $Q(w)\Psi(w)$ belongs to the class $K^{\ell(\cdot)}(\gamma;\rho)$, where ρ is a power function from $W^{\ell(\cdot)}(\gamma)$.

For this, we first note that $\Psi \in H^\varepsilon$ for some $\varepsilon > 0$, according to Corollary 9.

Further, taking into account (16) and the equality $M^+(\tau) = |z'(\tau)|^{-\frac{1}{\ell(\tau)}}M_0(\tau)$, $M_0 \in L^{\ell(\cdot)}(\gamma)$, that follows from Lemma 10, we obtain

$$\left|\frac{\Psi^+(\tau)}{X^+(\tau)}\right| = \frac{1}{|X_1^+(\tau)|}\left|\frac{M^+(\tau)}{\omega(z(\tau))r(\tau)}\right|$$

$$\sim \left|\prod_{k=1}^{\nu}(\tau-\tau_k)^{-\alpha_k\nu_k}\prod_{k=1}^{\lambda}(\tau-b_k)^{-u_k}\prod_{k=1}^{i}(\tau-a_k)^{-\frac{\nu_k-1}{\ell(\tau)}}\right|\widetilde{\Psi}_0(\tau), \quad \widetilde{\Psi}_0 \in L^{\ell(\cdot)}(\gamma).$$

We will need

Lemma 11. *If $\zeta \in \gamma$, $\alpha \in \mathbb{R}$, then*

$$0 < m \leq \frac{|\tau - \zeta|^{\frac{\alpha}{\ell(\tau)}}}{|\tau - \zeta|^{\frac{\alpha}{\ell(\zeta)}}} \leq M < \infty.$$

Proof. Indeed,

$$|\tau - \zeta|^{\frac{\alpha}{\ell(\tau)} - \frac{\alpha}{\ell(\zeta)}} = \exp\left\{\frac{\alpha(\ell(\zeta) - \ell(\tau))}{\ell(\zeta)\ell(\tau)} \ln|\tau - \zeta|\right\} = f(\tau, \zeta),$$

and since $\ell \in \mathcal{P}(\gamma)$ (see Chapter 3, Lemma 12), we see that $f(\tau, \zeta)$ is bounded and $\inf|f(\tau, \zeta)| > 0$. □

Now, by equalities (16) we conclude that

$$z'(w) \sim \prod_{k=1}^{i} (w - a_k)^{\nu_k - 1}.$$

By virtue of Lemma 11, from (5.4) it follows that

$$\left|\frac{\Psi^+(\tau)}{X^+(\tau)}\right| \sim \left|\prod_{k=1}^{\nu} (\tau - \tau_k)^{-\alpha_k \nu_k} \prod_{k=1}^{\lambda} (\tau - b_k)^{-u_k} \prod_{k=1}^{i} (\tau - a_k)^{-\frac{\nu_k - 1}{\ell(a_k)}}\right| R_0(\tau), \quad R_0 \in L^{\ell(\cdot)}(\gamma).$$

According to the adopted numbering of points τ_k, a_k, b_k (see (17)), we have

$$\frac{\Psi^+(\tau)}{X^+(\tau)} \sim \prod_{k=1}^{j} (\tau - w_k)^{-\delta_k} R_0(\tau), \quad R_0 \in L^{\ell(\cdot)}(\gamma), \tag{18}$$

where

$$\delta_k = \begin{cases} \alpha_k \nu_k + \frac{\nu_k - 1}{\ell(a_k)} + u_k, & k = \overline{1, \mu}, \\ \alpha_k \nu_k + \frac{\nu_k - 1}{\ell(a_k)}, & k = \overline{\mu+1, \mu+r}, \\ \alpha_k + u_{k-r}, & k = \overline{\mu+r+1, \mu+r+q}, \\ \frac{\nu_{k-q} - 1}{\ell(a_{k-q})} + u_{k-r}, & k = \overline{\mu+r+q+1, \mu+r+q+p}, \\ \alpha_{k-p}, & k = \overline{\mu+r+q+p+1, \mu+r+q+p+m}, \\ \frac{\nu_{k-q-m} - 1}{\ell(a_{k-q-m})}, & k = \overline{\mu+r+q+p+m+1, \mu+r+q+p+m+n}, \\ u_{k-r-m-n}, & k = \overline{\mu+r+q+p+m+n+1, \mu+r+q+p+m+n+s}. \end{cases} \tag{19}$$

For a real number x we assume $x = [x] + \{x\}$, where $[x]$ is integer and $0 \leq \{x\} < 1$. We require that

$$\{\delta_k\} \neq \frac{1}{\ell'(w_k)}, \quad \ell'(w_k) = \frac{\ell(w_k)}{\ell(w_k) - 1}, \tag{20}$$

for all k. Let

$$\gamma_k = \begin{cases} [\delta_k] & \text{if } \{\delta_k\} < \frac{1}{\ell'(w_k)}\,, \\ [\delta_k]+1 & \text{if } \{\delta_k\} > \frac{1}{\ell'(w_k)}\,. \end{cases}$$

Then

$$-\frac{1}{\ell(w_k)} < \delta_k - \gamma_k < \frac{1}{\ell'(w_k)}\,. \tag{21}$$

Setting

$$Q(w) = \prod_{k=1}^{j}(w-w_k)^{\gamma_k}, \;\; R(w) = Q(w)\Omega(w)[X(w)]^{-1}, \tag{22}$$

we have

$$R(\tau) = Q(\tau)\Psi^+(\tau)[X^+(\tau)]^{-1} \in L^{\ell(\cdot)}(\gamma;\rho), \tag{23}$$

where

$$\rho(\tau) = \prod_{k=1}^{j}(\tau-w_k)^{\delta_k-\gamma_k}. \tag{24}$$

By virtue of (21) we conclude that $\rho \in W^{\ell(\cdot)}(\gamma)$.

Remark 12. By the (18)

$$R(\tau) \sim Q(\tau)\prod_{k=1}^{j}(\tau-w_k)^{-\delta_k}R_0(\tau) = (\rho(\tau))^{-1}R_0(\tau) \tag{25}$$

and we obtain

$$\rho(\tau) \sim (R(\tau))^{-1}R_0(\tau) \Longrightarrow \rho(\tau) \sim w(z(\tau))r(\tau)|z'(\tau)|^{\frac{1}{\ell(\tau)}}[Q(\tau)]^{-1}R_0(\tau). \tag{26}$$

Since $\Psi \in H^{\eta_0}$, for some $\eta_0 > 0$ (see Corollary 9), it is obvious that $R(w) \in H^{\delta}$ for some $\delta > 0$. Furthermore, since $\rho^{-1} \in L^{\ell'(\cdot)}(\gamma)$ and $R(\tau) \in L^{\ell(\cdot)}(\gamma;\rho)$ (see (23)), by Lemma 4, we have

$$R(w) \in K^{\ell(\cdot)}(\gamma;\rho) \tag{27}$$

where ρ is defined by (24).

Since $\rho \in W^{\ell(\cdot)}(\gamma)$ and (27) are true, we can use following result from Theorem 2 of Chapter 4: if $\omega \in W^{p(\cdot)}(\Gamma)$ and $\Phi \in K^{p(\cdot)}(\Gamma;\omega)$ then $\Phi \in E^1(D)$. From this statement follows that

$$R(w) \in H^1. \tag{28}$$

Thus if $\phi \in K^{p(\cdot)}(D;\omega)$, then the inclusions (27) and (28) are valid.

Let us prove the converse statement, i.e. if (27) holds, then the function $\phi(z) = R(w(z))X(w(z))[Q(w(z)]^{-1}$ belongs to $K^{p(\cdot)}(D;\omega)$. In other words, we have to establish that $\phi_0(z) = \omega(z)\phi(z)(= \omega(z)R(w(z))X(w(z))[Q(w(z)]^{-1}))$ is representable by a Cauchy type integral with density from $L^{p(\cdot)}(\Gamma)$. For this, it is sufficient to show that $\phi_0 \in E^\delta(D)$, for some $\delta > 0$, and $\phi_0^+ \in L^{p(\cdot)}(\Gamma) \subset L(\Gamma)$ (whence it follows that $\phi_0 \in E^1(D)$).

If Γ_ρ is the image of circumference $|w| = \rho$ under a conformal mapping of U onto D, then

$$\begin{aligned}\int\limits_{\Gamma_\rho} |\phi_0(z)|^\delta |dz| &= \int\limits_{|w|=\rho} |\phi_0(z(w))|^\delta |z'(w)|\,|dw| \\ &= \int\limits_{|w|=\rho} \left|\omega(z(w))\,\frac{R(w)X(w)}{Q(w)}\right|^\delta |z'(w)|\,|dw| \\ &\le M_1 \int\limits_{|w|=\rho} \left|\omega(z(w))\,\frac{r(w)R(w)}{Q(w)}\right|^\delta |z'(w)|\,|dw| \qquad (29)\end{aligned}$$

where $M_1 = \max\limits_{\tau\in\gamma} |X_1(\tau)|^\delta$. According (28) $R \in H^1$. It is obvious that $[Q(w)]^{-1}$ and $r(w)$ belong to some Hardy class; the same property follows from (4) for $\omega(z(w))$. Therefore the function $F_1(w) = \omega(z(w))r(w)R(w)[Q(w)]^{-1}$ belongs to H^η for some $\eta > 0$. Assuming $0 < \delta < \eta = \min(1, \nu_1, \dots, \nu_i)$, (29) implies

$$\int\limits_{\Gamma_\rho} |\phi_0(z)|^\delta |dz| \le M_2 \Big(\int\limits_{|w|=\rho} |F_1(w)|^\eta |dw| \Big)^{\frac{\delta}{\eta}} \Big(\int\limits_{|w|=\rho} |z'(w)|^{\frac{\eta}{\eta-\delta}} |dw| \Big)^{\frac{\eta-\delta}{\eta}}. \qquad (30)$$

From (4) and the conditions $0 < \nu_k \le 2$ it follows that we can make the right-hand part of (30) be bounded with respect to ρ and thus $\phi_0 \in E^\delta(D)$. Further, we have

$$\int\limits_{\Gamma} |\phi_0(t)|^{p(t)} |dt| \le M_3 \int\limits_{\gamma} \left| \frac{\omega(z(\tau))r(\tau)R(\tau)}{Q(\tau)} |z'(\tau)|^{\frac{1}{\ell(\tau)}} \right|^{\ell(\tau)} |d\tau|. \qquad (31)$$

Applying (26) we find from (31) that

$$\int\limits_{\Gamma} |\phi_0(\tau)|^{p(t)} |dt| \le M_4 \int\limits_{\gamma} |R_0(\tau)|^{\ell(\tau)} |d\tau| < \infty.$$

Hence $\phi_0 \in L^{p(\cdot)}(\Gamma)$ and therefore $\phi \in K^{p(\cdot)}(D;\omega)$.

We have thus proved

Theorem 13. *If $\phi \in K^{p(\cdot)}(D;\omega)$ and $Q(w)$ is the rational function defined by equality (22), then inclusion (27) holds provided that conditions (20) are fulfilled. Conversely, if (27) holds, then $\phi \in K^{p(\cdot)}(D;\omega)$.*

We have studied above some properties of the function $R(w)$ in U. Now let us consider the properties of the same function in $U^- = \{w \in \mathbb{C} : \ |w| > 1\}$.

The function $R(w) = Q(w)\Omega(w)[X(w)]^{-1}$ is analytic in U^- everywhere except, perhaps, the point $z = \infty$ and has, at that point, order $\varkappa = \varkappa_0 + \varkappa_1$, where $\varkappa_0 = \sum\limits_{k=1}^{j} \gamma_k$ is the order at infinity of rational function Q (see (22)), while $\varkappa_1$ is the order of $[X(w)]^{-1}$. If $\varkappa < 0$, then $R(w)$ has at infinity zero of order $(-\varkappa)$. If $\varkappa \geq 0$, then since $\Omega(\infty) \neq 0$, there exists a polynomial $P(w)$ of order $\varkappa$ such that the function

$$N(w) = Q(w)\Omega(w)[X(w)]^{-1} - P(w) = Q(w)\overline{\Psi\left(\frac{1}{\overline{w}}\right)}[X(w)]^{-1} - P(w) \tag{32}$$

is analytic in U^- and $N(\infty) = 0$.

Let us show that $N \in E^1(U^-)$, i.e.

$$\sup_{\rho>1} \int\limits_{|w|=\rho} |N(w)|\,|dw| = \sup_{\rho>1} I_\rho < \infty.$$

Since $N(\infty) = 0$, it is sufficient to prove that $\sup\limits_{1<\rho\leq\rho_0} I_\rho < \infty$, $\rho_0 < \infty$. We have

$$\begin{aligned} I_\rho &\leq \int\limits_{|w|=\rho} \left|\frac{Q(w)\Omega(w)}{X(w)}\right| |dw| + 2\pi\rho_0 \max_{|w|\leq\rho_0} |P(w)| \\ &\leq M_5 \int\limits_{|w|=\rho} \left|\frac{Q(w)\Psi\left(\frac{1}{\overline{w}}\right)}{r(w)}\right| |dw| + M_6 = M_5 \mathcal{J}_\rho + M_6, \end{aligned} \tag{33}$$

where $M_5 = \max\limits_{|w|\leq\rho_0} \left|\frac{1}{X(w)}\right|$, $M_6 = 2\pi\rho_0 \max\limits_{|w|\leq\rho_0} |P(w)|$.

Next,

$$\begin{aligned} \mathcal{J}_\rho &= \int\limits_{|w|=\rho} \left|\frac{Q(w)\Psi\left(\frac{1}{\overline{w}}\right)}{r(w)}\right| |dw| = \rho^2 \int\limits_{|\zeta|=\frac{1}{\rho}} \left|\frac{Q\left(\frac{1}{\zeta}\right)\Psi(\zeta)}{r\left(\frac{1}{\zeta}\right)}\right| |d\zeta| \\ &= \rho^2 \int\limits_{|\zeta|=\frac{1}{\rho}} \left|\frac{Q\left(\frac{1}{\overline{\zeta}}\right)}{Q(\zeta)} \frac{r(\zeta)}{r\left(\frac{1}{\overline{\zeta}}\right)} \frac{Q(\zeta)\Psi(\zeta)}{r(\zeta)}\right| |d\zeta|. \end{aligned}$$

It is easy to verify that $\left|Q\left(\frac{1}{\overline{\zeta}}\right)\right|\left|[Q(\zeta)]^{-1}\right| = |\zeta|^{\varkappa_0}$, $\left|r\left(\frac{1}{\overline{\zeta}}\right)\right|\left|[r(\zeta)]^{-1}\right| = 1$ and therefore

$$\mathcal{J}_\rho \leq \frac{\rho^2}{\rho^{\varkappa_0}} \int\limits_{|\zeta|=\frac{1}{\rho}} \left|\frac{Q(\zeta)\Psi(\zeta)}{r(\zeta)}\right| |d\zeta| \leq \frac{M_7}{\rho^{\varkappa_0-2}} \int\limits_{|\zeta|=\frac{1}{\rho}} |R(\zeta)|\,|d\zeta|$$

where $M_7 = \max\limits_{|w|\leq\rho_0} |X_1(w)|$.

By virtue of (28) we have $\sup\limits_{\rho\leq\rho_0} \mathcal{J}_\rho < \infty$ and from (33) we conclude that $N(w) \in E^1(U^-)$. This and (32) imply that

$$R(w) = Q(w)\Omega(w)[X(w)]^{-1} = -(K_\gamma N^-)(w) + P(w), \quad |w| > 1, \tag{34}$$

where $N^- \in L^{\ell(\cdot)}(\gamma;\rho)$ (see (27)).

Note that by virtue of (28) we have

$$R(w) = (K_\gamma R^+)(w), \quad |w| < 1. \tag{35}$$

Definition 14. The set of functions F representable in the form

$$F(w) = \frac{1}{2\pi i}\int\limits_\gamma \frac{f(\tau)\,d\tau}{\tau - w} + P_n(w), \quad f \in L^{\ell(\cdot)}(\gamma;\rho), \quad |w| \neq 1,$$

where P_n is some polynomial of order n, is denoted by $K^{\ell(\cdot)}(\gamma;\rho;n)$. We consider this class also for negative n, assuming that in that case $P_n \equiv 0$ and $(K_\gamma f)(w)$ has zero of order $(-n)$ at infinite.

Now from (34) and (35) it follows

Theorem 15. *If $\phi \in K^{p(\cdot)}(D;\omega)$, $\Psi(w) = \phi(z(w))$, and the functions $X(w)$, $\Omega(w)$ and $Q(w)$ are defined in $\mathbb{C}\setminus\gamma$ by formulas (11), (12) and (22), then the function*

$$F(w) = Q(w)\Omega(w)[X(w)]^{-1}, \quad |w| \neq 1, \tag{36}$$

belongs to $K^{\ell(\cdot)}(\gamma;\rho;\varkappa)$.

5.5 Solution of the Riemann–Hilbert problem in the class $K^{p(\cdot)}(D;\omega)$ for a domain D with an arbitrary piecewise smooth boundary

Now we can go back to problem (13). Multiplying (13) by $Q(w)$ from (36) we obtain

$$F^+(\tau) - F^-(\tau) = c_2(\tau)Q(\tau), \quad \tau \in \gamma, \tag{37}$$

where

$$c_2(\tau) = 2C(\tau)\left[A(\tau) + iB(\tau)\right]^{-1}\left[X^+(\tau)\right]^{-1}.$$

The solution of this problem is to be sought for in the class $K^{\ell(\cdot)}(\gamma;\rho;\varkappa)$.

According to Theorem 13, if $F \in K^{\ell(\cdot)}(\gamma,\rho,\varkappa)$, then

$$\phi(z) = F(w(z))X(w(z))[Q(w(z))]^{-1}$$

is a function of the class $K^{p(\cdot)}(D;\omega)$. Thus we need a solution $\Omega(w)$ of problem (13) representable by the form $\Omega(w)=F(w)X(w)[Q(w)]^{-1}$, where $F(w)$ is the solution of (37) from $K^{\ell(\cdot)}(\gamma;\rho;\varkappa)$. The function $\Omega(w(z))$ is a solution of problem (1) if and only if $\Omega(w)$ satisfies the condition (14).

Thus problem (1) will be solved in the class $K^{p(\cdot)}(D;\omega)$ if we find all solutions of problem (37) in $K^{\ell(\cdot)}(\gamma;\rho;\varkappa)$ and choose among them solutions F for which

$$\left(\frac{FX}{Q}\right)_*(w)=\left(\frac{FX}{Q}\right)(w),\quad |w|\neq 1.$$

If this condition is fulfilled, then by the restricting of the function $\Omega(w)$ on U we find the function $\Psi(w)=\phi(z(w))=F(w)X(w)[Q(w)]^{-1}$ and, eventually, obtain

$$\phi(z)=F(w(z))X(w(z))[Q(w(z)]^{-1} \tag{38}$$

which is a solution of problem (1) in the class $K^{p(\cdot)}(D;\omega)$.

For problem (37) to be solvable it is necessary that the function

$$c_2(\tau)Q(\tau)=2C(\tau)[A(\tau)+iB(\tau)]^{-1}[X^+(\tau)]^{-1}Q(\tau)$$

belongs to $L^{\ell(\cdot)}(\gamma;\rho)$, which is equivalent to the requirement that $c\omega\in L^{p(\cdot)}(\Gamma)$. Under this condition we find

$$F(w)=(K_\gamma c_2)(w)+P_\varkappa(w)=F_c(w)+P_\varkappa(w),$$

where $F_c(w)=(K_\gamma c_2)(w)$ and $P_\varkappa(w)$ is an arbitrary polynomial of order $\varkappa$.

Note that condition (14) is satisfied not only by $\Omega(w)$, but also by $\Omega_*(w)$. Then taking

$$\widetilde{\Omega}_c(w)=\frac{1}{2}(\Omega_c(w)+(\Omega_c)_*(w)),\ |w|\neq 1,$$

where $\Omega_c(w)=F_c(w)X(w)[Q(w)]^{-1}$, we obtain a solution of the nonhomogeneous problem that satisfies the condition $(\widetilde{\Omega}_c)_*(w)=\widetilde{\Omega}_c(w)$, $|w|\neq 1$.

Let us write a general solution of problem (13) in the form

$$\Omega(w)=\widetilde{\Omega}_c(w)+X(w)[Q(w)]^{-1}P_\varkappa(w)=\widetilde{\Omega}_c(w)+\Omega_0(w). \tag{39}$$

In order that $\Omega(w)$ gives the desired solution, we must have $(\Omega_0)_*(w)=\Omega_0(w)$, $|w|\neq 1$. This imposes additional restrictions on $P_\varkappa(w)$.

So, $P_\varkappa$ must be such that

$$\frac{X_*(w)(P_\varkappa)_*(w)}{Q_*(w)}=\frac{X(w)}{Q(w)}P_\varkappa(w),\ |w|\neq 1,$$

i.e.

$$r_*(w)(X_1)_*(w)Q(w)(P_\varkappa)_*(w)=r(w)X_1(w)P_\varkappa(w)Q_*(w). \tag{40}$$

By the construction of the function $r(w)$ (see (9)) it is easy to verify that $r_*(w) = r(w)$, while by an appropriate choice of the constant C in equality (10) we can obtain $(X_1)_*(w) = w^{\varkappa_1} X_1(w)$ (see [98, p. 147]). Hence (40) implies

$$w^{\varkappa_1} Q(w) \overline{P_\varkappa\left(\frac{1}{\overline{w}}\right)} = Q_*(w) P_\varkappa(w). \tag{41}$$

Further,

$$\begin{aligned} Q_*(w) &= \overline{\prod_{k=1}^{j}\left(\frac{1}{\overline{w}} - w_k\right)^{\gamma_k}} = \prod_{k=1}^{j}\left(\frac{1}{w} - \overline{w}_k\right)^{\gamma_k} = \prod_{k=1}^{j}\left(\frac{1}{w} - \frac{1}{w_k}\right)^{\gamma_k} \\ &= w^{-\varkappa_0}(-1)^{\varkappa_0} \prod_{k=1}^{j} w_k^{-\gamma_k} Q(w) = A w^{-\varkappa_0} Q(w), \end{aligned} \tag{42}$$

$$A = (-1)^{\varkappa_0} \prod_{k=1}^{j} w_k^{-\gamma_k} \quad (|A| = 1, \text{ since } w_k \in \gamma). \tag{43}$$

Remind that $\varkappa_0 = \operatorname{ord}[Q(w), \infty] = \sum_{k=1}^{j} \gamma_k$.

Now from (41)–(43) we obtain

$$w^{\varkappa_0 + \varkappa_1} \overline{P_*\left(\frac{1}{\overline{w}}\right)} = A P_\varkappa(w).$$

Let æ ≥ 0 and

$$P(w) = h_0 + h_1 w + \cdots + h_\varkappa w^\varkappa. \tag{44}$$

Then

$$w^\varkappa\left(\overline{h}_0 + \overline{h}_1 \frac{1}{w} + \cdots + \frac{\overline{h}_{æ}}{w^\varkappa}\right) = A(h_0 + h_1 w + \cdots + h_\varkappa w^\varkappa),$$

whence

$$\overline{h}_k = A h_{\varkappa - k}, \quad k = \overline{0, \varkappa}. \tag{45}$$

Thus if for $\varkappa \geq 0$ the coefficients h_k in polynomial (44) satisfy condition (45), then the functions given by equality (39) are the desired solutions of problem (13). Then the set of functions

$$\Omega(w(z)) = \widetilde{\Omega}_c(w(z)) + X(w(z))[Q(w(z))]^{-1} P_\varkappa(w(z)) \tag{46}$$

contains all solutions of problem (1).

If $\varkappa \geq 0$, then these functions give solutions, since, by virtue of Theorems 13 and 15, the functions $\Omega(w(z))$ belong to $K^{p(\cdot)}(D; \omega)$, equality (14) is fulfilled and the order of $\Omega(w)$ at infinity is equal to $\varkappa$.

If $\varkappa < 0$, then $P_\varkappa \equiv 0$ and for $\Omega(w(z))$ to be a solution of problem (1), it is necessary that $\widetilde{\Omega}_c(w)$ have zero of order $(-\varkappa)$ at infinity. For this it is necessary and sufficient that the function $K_\gamma c_2$ have zero of order $(-\varkappa)$ at infinity, i.e.

$$\int_\Gamma \frac{c(t)Q(w(t))w^k(t)w'(t)}{X^+(w(\tau))(a(t)+ib(t))}\,dt = 0, \quad k = \overline{0,|\varkappa|}, \tag{47}$$

and, upon the fulfillment of this condition, the problem has a unique solution given by equality (46) for $P_\varkappa \equiv 0$.

Let us summarize the obtained results.

Let 1) D be a domain bounded by a curve $\Gamma \in C^{1,L}(A_1,\dots,A_i;\nu_1,\dots,\nu_i)$, $0 < \nu_k \le 2$; 2) $\omega(z) = \prod\limits_{k=1}^{\nu}(z-t_k)^{\alpha_k}$, $t_k \in \Gamma$, $\alpha_k \in \mathbb{R}$; 3) $a(t)$, $b(t)$ be piecewise Hölder real functions with the condition $\inf(a^2(t)+b^2(t)) > 0$ such that $G(t) = -[a(t)-ib(t)][a(t)+ib(t)]^{-1}$ has discontinuity points B_k, $k = \overline{1,\lambda}$, and also $G(B_k-)[G(B_k+)]^{-1} = \exp(2\pi i u_k)$, $u_k \in \mathbb{R}$; 4) $p(t)$ be a function from the class $\mathcal{P}(\Gamma)$ and $\ell(\tau) = p(z(\tau))$.

Let $\tau_k = w(t_k)$, $a_k = w(A_k)$, $b_k = w(B_k)$. Thus we proved

Theorem 16. *Let*

i) *the points τ_k, a_k, b_k be numbered according to* (17)*, the numbers δ_k be defined by equalities* (19) *and $\{\delta_k\} \ne [\ell'(w_k)]^{-1}$;*

ii) *$Q(w)$ be a rational function defined by equality* (22) *and the order of $Q(w)$ at infinity be equal to $\varkappa_0$;*

iii) *the functions $r(w)$ and $X_1(w)$ be given by equalities* (9), (10)*; $X(w) = r(w)X_1(w)$ so that $X(w)$ has order $(-\varkappa_1)$ at infinity;*

iv) $c(t) \in L^{p(\cdot)}(\Gamma;\omega)$.

Assume that $\varkappa = \varkappa_0 + \varkappa_1$.

Then

a) *if $\varkappa < 0$, then for problem* (1) *to be solvable in the class $K^{p(\cdot)}(D;\omega)$ it is necessary and sufficient that conditions* (47) *be fulfilled and, upon their fulfillment,* (1) *has a unique solution given by equality* (46) *with $P_\varkappa = 0$;*

b) *for $\varkappa \ge 0$, problem* (1) *is unconditionally solvable and all its solutions are given by equality* (46)*, where $P_\varkappa$ is an arbitrary polynomial* (44) *whose coefficients satisfy conditions* (45).

5.6 Some particular cases

a) The Riemann–Hilbert problem with Hölder coefficients $a(t)$, $b(t)$ in the class $K^{p(\cdot)}(D;\omega)$ for $\omega^+ = \omega \in W^{p(\cdot)}(\Gamma)$

In this case there are no points b_k and $-\frac{1}{p(t_k)} < \alpha_k < \frac{1}{p'(t_k)}$. Only the points τ_k and a_k may coincide; to simplify numeration (17) it is assumed that $w_1 = \tau_1, \dots, w_m = \tau_m$,

$w_{m+1} = a_1, \ldots, w_{m+n} = a_n$, $w_{m+n+1} = \tau_{m+1} = a_{n+1}, \ldots, w_{m+n+r} = \tau_{m+r} = a_{n+r}$. Accordingly, $\delta_1 = \cdots = \delta_m = 0$, $\delta_{m+k} = \frac{\nu_k - 1}{p(A_k)}$, $k = \overline{1,n}$, $\delta_{m+n+k} = \alpha_{n+k}\nu_{n+k} + \frac{\nu_{n+k}-1}{p(A_{n+k})}$, $k = \overline{1,r}$. Therefore $\gamma_1 = \gamma_2 = \cdots = \gamma_m = 0$,

$$\gamma_{m+k} = \begin{cases} 0 & \text{if } \nu_k < \frac{1}{p'(A_k)}, \\ 1 & \text{if } \nu_k > \frac{1}{p'(A_k)}, \end{cases} \qquad k = \overline{1,n},$$

$$\gamma_{m+n+k} = \tag{48}$$

$$\begin{cases} \left[\alpha_{n+k}\nu_{n+k} + \frac{\nu_{n+k}-1}{\ell(a_{n+k})}\right] & \text{if } \left\{\alpha_{n+k}\nu_{n+k} + \frac{\nu_{n+k}-1}{\ell(a_{n+k})}\right\} < \frac{1}{p'(A_{n+k})}, \\ \left[\alpha_{n+k}\nu_{n+k} + \frac{\nu_{n+k}-1}{\ell(a_{n+k})}\right] + 1 & \text{if } \left\{\alpha_{n+k}\nu_{n+k} + \frac{\nu_{n+k}-1}{\ell(a_{n+k})}\right\} > \frac{1}{p'(A_{n+k})}, \quad k = \overline{1,r}. \end{cases}$$

The latter numbers can be considered in more detail: let $\beta = \gamma_{m+n+k}$, $\alpha \in \{\alpha_{n+1}, \ldots, \alpha_{n+r}\}$, $\nu \in \{\nu_{n+1}, \ldots, \nu_{n+r}\}$, $\ell = p(A_{n+k}) = \ell(a_{n+k})$, $\ell' = p'(A_{n+k}) = \ell'(a_{n+k})$ and $v = \alpha\nu + \frac{\nu-1}{\ell}$. Then (48) takes the form

$$\beta = \begin{cases} [v] & \text{if } \{v\} < \frac{1}{\ell'}, \\ [v] + 1 & \text{if } \{v\} > \frac{1}{\ell'}. \end{cases}$$

Since $-\frac{1}{\ell} < \alpha < \frac{1}{\ell'}$, $\frac{1-\ell-\nu}{\ell\nu} < -\frac{1}{\ell}$ and $\frac{2\ell-\nu}{\ell\nu} \geq \frac{1}{\ell'}$, it is sufficient to consider the following possible cases:

i) $\frac{1-\ell-\nu}{\ell\nu} < \alpha \leq \frac{1-\nu}{\ell\nu}$; **ii)** $\frac{1-\nu}{\ell\nu} < \alpha < \frac{\ell-\nu}{\ell\nu}$; **iii)** $\frac{\ell-\nu}{\ell\nu} < \alpha < \frac{1+\ell-\nu}{\ell\nu}$; **iv)** $\frac{1+\ell-\nu}{\ell\nu} \leq \alpha < \frac{2\ell-\nu}{\ell\nu}$.

Case i) We have $-1 < v \leq 0$. If $v = 0$, then $[v] = \{v\} = 0$ and therefore $\beta = 0$. If $-1 < v < 0$, then $[v] = -1$, $\{v\} = 1 + v = \alpha\nu + \frac{\nu-1}{\ell} + 1 = \alpha\nu + \frac{\nu}{\ell} + \frac{1}{\ell'} > (-\frac{1}{\ell})\nu + \frac{\nu}{\ell} + \frac{1}{\ell'} = \frac{1}{\ell'}$ and therefore $\beta = -1 + 1 = 0$.

Case ii) We have $0 < v < \frac{1}{\ell'}$. Hence $[v] = 0$, $\{v\} = v = \alpha\nu + \frac{\nu-1}{\ell} < \frac{\ell-\nu}{\ell} + \frac{\nu-1}{\ell} = \frac{1}{\ell'}$; thus $\beta = 0$.

Case iii) We have $\frac{1}{\ell'} < v < 1$, i.e. $[v] = 0$, $\{v\} = v = \alpha\nu + \frac{\nu-1}{\ell} > \frac{1}{\ell'}$ and therefore $\beta = [v] + 1 = 1$.

Case iv) We have $1 \leq v < 2 - \frac{1}{\ell'}$ (< 2). If $v = 1$, then $[v] = 1$, $\{v\} = 0$, and therefore $\beta = 1$. If $v > 1$, then $[v] = 1$, $\{v\} = v - 1 = \alpha\nu + \frac{\nu-1}{\ell} - 1 < \frac{2\ell-\nu}{\ell} + \frac{\nu-1}{\ell} - 1 = \frac{1}{\ell'}$. Therefore $\beta = [v] = 1$.

Thus

$$\begin{cases} \beta = 0 & \text{if } \dfrac{1-\ell-\nu}{\ell\nu} < \alpha < \dfrac{\ell-\nu}{\ell\nu}, \\ \beta = 1 & \text{if } \dfrac{\ell-\nu}{\ell\nu} < \alpha < \dfrac{2\ell-\nu}{\ell\nu}. \end{cases} \tag{49}$$

Equalities (49) can be written in the form

$$\begin{cases} \beta = 0 & \text{if } 0 < \nu < \dfrac{\ell}{1+\alpha\ell}, \\ \beta = 1 & \text{if } \dfrac{\ell}{1+\alpha\ell} < \nu < \dfrac{2\ell}{1+\alpha\ell}. \end{cases}$$

By virtue of (49), it follows from (22) that $Q(\tau)$ is a polynomial and its order is equal to

$$\varkappa_0 = \mathcal{N}\{w_k : \nu_k > \ell(w_k)\} + \mathcal{N}\left\{\tau_k = a_k : \frac{\ell(a_k)}{1+\alpha_k\ell(a_k)} < \nu_k < \frac{2\ell(a_k)}{1+\alpha_k\ell(a_k)}\right\} \tag{50}$$

where $\mathcal{N}(E)$ denotes the number of elements of a set E.

Thus we derive the following

Proposition 17. *If problem* (1) *is considered in the class* $K^{p(\cdot)}(D;\omega)$, $\omega \in W^{p(\cdot)}(\Gamma)$ *and* $a(t)$, $b(t)$ *belong to the Hölder class, then Theorem* 16 *is valid where number* $\varkappa_0$ *is calculated by equality* (50).

b) The Dirichlet problem in the class $\operatorname{Re} K^{p(\cdot)}(D;\omega)$

In what follows, the set of functions $u(z)$ =Re$\phi(z)$ where $\phi \in K^{p(\cdot)}(D;\omega)$ is denoted by $k^{p(\cdot)}(\omega)$. Let $a(t)=1$, $b(t)=0$, ω is defined by (2), $\omega \in W^{p(\cdot)}(\Gamma)$, $c \in L^{p(\cdot)}(\Gamma;\omega)$. Problem (1) is posed as follows: find a function $\phi(z) \in K^{p(\cdot)}(D;\omega)$, that satisfies the condition $\operatorname{Re}\phi^+(t) = c(t)$ a.e. on Γ, i.e. we deal with the Dirichlet problem: find a function u for which

$$\begin{cases} \Delta u = 0, \quad u = \operatorname{Re}\phi, \quad \phi \in K^{p(\cdot)}(D;\omega), \\ u^+(t) = c(t), \quad t \in \Gamma, \quad c(t) \in L^{p(\cdot)}(\Gamma;\omega). \end{cases} \tag{51}$$

Then $r(w)=1$,

$$X_1(w) = \begin{cases} -i, & |w|<1, \\ i, & |w|>1 \end{cases}$$

(we need this to have $(X_1)_*(w) = X_1(w)$). Thus $\varkappa_1 = 0$, i. e. $\varkappa = \varkappa_0$, where $\varkappa_0$ is calculated by formula (50) and therefore $\varkappa \geq 0$.

c) The Dirichlet problem in the class $k^{p(\cdot)}(D)$

In that case, r, X_1 and $\varkappa_1$ are calculated as in b). From condition (20) we obtain $\nu_k \neq p(A_k)$, $k=\overline{1,m}$. The order $\varkappa_0$ of $Q(w)$ at infinity is equal to the number of angular points for which $\nu_k > p(A_k)$.

Let us find the solution when $m=1$, $\nu > p(A_1) = \ell(a_1)$ and $c(t)=0$.

From (46) we have

$$\Psi(w) = i\,\frac{P_1(w)}{w-w_1} = i\,\frac{h_0+h_1 w}{w-w_1}, \quad w_1 = w(A_1),$$

where the coefficient $h_1 = is_1$ with $s_1 \in \mathbb{R}$ (this fact follows from equality $\Omega(\infty) = \widetilde{\Omega}(\infty) = \frac{1}{2}(\Psi(0)+\overline{\Psi(0)})$). Formula (45) gives $\overline{h_0} = -Ah_1$, where $A = -\frac{1}{w_1}$ (see

(43)). Hence $h_0 = \frac{is_1}{\overline{w}_1} = is_1 w_1$ and therefore

$$\phi(w) = i\,\frac{is_1 w_1 + is_1 w}{w - w_1} = -s_1\,\frac{w + w_1}{w - w_1}, \qquad s_1 \in \mathbb{R}.$$

Thus for $\nu > p(A_1)$ the problem

$$\begin{cases} \Delta u = 0, \quad u \in k^{p(\cdot)}(D), \quad p \in \mathcal{P}(\Gamma), \quad \Gamma \in C^{1,L}(A_1, \nu), \\ u^+(t) = 0, \quad t \in \Gamma, \end{cases}$$

has a solution

$$u(z) = s_1 \operatorname{Re} \frac{w(z) + w(A_1)}{w(z) - w(A_1)}$$

depending on one real parameter.

If $\nu < p(A_1)$, then the problem has only a trivial solution.

Part II

The case of domains with piecewise smooth boundaries

In this Part, the investigation of problem (1) is carried out under the following assumptions:

i) Γ is a closed piecewise smooth curve from the class $C^1(A_1,\ldots,A_i;\nu_1,\ldots,\nu_i)$, $0 < \nu_k \le 2$;

ii) the coefficients $a(t)$, $b(t)$ are piecewise continuous; $c \in L^{p(\cdot)}(\Gamma;\omega)$;

iii) a function $p(t)$ belongs to the class

$$\widetilde{\mathcal{P}}(\Gamma) = \bigcup_{\varepsilon>0} \mathcal{P}_{1+\varepsilon}(\Gamma), \tag{52}$$

where $\mathcal{P}_{1+\varepsilon}(\Gamma)$ denotes the set of those real functions p, for which $\underline{p} > 1$ and there exists positive numbers A and ε such that

$$|p(t_1) - p(t_2)| < \frac{A}{|\ln|t_1 - t_2||^{1+\varepsilon}} \tag{53}$$

for arbitrary points t_1 and t_2 on Γ.

When investigating problem (1) in the case of a piecewise smooth boundary, we have to deal with the following problems:

1) As is known, if $-[\ell(\tau_k)]^{-1} < \alpha_k < [\ell'(\tau_k)]^{-1}$, $\ell \in \mathcal{P}(\gamma)$, and φ is a continuous real function on γ, then

$$\rho(\tau) = \prod_{k=1}^{\nu} (\tau - \tau_k)^{\alpha_k} \exp\left(\frac{1}{\pi}\int_{\gamma} \frac{\varphi(\zeta)\,d\zeta}{\zeta - \tau}\right) \in W^{\ell(\cdot)}(\gamma) \tag{54}$$

(this follows from Corollary 11 of Chapter 4).

If however Γ is a piecewise smooth curve, then for the investigation we need to know whether the function

$$\rho(\tau) = \prod_{k=1}^{\nu} (\tau - \tau_k)^{\alpha_k} \exp\left(\frac{1}{\ell(\tau)}\frac{1}{\pi}\int_{\gamma}\frac{\varphi(\zeta)\,d\zeta}{\zeta-\tau}\right), \quad -\frac{1}{\ell(\tau_k)} < \alpha_k < \frac{1}{\ell'(\tau_k)} \tag{55}$$

possesses the same property.

2) If $X(w)$ is a canonical function for a piecewise continuous function, then we do not know whether the functions $X^+(\tau)$ and $\rho(\tau)X^+(\tau)$, where $\rho(\tau)$ is given by equality (54), belong to the class $W^{\ell(\cdot)}(\gamma)$.

3) When Γ is a piecewise smooth curve from $C^1(A_1,\dots,A_i;\nu_1,\dots,\nu_i)$, and $z(a_k) = A_k$, where $z = z(w)$ is the conformal mapping of the circle U on D, then we must know the weight properties of the function $z(\tau) - z(a_k)$.

4) We must know whether the log-Hölder condition holds for the function $\ell(\tau) = p(z(\tau))$ on γ when $p(t)$ satisfies the log-Hölder condition on Γ.

The reasoning in Part I clearly implies that positive answers to the above four questions make if possible to establish the validity of the results obtained there also for the case where Γ is piecewise smooth, while $a(t)$, $b(t)$ are piecewise continuous.

In this Part we first investigate the above questions and then proceed to the investigation of problem (1) under the given assumptions.

As to questions 1) and 2), we succeeded in showing that they both have a positive answer if $p \in \widetilde{\mathcal{P}}(\Gamma)$.

As to question 3), the following was clarified: if we follows the method of investigation of the functions z' and z, which is used in [40, Ch. III] and apply inclusion (54) given above, then it can be shown that

$$\prod_{k=1}^{i} (z(\tau) - z(a_k)) \in W^{\ell(\cdot)}(\gamma) \tag{56}$$

for $0 < \nu_k < \frac{1}{\ell'(a_k)}$.

As to question 4), we can state that the function $\ell(\tau) = p(z(\tau))$ belongs to the class $\widetilde{\mathcal{P}}(\gamma)$ if so does the function $p(t)$ on $\Gamma \in C^1(A_1,\dots,A_i;\nu_1,\dots,\nu_i)$.

Having these results and following to reasoning of Part I, it is not difficult to justify the results obtained in that Part I are true also for the Riemann–Hilbert problem under the assumptions i)–iii).

5.7 Invariance of the classes $\mathcal{P}_{1+\varepsilon}(\Gamma)$ under a conformal mapping

Lemma 18. *If $\Gamma \in C^1(A_1,\dots,A_i;\nu_1,\dots,\nu_i)$, $0 < \nu_k \le 2$, $k = \overline{1,i}$, is the closed curve bounding the domain D, $p \in \mathcal{P}_{1+\varepsilon}(\Gamma)$, $\varepsilon > 0$, and $z = z(w)$ is the conformal* *mapping of the circle U onto D, then the functions $\ell(\tau) = p(z(\tau))$ and $\frac{1}{\ell(\tau)}$ belong to $\mathcal{P}_{1+\varepsilon}(\gamma)$.*

This Lemma is proved exactly in the same manner as Lemma 13 in Chapter 3.

Corollary 19. *If $p \in \widetilde{\mathcal{P}}(\Gamma)$, then $\ell \in \widetilde{\mathcal{P}}(\gamma)$.*

5.8 On the belonging of a function $\prod_{k=1}^{i}(z(\tau) - z(a_k))$ to the class $W^{\ell(\cdot)}(\gamma)$

Lemma 20. *If $\Gamma \in C^1(A;\nu)$, $0 < \nu < 1$, $a \in \gamma$, and $z(a) = A$, then*

$$z(w) - z(a) = (w-a)^{\nu} z_0(w) \exp\left(\frac{1}{2\pi}\int_{\gamma} \frac{\mu(\zeta)\, d\zeta}{\zeta - w}\right), \tag{57}$$

where $z_0^{\pm 1}$ are bounded functions, and $\mu(\zeta)$ is a real continuous function on γ.

Proof. Let us consider the function

$$\Omega(w) = \begin{cases} z(w) - z(a), & |w| < 1, \\ \overline{z(\frac{1}{\overline{w}}) - z(a)}, & |w| > 1, \end{cases} \tag{58}$$

on the plane cut along γ. This function is analytic outside γ and satisfies the boundary condition

$$\Omega^+(\tau) = [z(\tau) - z(a)]\,\overline{[z(\tau) - z(a)]}^{-1}\Omega^-(\tau), \quad \Omega(\infty) = const. \tag{59}$$

Also,

$$\Omega(w) = \frac{1}{2\pi i}\int_{\gamma} \frac{\Omega^+(\zeta) - \Omega^-(\zeta)}{\zeta - w}\, d\zeta + \Omega(\infty),$$

where $\Omega^{\pm}(\zeta)$ are continuous functions on γ, and therefore $\Omega \in \widetilde{K}_0^q(\gamma)$ for arbitrary q, $1 < q < \infty$.

For the coefficient of problem (59) $G(\tau) = [z(\tau) - z(a)]\,\overline{[z(\tau) - z(a)]}^{-1} = \exp i\, 2\lambda(\tau)$, we choose a branch of the function $\lambda(\tau) = \arg(z(\tau) - z(a))$, which varies continuously on $\gamma \setminus \{a\}$. It is piecewise continuous on γ with one discontinuity point $\tau = a$ and at this point has a discontinuity of the value $\pi\nu$.

Let us consider the function

$$X_0(w) = \exp\left\{\frac{1}{2\pi i}\int_{\gamma} \frac{\ln G(\zeta)\, d\zeta}{\zeta - w}\right\} = \exp\left\{\frac{1}{\pi}\int_{\gamma} \frac{\lambda(\zeta)\, d\zeta}{\zeta - w}\right\}.$$

Since $\lambda(\zeta)$ is bounded, $X_0(w)$ belongs to some Hardy class H^{δ}, $\delta > 0$ [40, p. 97].

Let $\lambda_a(\zeta) = \nu \arg(\zeta - a)$, then the function $\lambda(\zeta) - \lambda_a(\zeta)$ is continuous on γ and therefore

$$X_0(w) = \exp\left\{\frac{1}{\pi}\int\limits_{\gamma}\frac{\nu\arg(\zeta-a)\,d\zeta}{\zeta-w}\right\}\exp\left\{\frac{1}{\pi}\int\limits_{\gamma}\frac{\mu(\zeta)\,d\zeta}{\zeta-w}\right\}, \tag{60}$$

where $\mu(\zeta) = \lambda(\zeta) - \lambda_a(\zeta)$.

Since the function $\nu\arg(\tau - a)$ has on γ one discontinuity point a, and, in the unilateral neighborhoods of this point, satisfies the Hölder condition, we have near γ

$$\frac{1}{\pi}\int\limits_{\gamma}\frac{\nu\arg(\zeta-a)}{\zeta-w}\,d\zeta = \nu\ln(w-a) + \Phi_0(w),$$

where $\Phi_0(w)$ is a bounded analytic function (see [98, p. 75]). Therefore

$$X_0(w) = (w-a)^{\nu}\exp\Phi_0(w)\exp\left(\frac{1}{\pi}\int\limits_{\gamma}\frac{\mu(\zeta)\,d\zeta}{\zeta-w}\right). \tag{61}$$

Since $0 < \nu < 1$, it is not difficult to verify that $[X_0(w)]^{-1}$ belongs to $\widetilde{K}_0^{\lambda}(\gamma)$ for some $\lambda > 1$. By virtue of this fact we have that $\Omega(w)[X_0(w)]^{-1} \in \widetilde{K}_0^{1}(\gamma)$ is the solution of problem (59) and $\Omega(\infty)X_0(\infty) = const.$ Thus $\Omega(w)[X_0(w)]^{-1} = const.$ Hence it follows that all solutions of problem (59) lie in the set of functions

$$\Omega(w) = CX_0(w), \tag{62}$$

where C is an arbitrary complex constant.

Since $\nu > 0$, we have $X_0 \in \bigcap\limits_{q>1}\widetilde{K}^q$. Therefore equality (62) gives a general solution of problem (59). Now, from (58) it follows that there exists a constant C_0 such that

$$z(w) - z(a) = C_0X_0(w).$$

Hence, by virtue of (61), we obtain representation (57) with $z_0(w) = C_0\exp\Phi_0(w)$. □

From (57) we have

$$\begin{aligned} z(\tau) - z(a) &\sim (\tau-a)^{\nu}z_0^{+}(\tau)\exp\frac{i\mu(\tau)}{2}\exp\left(\frac{1}{2\pi}\int\limits_{\gamma}\frac{\mu(\zeta)\,d\zeta}{\zeta-\tau}\right)\\ &= (\tau-a)^{\nu}\,\widetilde{z}_0(\tau)\exp\left(\frac{1}{2\pi}\int\limits_{\gamma}\frac{\mu(\zeta)\,d\zeta}{\zeta-\tau}\right) = \rho(\tau)\widetilde{z}_0(\tau), \quad \tau\in\gamma, \end{aligned} \tag{63}$$

where $\widetilde{z}_0 = \left[z_0^{+}(\tau)\exp\frac{i\mu(\tau)}{2}\right]$ and therefore $\widetilde{z}_0^{\pm 1}$ are bounded functions, and $\rho(\tau) = (\tau-a)^{\nu}\exp\frac{1}{2\pi}\int\limits_{\gamma}\frac{\mu(\zeta)d\zeta}{\zeta-\tau}$.

In representation (63), the functions $\widetilde{z}_0$, $\frac{1}{\widetilde{z}_0}$ are bounded, but $\rho(\tau)$ belongs to $W^{\ell(\cdot)}(\gamma)$ if and only if the function $\rho_0(\tau) = (\tau - a)^{\nu}$ belongs to the class $W^{\ell(\cdot)}$. If $\ell \in \mathcal{P}(\gamma)$, then for this it is necessary and sufficient that the condition

$$-\frac{1}{\ell(a)} < \nu < \frac{1}{\ell'(a)},$$

would be fulfilled. It is not difficult to verify that if this condition is fulfilled, then $\rho_0^{-1} = (\tau - a)^{-\nu}$ (and thereby $\rho^{-1}(\tau)$, too) belongs to $L^{\ell'(\cdot)+\varepsilon}(\gamma)$. Therefore we have a right to apply Corollary 11 from Chapter 4. As a result we conclude that the following lemma is valid.

Lemma 21. *If $\Gamma \in C^1(A;\nu)$, $0 < \nu < \frac{1}{\ell'(a)}$, $\ell \in \mathcal{P}(\gamma)$, then $[z(\tau) - z(a)] \in W^{\ell(\cdot)}(\gamma)$.*

Corollary 22. *If $\Gamma \in C^1(A_1, \dots, A_i; \nu_1, \dots, \nu_i)$, $0 < \nu_k < \frac{1}{\ell'(a_k)}$, then $\left[\prod_{k=1}^{i} (z(\tau) - z(a_k))\right] \in W^{\ell(\cdot)}(\gamma)$.*

5.9 One weight function

Lemma 23. *If $\ell \in \widetilde{\mathcal{P}}(\gamma)$, φ is a real continuous function on γ and $-\frac{1}{\ell(\tau_k)} < \alpha_k < \frac{1}{\ell'(\tau_k)}$, then the function*

$$\rho(\tau) = \prod_{k=1}^{i} (\tau - \tau_k)^{\alpha_k} \exp\left(\frac{1}{\ell(\tau)} \int_{\gamma} \frac{\varphi(\zeta)\, d\zeta}{\zeta - \tau}\right)$$

belongs to $W^{\ell(\cdot)}(\gamma)$.

Proof. Since $\ell \in \widetilde{\mathcal{P}}$, there exist positive constants ε and B such that

$$|\ell(\tau_1) - \ell(\tau_2)| < \frac{B}{|\ln|\tau_1 - \tau_2||^{1+\varepsilon}} \tag{64}$$

for arbitrary τ_1 and τ_2 on the circumference γ. Furthermore,

$$\frac{1}{\ell(\tau)} \int_{\gamma} \frac{\varphi(\zeta)\, d\zeta}{\zeta - \tau} = \int_{\gamma} \frac{\varphi(\zeta)}{\ell(\zeta)} \frac{d\zeta}{\zeta - \tau} + \int_{\gamma} \frac{\varphi(\zeta)[\ell(\zeta) - \ell(\tau)]}{\ell(\zeta)\ell(\tau)(\zeta - \tau)}\, d\zeta$$

$$= \int_{\gamma} \frac{\varphi(\zeta)}{\ell(\zeta)} \frac{d\zeta}{\zeta - \tau} + M(\tau), \quad M(\tau) = \int_{\gamma} \frac{\varphi(\zeta)[\ell(\zeta) - \ell(\tau)]}{\ell(\zeta)\ell(\tau)(\zeta - \tau)} d\zeta, \quad \frac{\varphi}{\ell} \in C(\gamma).$$

By virtue of (64) we have

$$|M(\tau)| \le \max_{\zeta \in \gamma} \left|\frac{\varphi(\zeta)}{\ell(\zeta)}\right| \int_0^{2\pi} \frac{B\, ds}{|s - s_\tau|\, |\ln|s - s_\tau||^{1+\varepsilon}} \le M_0 < \infty. \tag{65}$$

Then

$$\rho(\tau)=\prod_{k=1}^{i}(\tau-\tau_k)^{\alpha_k}\exp\left\{\int\limits_{\gamma}\frac{\varphi(\zeta)}{\ell(\zeta)}\frac{d\zeta}{\zeta-\tau}\right\}\exp M(\tau),\quad \frac{\varphi}{\ell}\in C(\gamma).$$

Since $-\ell^{-1}(\tau_k)<\alpha_k<[\ell'(\tau_k)]^{-1}$, then, by (54) and (65), we have $\rho\in W^{\ell(\cdot)}(\gamma)$. □

Corollary 24. *Let the assumptions of Lemma 23 are valid. If*

$$X_1(w)=\exp\left(\frac{1}{2\pi}\int\limits_{\gamma}\frac{\ln G_1(\zeta)\,d\zeta}{\zeta-w}\right),$$

where $G_1\in C(\gamma)$, $G_1\neq 0$ *and* $\operatorname{ind}G_1(\tau)=0$, *then*

$$\rho(\tau)X_1^+(\tau)\in W^{\ell(\cdot)}(\gamma). \tag{66}$$

Indeed, it is easy to verify that

$$\rho(\tau)X_1^+(\tau)\sim\prod_{k=1}^{i}(\tau-\tau_k)^{\alpha_k}\exp\left\{\int\limits_{\gamma}\left(\frac{\varphi(\zeta)}{\ell(\zeta)}+\frac{1}{2}\arg G_1(\zeta)\right)\frac{d\zeta}{\zeta-\tau}\right\}M_1(\tau),$$

where $M_1(\tau)=\sqrt{G_1(\tau)}\,M(\tau)$. Since $\frac{\varphi(\zeta)}{\ell(\zeta)}+\frac{1}{2}\arg G(\zeta)$ is continuous on γ, the validity of (66) follows from inequality (65) and the assumptions of lemma 23. □

5.10 Reduction of the Riemann–Hilbert problem to a linear conjugation problem

Let $c\in L^{p(\cdot)}(\Gamma,\omega)$, $a(t)$ and $b(t)$ be piecewise continuous function on Γ, with the condition $\inf\limits_{t\in\Gamma}(a^2(t+b^2(t))>0$, and B_k, $k=\overline{1,\lambda}$, be the discontinuity points of the functions $\widetilde{G}(t)=-[a(t)-ib(t)]\,[a(t)+ib(t)]^{-1}$.

Let us consider the Riemann–Hilbert problem: find a function $\Phi\in K^{p(\cdot)}(D;\omega)$ satisfying the boundary condition (1).

Let $\Psi(w)=\Phi(z(w))$. Then

$$\begin{aligned}\Psi^+(\tau)&=-[A(\tau)-iB(\tau)]\,[A(\tau)+iB(\tau)]^{-1}\overline{\Psi^+(\tau)}\\&\quad+2C(\tau)[A(\tau)+iB(\tau)]^{-1},\\ A(\tau)&=a(z(\tau)),\quad B(\tau)=b(z(\tau)),\quad C(\tau)=c(z(\tau)).\end{aligned} \tag{67}$$

Let us introduce the new unknown function

$$\Omega(w)=\begin{cases}\Psi(w), & |w|<1,\\ \overline{\Psi\left(\frac{1}{\overline{w}}\right)}, & |w|>1.\end{cases} \tag{68}$$

Then $\Omega_*(w) = \Omega(w)$, $|w| \neq 1$ and (67) takes the form

$$\Omega^+(\tau) = G(\tau)\Omega^-(\tau) + C_1(\tau),$$
$$G(\tau) = -[A(\tau) - iB(\tau)]\,[A(\tau) + iB(\tau)]^{-1}, \quad C_1(\tau) = C(\tau)[A(\tau) + iB(\tau)]^{-1},$$

and thereby we have to find a solution of the problem

$$\begin{cases} \Omega^+(\tau) = G(\tau)\Omega^-(\tau) + C_1(\tau), & \tau \in \gamma, \quad C_1 \in L^{p(\cdot)}(\gamma), \\ \Omega_*(w) = \Omega(w), & |w| \neq 1, \quad \Omega(\infty) = const. \end{cases} \tag{69}$$

Solution of problem (69)

In Problem (69), for the time being the question in what class the function Ω should be sought for remains open.

Let $G(B_k^-)/G(B_k^+) = \exp 2\pi i u_k$, $k = \overline{1, \lambda}$ ($u_k \in \mathbb{R}$ since $|G| = 1$).

The scheme of our investigation will be as follows.

Step I. Write G in the form $G(\tau) = X^+(\tau)[X^-(\tau)]^{-1}$, where

$$\begin{aligned} &X(w) = r(w)X_1(w), \quad r(w) = \prod_{k=1}^{\lambda} r_k(w), \\ &r_k(w) = \begin{cases} (w - b_k)^{u_k}, & |w| < 1, \\ \overline{\left(\frac{1}{\overline{w}} - b_k\right)^{u_k}}, & |w| > 1, \end{cases} \end{aligned} \tag{70}$$

$$X_1(w) = \begin{cases} C \exp\left\{ \dfrac{1}{2\pi i} \displaystyle\int_\gamma \dfrac{\ln G_1(\tau)\tau^{-\varkappa_1} d\tau}{\tau - w} \right\}, & |w| < 1, \\ C w^{-\varkappa_1} \exp\left\{ \dfrac{1}{2\pi i} \displaystyle\int_\gamma \dfrac{\ln G_1(\tau)\tau^{-\varkappa_1} d\tau}{\tau - w} \right\}, & |w| > 1. \end{cases} \tag{71}$$

Here

$$G_1(\tau) = \prod_{k=1}^{\lambda} r_k^+(\tau)[r_k(\tau)]^{-1} G(\tau) \in C(\gamma) \tag{72}$$

$\varkappa_1 = \operatorname{ind} G_1(\tau)$, $C = const$.

Step II. Rewrite the boundary condition from (69) in the form

$$\frac{\Omega^+(\tau)}{X^+(\tau)} = \frac{\Omega^-(\tau)}{X^-(\tau)} + \frac{C_1(\tau)}{X^+(\tau)}.$$

Step III. Construct a rational function $Q(w)$ with zeros and poles on γ, such that the function $F(w) = Q(w)\Omega(w)[X(w)]^{-1}$ belongs to $K^{\ell(\cdot)}(\gamma; \rho)$, where $\ell(\tau) = p(z(\tau))$ and $\rho(\tau)$ is a power function from $W^{\ell(\cdot)}(\gamma)$. For this construction, the numbers α_k, ν_k, u_k together must satisfy a certain condition (see conditions (74)–(76) below).

To construct $Q(w)$, we proceed as follows. Let

$$T=\{\tau_k:\ \tau_k=w(t_k)\},\quad A=\{a_k:\ a_k=w(A_k)\},\quad B=\{b_k:\ b_k=w(B_k)\}.$$

Let us enumerate the points of the set $T\cup A\cup B$ such in (17).

In the circle U the functions $\Omega(w)[X(w)]^{-1}$ and $\omega^{-1}(z)\Omega(w)[X(w)]^{-1}$ belong to the Hardy class H^δ for some $\delta>0$. The proof of these facts given in Part I is consistent because in representations of z' (see (4)) and (57) the multipliers of power functions belong to $\bigcap\limits_{q>0} H^q$.

Besides, if $M(w)=(K_\Gamma\varphi)(z(w))$, $\varphi\in L^{p(\cdot)}(\Gamma)$, then $M^+(\tau)=|z'(\tau)|^{-\frac{1}{\ell(\tau)}}M_0(\tau)$, $M_0(\tau)\in L^{\ell(\cdot)}(\gamma)$ (see Lemma 10).

But

$$[z'(\tau)]^{\frac{1}{\ell(\tau)}}\sim\prod_{k=1}^{i}(\tau-a_k)^{\frac{\nu_k-1}{\ell(\tau)}}\exp\left\{\frac{1}{\ell(\tau)}\int_\gamma\frac{\beta_k(\zeta)\,d\zeta}{\zeta-\tau}\right\}$$

$$\sim\prod_{k=1}^{i}(\tau-a_k)^{\frac{\nu_k-1}{\ell(\tau_k)}}\exp\left\{\frac{1}{\ell(\tau)}\int_\gamma\frac{\beta(\zeta)\,d\zeta}{\zeta-\tau}\right\},\quad \beta(\zeta)=\sum_{k=1}^{i}\beta_k(\zeta),\ \ \beta\in C(\gamma).$$

Here we apply the relation $(\tau-a_k)^{\frac{1}{\ell(\tau)}}\sim(\tau-a_k)^{\frac{1}{\ell(\tau_k)}}$, (see Lemma 11). Therefore

$$M^+(\tau)\sim\prod_{k=1}^{i}(\tau-a_k)^{\frac{\nu_k-1}{\ell(a_k)}}\exp\left\{\frac{1}{\ell(\tau)}\int_\gamma\frac{\beta(\zeta)\,d\zeta}{\zeta-\tau}\right\}\cdot M_0(\tau),\quad M_0\in L^{\ell(\cdot)}(\gamma).\tag{73}$$

Let δ_k are numbers defined by equalities (19).

Assume like to (20)–(21)

$$\{\delta_k\}\neq\frac{1}{\ell'(w_k)},\tag{74}$$

$$\gamma_k=\begin{cases}[\delta_k], & \text{if } \{\delta_k\}<\frac{1}{\ell'(w_k)},\\ [\delta_k]+1, & \text{if } \{\delta_k\}>\frac{1}{\ell'(w_k)},\end{cases}\qquad k=\overline{1,j}.\tag{75}$$

Then

$$-\frac{1}{\ell(w_k)}<\delta_k-\gamma_k<\frac{1}{\ell'(w_k)},\quad k=\overline{1,j}.\tag{76}$$

Let

$$Q(w)=\prod_{k=1}^{j}(w-w_k)^{\gamma_k}.\tag{77}$$

Lemma 25. *If p belongs to $\widetilde{\mathcal{P}}(\gamma)$, then*

$$F=Q\Omega/X^+\in L^{\ell(\cdot)}(\gamma,\rho),$$

where

$$\rho(\tau)=\prod_{k=1}^{j}(\tau-w_k)^{\delta_k-\gamma_k}\exp\left(\frac{1}{\ell(\tau)}\frac{1}{\pi}\int\limits_{\gamma}\frac{v(\zeta)\,d\zeta}{\zeta-\tau}\right),\quad v\in C(\gamma),\quad \operatorname{Im}v=0. \qquad (78)$$

Proof. We have

$$\Omega^+(\tau)=\Psi^+(\tau)=\Phi^+(z(\tau))=[\omega(z(\tau))]^{-1}(K_\Gamma\varphi)^+(z(\tau)).$$

This means that

$$\Omega^+(\tau)=[\omega(z(\tau))]^{-1}M^+(\tau),$$

where $M(w)=(K_\Gamma\varphi)(z(w))$, $\varphi\in L^{p(\cdot)}(\Gamma)$.

Setting $\nu_k=1$ if $\tau_k\notin A$ and taking (63) and (73), we obtain

$$F(\tau)\sim Q(\tau)\omega^{-1}(z(\tau))M^+(\tau)[X^+(\tau)]^{-1}$$

$$\sim\prod_{k=1}^{j}(\tau-w_k)^{\gamma_k}\prod_{k=1}^{\nu}(\tau-\tau_k)^{-\alpha_k\nu_k}\exp\left(\frac{-\alpha_k}{\pi}\int\limits_{\gamma}\frac{\mu_k(\zeta)\,d\zeta}{\zeta-\tau}\right)\prod_{b_k\in B}(\tau-b_k)^{-u_k}$$

$$\times\prod_{a_k\in A}(\tau-a_k)^{\frac{\nu_k-1}{\ell(a_k)}}\exp\left(\frac{1}{\ell(\tau)\pi}\int\limits_{\gamma}\frac{\beta(\zeta)\,d\zeta}{\zeta-\tau}\right)M_0(\tau)$$

$$=\prod_{k=1}^{j}(\tau-w_k)^{\gamma_k-\delta_k}\exp\left(\frac{1}{\pi}\int\limits_{\gamma}\frac{\mu(\zeta)+\beta(\zeta)[\ell(\tau)]^{-1}}{\zeta-\tau}\,d\zeta\right)M_1(\tau)M_0(\tau),$$

where $\mu(\zeta)=-\sum\limits_{k=1}^{\nu}\alpha_k\mu_k(\zeta)$ is a continuous function, M_1 and $1/M_1$ are bounded functions and $M_0\in L^{\ell(\cdot)}(\gamma)$. Hence, by virtue of (76) we obtain the validity of Lemma 25 with $v(\zeta)=\mu(\zeta)+\beta(\zeta)/\ell(\zeta)$. □

Note that, by virtue of (76) and according to Lemma 23, the function ρ given by (78) belongs to $W^{\ell(\cdot)}(\gamma)$.

Let

$$F(w)=\frac{\Omega(w)Q(w)}{X(w)}.$$

Based on Lemma 11, the next lemma is proved in the same manner as Theorem 15.

Lemma 26. *Let $\Phi\in K^{p(\cdot)}(D;\omega)$, $\Psi(w)=\Phi(z(w))$, and $\Omega(w)$, $X(w)$ and $Q(w)$ be defined in $\mathbb{C}\setminus\gamma$ by formulas (68), (70) and (77), $\varkappa=\varkappa_0+\varkappa_1$, where $\varkappa_0$ is the order of the function $Q(w)$ at infinity, and $\varkappa_1$ is the index of the functions $G_1(w)$ giving by (72); therefore $-\varkappa_1$ is the order of the function $X(w)$ at infinity. Then*

$$F\in K^{\ell(\cdot)}(\gamma;\rho;\varkappa),$$

where $\rho(\tau)$ is a function of the class $W^{\ell(\cdot)}(\gamma)$ given by equality (78) and

$$K^{\ell(\cdot)}(\gamma;\rho;\varkappa)=\left\{F:\ F(w)=\frac{1}{2\pi i}\int\limits_{\gamma}\frac{\varphi(\zeta)\,d\zeta}{\zeta-w}+P_\varkappa(w),\ \varphi\in L^{\ell(\cdot)}(\gamma;\rho)\right\},$$

where $P_\varkappa$ is some polynomial of order $\varkappa$ for $\varkappa\geq 0$. If $\varkappa<0$, then $P_\varkappa\equiv 0$ and it is assumed that $(K_\gamma\varphi)(w)$ has zero of order $-\varkappa$ at the point $w=\infty$.

Now problem (69) can be formulated as follows: let ρ be given by equality (78). Find a function $\Omega(w)=X(w)[Q(w)]^{-1}F(w)$, where $F\in K^{\ell(\cdot)}(\gamma;\rho;\varkappa)$ and

$$\begin{cases} F^+(\tau)=F^-(\tau)+C_2(\tau),\\ C_2(\tau)=C_1(\tau)Q(\tau)[X^+(\tau)]^{-1}=\dfrac{c(z(\tau))}{a(z(\tau))+ib(z(\tau))}\,Q(\tau)[X^+(\tau)]^{-1},\\ \left(\dfrac{FX}{Q}\right)_*(w)=\dfrac{F(w)X(w)}{Q(w)},\quad |w|\neq 1. \end{cases}$$

The further reasoning given in Part I for problem (1) does not need additional comments.

5.11 Main result

Let: 1) D be an internal domain bounded by a curve $\Gamma\in C^1(A_1,\dots,A_i;\nu_1,\dots,\nu_i)$, $0<\nu_k\leq 2$; 2) $p\in\widetilde{\mathcal{P}}(\Gamma)$; 3) $\omega(z)=\prod\limits_{k=1}^{\nu}(z-t_k)^{\alpha_k}$, $-\frac{1}{p(t_k)}<\alpha_k<\frac{1}{p'(t_k)}$; 4) $a(t)$, $b(t)$ be piecewise continuous real functions with the condition $\inf(a^2(t)+b^2(t))>0$ such that the function $G(t)=-[a(t)-ib(t)][a(t)+ib(t)]^{-1}$ has discontinuity points B_k, $G(B_k^-)/G(B_k^+)=\exp 2\pi i u_k$, $u_k\in\mathbb{R}$; 5) $c(t)\in L^{p(\cdot)}(\Gamma;\omega)$; 6) $z=z(w)$ be the conformal mapping of the circle U onto D and let $w=w(z)$ be the inverse function and $\tau_k=w(t_k)$, $a_k=w(A_k)$, $b_k=w(B_k)$.

Theorem 27. *Let* i) *the points τ_k, a_k, b_k be enumerated according to (17); the numbers δ_k be defined by equalities (19) and $\{\delta_k\}\neq\frac{1}{\ell'(w_k)}$, and the integer numbers γ_k be chosen according to (75);* ii) *$Q(w)$ be the rational function defined by equality (77), and $\varkappa_0$ be its order at the point $w=\infty$;* iii) *the functions $r(w)$, $X_1(w)$ and $G_1(\tau)$ be defined by equalities (70), (71) and (72); therefore X_1 have order $(-\varkappa_1)$ at infinity.*

Assume that $\varkappa=\varkappa_0+\varkappa_1$. Then:

a) *if $\varkappa<0$, then for problem (1) to be solvable in the class $K^{p(\cdot)}(D;\omega)$ it is necessary and sufficient that the conditions*

$$\int\limits_{\gamma}\frac{c(z(\tau))Q(\tau)}{X^+(\tau)[a(z(\tau))+ib(z(\tau))]}\,\tau^k d\tau=0,\quad k=\overline{0,\varkappa},\tag{79}$$

would be fulfilled; under these conditions problem (1) *has a unique solution*

$$\Phi_0(z) = \Omega(w(z)) = \widetilde{\Omega}_c(w(z)) = \frac{1}{2}\left(\Omega_c(w(z)) + (\Omega_c)_*(w(z))\right), \tag{80}$$

where

$$\Omega_c(w) = \frac{X(w)}{Q(w)}\frac{1}{2\pi i}\int\limits_{\gamma}\frac{c(z(\tau))Q(\tau)}{X^+(\tau)[a(z(\tau))+ib(z(\tau))]}\frac{d\tau}{\tau - w};$$

b) *if* $\varkappa \geq 0$, *then problem* (1) *is solvable unconditionally and all its solutions are given by the equality*

$$\Phi(z) = \Phi_0(z) + X(w(z))Q^{-1}(w(z))P_\varkappa(w(z)), \tag{81}$$

where $P_\varkappa(w)$ *is an arbitrary polynomial* $P_\varkappa(w) = \sum\limits_{k=0}^{\varkappa} h_k w^k$ *whose coefficients satisfy the condition*

$$\overline{h_k} = Ah_{\varkappa - k}\,, \quad k = \overline{0, \varkappa}, \quad A = (-1)^{\varkappa_0}\prod_{k=1}^{\varkappa_j} w_k^{-\gamma_k}.$$

5.12 Application to the Neumann problem

All the assertions from Part I (see Section 5.6) concerning the particular cases considered there also remain in force. We are not going to dwell on them here, but will give a supplement to the Neumann problem formulated as follows:

Define a harmonic function $u = \operatorname{Re}\Phi$, where Φ is an analytic function such that $\Phi' \in K^{p(\cdot)}(D)$, which satisfies the condition

$$\left(\frac{\partial u}{\partial n}\right)(t) = f(t), \quad f \in L^{p(\cdot)}(\Gamma),$$

a.e. on Γ.

Following [98, pp. 243–248], we come to the problem in the class $K^{p(\cdot)}(D)$

$$\operatorname{Re}\left[ie^{-2i\vartheta}\Psi^+(t)\right] = f(t), \quad \Psi = \Phi', \tag{82}$$

where ϑ is the angle formed by the tangent to Γ at the point t and the abscissa axis. Here the sets of points $\{a_k\}$ and $\{b_k\}$ coincide, while the set $\{\tau_k\}$ is empty. Furthermore, $u_k = 1 - \nu_k$ and thus condition (74) takes the form $\left\{\frac{\nu_k - 1}{p(A_k)}\right\} \neq \frac{1}{p'(A_k)}$. It is not difficult to verify that this is equivalent to the condition $\nu_k \neq p'(A_k)$. For problem (82) to be solvable it is necessary that $f(t)$ would be orthogonal to them solutions of the class $\widetilde{K}_0^{p'(\cdot)}(\Gamma)$ of the problem

$$F^+ = e^{2i\vartheta}F^-$$

for which $F(\infty) = \text{const}$. (This is the conjugate problem to the Riemann problem corresponding to problem (82)). It has $\varkappa(p') + 1$ solutions, where

$$\varkappa(p') = \mathcal{N}\{A_k : \nu_k > p'(A_k)\}$$

(see (50)). Thus we have the following

Statement 28. *If $0 < \nu_k \le 2$, $\nu_k \ne p'(A_k)$, then for the posed Neumann problem to be solvable it is necessary that $\varkappa(p') + 1$ above mentioned orthogonality conditions would be fulfilled. If these conditions are fulfilled, then the problem has solutions depending on $\varkappa(p) + 1 = \mathcal{N}\{A_k : \nu_k > p(A_k)\} + 1$ arbitrary constants. To find them we proceed as follows: using the formulas from Theorem 27, we find Φ', then integrate it and separate the real part.*

5.13 Non-Fredholmian case

In the assumptions we have considered above, problem (1) has turned out to be Fredholmian in the sense that the homogeneous problem has a finite number of linearly independent solutions, while the set of functions $c(t)$, for which it is solvable, is closed in $L^{p(\cdot)}(\Gamma;\omega)$. This together with conditions 1)–6) was guaranteed by condition (74).

Let us now consider the case where condition (74) is violated at individual points w_k and assume that $a(t)$, $b(t)$ are piecewise Hölder functions.

Let

$$\{\delta_{k_i}\} = \frac{1}{\ell'(w_{k_i})}. \tag{83}$$

In that case, we choose integer numbers γ_{k_i} such that we have

$$\delta_{k_i} - \gamma_{k_i} = \frac{1}{\ell'(w_{k_i})}. \tag{84}$$

Then it is not difficult to verify that all possible solutions of the considered problem lie in the set of functions given by equality (80) (for $\varkappa < 0$) and by equality (81) (for $\varkappa \ge 0$). In order that the functions defined by these equalities would indeed be solutions, it is necessary and sufficient that their boundary functions would belong to the class $L^{p(\cdot)}(\Gamma;\omega)$. This is equivalent to the requirement that the function

$$(M_c)(t) = \frac{X^+(t)}{Q(t)} \int\limits_{\gamma} \frac{c(z(\tau))Q(\tau)}{X^+(\tau)(a(z(\tau)) + ib(z(\tau)))} \frac{d\tau}{\tau - t}, \quad t \in \gamma, \tag{85}$$

would belong to the class $L^{\ell(\cdot)}(\gamma;\rho)$, where ρ defined by (26). Under our assumptions we have

$$\rho(\tau) \sim \prod_{k=1}^{j} (\tau - w_k)^{\beta_k}, \quad \beta_k = \delta_k - \gamma_k, \quad \beta_{k_i} = \frac{1}{\ell'(w_{k_i})}. \tag{86}$$

Thus it is required that the condition $\rho(t)(M_c)(t) \in L^{\ell(\cdot)}(\gamma)$ or, which is same, the condition

$$\frac{\omega(z(t))r(t)|z'(t)|^{\frac{1}{\ell(t)}}}{Q(t)} \int_\gamma \frac{c(z(\tau))Q(\tau)}{X^+(\tau)(a(z(\tau)) + ib(z(\tau)))} \frac{d\tau}{\tau - t} \in L^{\ell(\cdot)}(\gamma)$$

be fulfilled.

Since in the case of piecewise Hölder coefficients we have $X^+(t) \sim r(t)$, the latter condition can be written in the form

$$\frac{R_0(t)\omega(z(t))r(t)|z'(t)|^{\frac{1}{\ell(t)}}}{Q(t)} \int_\gamma \frac{c(z(\tau))\omega(z(\tau))|z'(\tau)|^{\frac{1}{\ell(\tau)}} Q(\tau)}{\omega(z(\tau))|z'(\tau)|^{\frac{1}{\ell(\tau)}} X^+(\tau)(a(z(\tau)) + ib(z(\tau)))} \frac{d\tau}{\tau - t}$$

$$= \rho(t) \int_\gamma \frac{g(\tau)}{\rho(\tau)} \frac{d\tau}{\tau - t} \in L^{\ell(\cdot)}(\gamma),$$

where

$$g(\tau) = c(z(\tau))\omega(z(\tau))|z'(\tau)|^{\frac{1}{\ell(\tau)}} \left(a(z(\tau)) + ib(z(\tau))\right)^{-1} R_0(\tau) \in L^{\ell(\cdot)}(\gamma). \tag{87}$$

Assume that

$$(Tg)(t) = \rho(t) \int_\gamma \frac{g(\tau)}{\rho(\tau)} \frac{d\tau}{\tau - t}, \quad t \in \gamma.$$

Finally, we have

Lemma 29. *Let the equality* (83) *holds. For problem* (1) *to be solvable in the class* $K^{p(\cdot)}(D;\omega)$ *it is necessary and sufficient that the function* Tg *would belong to the class* $L^{\ell(\cdot)}(\gamma)$, *where* ρ *is given by equality* (86), *and* g *by equality* (87).

Under assumptions (83)–(84) this condition is fulfilled not for all $c \in L^{p(\cdot)}(\Gamma;\omega)$; otherwise the operator S_γ would turn out to be continuous in $L^{\ell(\cdot)}(\gamma;\rho)$, where ρ is given by equality (86), which is impossible because condition (74) is violated.

We naturally pose the problem of indicating wide subclasses of a function $g \in L^{p(\cdot)}(\gamma)$, for which $Tg \in L^{\ell(\cdot)}(\gamma)$.

If $p(t) = p = const > 1$, then one of such possible classes is a family of those functions g, for which

$$g(\tau) \ln|\tau - \tau_{k_i}| \in L^{\ell(\cdot)}(\gamma)$$

(see [40, p. 163]).

If on Γ there is a point t_0, for which $\delta(w_0) = \frac{1}{\ell'(w_0)}$ $(w_0 = w(t_0))$ and also $p(t_0) = \underline{p} = \min_{t\in\Gamma} p(t)$, then, using the condition

$$g(\tau)\ln|\tau - w_0| \in L^{\ell(\cdot)}(\gamma) \tag{88}$$

we have $Tg \in L^{\ell(\cdot)}(\gamma)$ (see [22]).

As a result of the above consideration we come to

Theorem 30. *If $a(t)$, $b(t)$ are piecewise Hölder functions and the conditions of Theorem 27, except condition (74), are fulfilled, then for problem (1) to be solvable in the class $K^{p(\cdot)}(D;\omega)$ it is necessary and sufficient that, in addition to conditions (79) (for $\varkappa < 0$), the condition $M_c \in L^{\ell(\cdot)}(\gamma;\rho)$ would be fulfilled, where M_c is the function given by equality (85), and ρ by equality (86).*

If $p(t)$ attains a minimum at the point t_0, at which $\delta(w_0) = \frac{1}{\ell'(w_0)}$, $w_0 = w(t_0)$ and if $c(t)$ is such a function that the function g given by formula (87) satisfies condition (88), then for this function $c(t)$, problem (1) is solvable.

Remark 31. Condition (88) is equivalent to the condition

$$c(t)\omega(t)\ln|w(t) - w(t_0)| \in L^{p(\cdot)}(\Gamma),$$

and if $\Gamma \in C^{1,L}(A_1,\dots,A_i;\nu_1,\dots,\nu_i)$, $0 < \nu_k < 2$, then to the condition

$$c(t)\omega(t)\ln|t - t_0| \in L^{p(\cdot)}(\Gamma).$$

5.14 Notes and comments on Chapter 5

The problem studied in this Chapter is a particular case of the quite general problem posed by the B. Riemann [111] and discussed by the D. Hilbert [33]. The survey of the results of this problem can be found in the monographs by N. I. Muskhelishvili [96], [97] and I. N. Vekua [128].

Muskhelishvili created an effective method of solving of the Riemann-Hilbert problem by reducing it to the problem of linear conjugation (the Riemann problem). Later on, this method was extended in [42] to the case when a set of sought functions is the class of Cauchy type integrals with density from the Lebesgue class L^p, $1 < p < \infty$. In the above-mentioned paper the Riemann-Hilbert problem was studied in the domain with Lyapunov boundary. The similar problem in the domain with a non-smooth boundary was explored in [54].

The Riemann-Hilbert problem in a simply-connected domain with a piecewise smooth boundary in general setting, when the set of sought functions is the weighted class of Cauchy type integrals with a density from variable exponent Lebesgue space was investigated in [56], [59], [60]. In this papers we heavily rely on the Muskhelishvili's mentioned method.

The another important tool solving the Riemann-Hilbert problem in the domain with non-smooth boundary is a behavior of derivative of conformal mapping in a neighborhood of a angular points. For the extension of classical results in this direction we refer the reader to the papers [38], [40], (see also [1]).

The material of Chapter 5 is taken from the papers [59], [60].

Chapter 6

The Dirichlet Problem for Variable Exponent Smirnov Classes Harmonic Functions in Domains with Arbitrary Piecewise Smooth boundaries

This Chapter contains four parts.

In Part I we consider the Dirichlet problem in a simply connected domain bounded by a piecewise Lyapunov boundary for the Laplace equation, assuming that the unknown function, being the real part of an analytic function representable by Cauchy type integral with density in $L^{p(\cdot)}(\Gamma)$. In that case, the considered problem reduces to the Dirichlet problem for a circle. The Muskhelishvili's method played an important role in reducing this problem to the Riemann problem ([98, §§40–41]), in the class $h^{p(\cdot)}(\omega)$, where ω is power weight.

For $p(t) = const > 1$, the Cauchy type integral with a density from $L^p(\Gamma)$, where Γ is a simple closed Carleson curve bounding the domain D, belongs to the Smirnov class $E^p(D)$ ([40, p. 29]) and thus we have additional useful information on the problem solution.

In Chapter 3, the variable exponent Hardy and Smirnov classes were introduced and studied in both simply and doubly-connected domains. It turned out, in particular, that if Γ is a simple closed piecewise Lyapunov curve bounding the domain D, then the Cauchy type integral with a density from $L^{p(\cdot)}(\Gamma)$ belongs to the class $E^{p(\cdot)}(D)$. The Dirichlet problem as it is formulated above is thereby solved in the class $e^{p(\cdot)}(D) = \operatorname{Re} E^{p(\cdot)}(D)$, too. However, the important and interesting case of domains with arbitrary piecewise smooth boundaries has so far been remaining uninvestigated.

In Part II, for such domains we study the Dirichlet problem for harmonic functions from $e^{p(\cdot)}(D)$. By the conformal mapping it again reduces to a problem for the circle, but this time in the weight class $h^{p(\cdot)}(\omega) = \operatorname{Re} H^{p(\cdot)}(\omega)$, where the weight ω is more general than the power weight. After solving the problem in the class $h^{p(\cdot)}(\omega)$, we proceed to our main goal, which is to study the case of "bad domains".

In Part III we consider the Dirichlet problem in doubly-connected domain with an arbitrary piecewise smooth boundary Γ in the class of those harmonic functions which are real parts of analytic in D functions of Smirnov class $E^{p_1(\cdot),p_2(\cdot)}(D)$.

Part IV devoted to investigation of the Dirichlet problem with data from grand Lebesgue spaces.

Depending on the boundary geometry and the values of the exponents at angular points, the Dirichlet problem may turn out unsolvable, solvable uniquely or non-uniquely. For the unsolvable case, a necessary and sufficient condition is found for the boundary function, which provides the problem solvability. For all solvability cases, solutions are constructed in explicit form.

Part I

The Dirichlet problem in class $k^{p(\cdot)}(D)$ for domain D with arbitrary piecewise Lyapunov boundary

Let $\Gamma \in C^{1,L}(A_1,\dots,A_n;\nu_1,\dots,\nu_n)$, $0 \le \nu_k \le 2$, be a simply closed piecewise Lyapunov curve bounding the finite domain D. For $n = 1$ we assume $A_1 = A$, $\nu_1 = \nu$. Let $p \in \mathcal{P}(\Gamma)$ and $z = z(w)$ is conformal mapping of U onto D and $r(\tau) = p(z(\tau))$, $\mu = \min\limits_{t\in\Gamma} p(t)$, $A = z(a)$.

We set

$$k^{p(\cdot)}(D) = \Big\{u:\ u = \operatorname{Re}\Phi,\ \ \Phi \in K^{p(\cdot)}(D)\Big\}.$$

(Definition of $K^{p(\cdot)}(D)$ see in section 5.1).

Let us consider the following Dirichlet problem:

Define a function u satisfying the conditions

$$\begin{cases} \Delta u = 0, \quad u \in k^{p(\cdot)}(D), \quad p \in \mathcal{P}(\Gamma), \\ u^+(t) = f(t), \quad f \in L^{p(\cdot)}(\Gamma), \end{cases} \tag{1}$$

where $u^+(t)$ denotes an angular boundary value of the function $u(z)$ at the point t; it exists almost everywhere on Γ and the equality in (1) is understood also almost everywhere.

Let now Φ be function from $K^\mu(D)$ then $\Phi = K_\Gamma\varphi$, where $\varphi \in L^\mu(\Gamma)$. Hence it follows $\Phi \in E^\mu(D)$.

6.1 Solution of the Dirichlet problem in $k^{p(\cdot)}(D)$ when boundary Γ of D belongs to $C^{1,L}(A;\nu)$

Since $p(t) \in \mathcal{P}(\Gamma)$, we have $L^{p(\cdot)}(\Gamma) \subset L^{\mu}(\Gamma)$, $\mu > 1$, and therefore $K^{p(\cdot)}(D) \subset K^{\mu}(D)$. Let now Φ be a function from $K^{\mu}(D)$ then $\Phi = K_{\Gamma}\varphi$ where $\varphi \in L^{\mu}(\Gamma)$. Hence it follows that $\Phi \in E^{\mu}(D)$ (see 0.20). Hence it follows that if a function u is a solution of problem (1), then it is also a solution of the problem

$$\begin{cases} u \in e^{\mu}(D), \quad \mu = \min\limits_{t\in\Gamma} p(t), \quad \mu > 1, \\ u^{+}(t) = f(t), \quad f(t) \in L^{p(\cdot)}(\Gamma), \end{cases} \tag{2}$$

which is well studied (see [54] and also [40], Ch. IV).

The scheme of our further consideration is as follows: when problem (2) is solvable, having a solution u from the wide class $e^{\mu}(D)$ and knowing that f belongs to the class $L^{p(\cdot)}(\Gamma)$, we will establish that u belongs to the narrow class $k^{p(\cdot)}(D)$. If problem (2) is unsolvable in $e^{\mu}(D)$ for arbitrary f's from $L^{\mu}(D)$, then we will show that the belonging of f's to $L^{p(\cdot)}(\Gamma)$ does not change the situation. In doing so, it is important for us to use results from Chapter 4 and the already known formulas giving solutions of problem (2) or the sets containing all possible solutions.

Let us begin by assuming for the sake of simplicity that $\Gamma \in C^{1}(A;\nu)$, $0 \le \nu \le 2$, $r(\tau) = p(z(\tau))$ and consider the following cases:

I. $0 < \nu < p(A)$; II. $\nu > p(A)$; III. $\nu = p(A)$; IV. $\nu = 0$.

I. $0 < \nu < p(A) = r(a)$.

The consideration of this case is divided into two parts: I_i – $0 < \nu < \mu$ and I_{ii} – $\mu \le \nu < p(A)$.

I_i. $0 < \nu < \mu$. In this case the solution of problem (2) is unique and given by the equality (see, e.g., [40, pp. 197–198])

$$u = u_f(w(z)), \tag{3}$$

where

$$u_f(w) = \frac{1}{2}\,\mathrm{Re}\left[\frac{1}{\sqrt[\mu]{z'(w)}}\left(\Phi(w) + \overline{\Phi\left(\frac{1}{\overline{w}}\right)}\right)\right], \tag{4}$$

$$\Phi(w) = \frac{\sqrt[\mu]{z'(w)}}{2\pi i}\int\limits_{\gamma} \frac{g(\zeta)}{\sqrt[\mu]{z'(\zeta)}}\,\frac{d\zeta}{\zeta - w}, \quad w \in U, \tag{5}$$

$$g(\zeta) = f(z(\zeta))\,\sqrt[\mu]{z'(\zeta)}\,. \tag{6}$$

Let us show that under the assumption $f \in L^{p(\cdot)}(\Gamma)$, the function u_f given by equalities (3)–(6) is a solution of problem (1), i.e., u_f belongs to $k^{p(\cdot)}(D)$. For this, since the integrability properties of the functions $\overline{\Phi(\frac{1}{\overline{w}})}$ and $\Phi(\frac{1}{\zeta})$ are analogous to those of the functions $\Phi(w)$ and $\Phi(\zeta)$, it is sufficient to prove that a boundary function $\Psi^+(\tau)$ of the function $\Psi(w)=\frac{\Phi(w)}{\sqrt[\mu]{z'(w)}}$ possesses the property

$$|\Psi^+(\tau)|\,|z'(\tau)|^{\frac{1}{r(\tau)}}\in L^{r(\cdot)}(\gamma),\quad r(\tau)=p(z(\tau)) \tag{7}$$

From (5)–(7) it follows that

$$\Psi^+(\tau)=\frac{1}{2}\,f(z(\tau))+\frac{1}{2\pi i}\int\limits_{\gamma}\frac{f(z(\zeta))\,d\zeta}{\zeta-\tau}. \tag{8}$$

Since $f(t)\in L^{p(\cdot)}(\Gamma)$, we have $|f(z(\zeta))|\,|z'(\zeta)|^{\frac{1}{r(\tau)}}\in L^{r(\cdot)}(\gamma)$ and therefore $f(z(\tau))=\varphi(\tau)|z'(\tau)|^{-\frac{1}{r(\tau)}}$, where $\varphi\in L^{r(\cdot)}(\gamma)$. Now (8) implies

$$|z'(\tau)|^{\frac{1}{r(\tau)}}|\Psi^+(\tau)|\le|\varphi(\tau)|+\left|\frac{|z'(\tau)|^{\frac{1}{r(\tau)}}}{2\pi i}\int\limits_{\gamma}\frac{\varphi(\zeta)}{|z'(\zeta)|^{\frac{1}{r(\zeta)}}}\,\frac{d\zeta}{\zeta-\tau}\right|$$

$$=|\varphi(\tau)|+|(N\varphi)(\tau)|,\ \ (N\varphi)(\tau)=\frac{|z'(\tau)|^{\frac{1}{r(\tau)}}}{2\pi i}\int\limits_{\gamma}\frac{\varphi(\zeta)}{|z'(\zeta)|^{\frac{1}{r(\zeta)}}}\,\frac{d\zeta}{\zeta-\tau}. \tag{9}$$

Under our assumptions with regard to Γ, by Warschawski's theorem $z'(w)\sim(w-a)^{\nu-1}z_0(w)$, where $z_0(w)$ is nonzero continuous function in $\overline{U}$ (see 0.25), therefore $0<m\le\frac{|z'(\tau)|}{|\tau-a|^{\nu-1}}\le M<\infty$. From here follows that the second summand from (9) will belong to $L^{r(\cdot)}(\gamma)$ if for any function $\psi\in L^{r(\cdot)}(\gamma)$ the function

$$(T\psi)(\tau)=\frac{\rho(\tau)}{\pi i}\int\limits_{\gamma}\frac{\psi(\zeta)}{\rho(\zeta)}\,\frac{d\zeta}{\zeta-\tau},\quad \rho(\tau)=|\tau-a|^{\frac{\nu-1}{r(a)}}, \tag{10}$$

belongs to $L^{r(\cdot)}(\gamma)$.

For this, it is sufficient to have

$$-\frac{1}{r(a)}<\frac{\nu-1}{r(a)}<\frac{r(a)-1}{r(a)}, \tag{11}$$

i.e., $0<\nu<r(a)=p(A)$. But this condition is fulfilled due to the assumption $\nu<\mu$ ($\mu\le p(A)$) and therefore $T\psi\in L^{r(\cdot)}(\gamma)$ for any $\psi\in L^{r(\cdot)}(\gamma)$. Thereby (9) implies inclusion (7). This means that the function u_f given by equalities (3)–(6) is a solution of problem (1).

I$_{ii}$. $\mu\le\nu<p(A)$. Then alls solutions of problem (2) are contained in the set of functions

$$u(z)=u_0(w(z))+u_f(w(z)), \tag{12}$$

where

$$u_0(w) = c\,\mathrm{Re}\,\frac{w+a}{w-a} = c\,\mathrm{Re}\,\Phi_0, \quad \Phi_0(w) = \frac{w+a}{w-a}, c \in \mathbb{R}, \tag{13}$$

while u_f is again given by equality (4), where this time Φ is some function from the set

$$\Phi(w) = \frac{\sqrt[\mu]{z'(w)}}{(w-a)2\pi i}\int_\gamma \frac{g(\zeta)(\zeta-a)}{\sqrt[\mu]{z'(\zeta)}}\frac{d\zeta}{\zeta-w} + B\frac{\sqrt[\mu]{z'(w)}}{w-a} + D\sqrt[\mu]{z'(w)}, \tag{14}$$

where B, D are constants (see [40], p. 159).

According to Warshawskiĭ's Theorem we have

$$(w(z) - w(A)) \sim (z-A)^{\frac{1}{\nu}} w_0(z) \tag{15}$$

where $w_0(z)$ is nonzero continuous function on $\overline{D}$, (see 0.25). From here we conclude that in the neighborhood of the point A, $u_0(w(z)) = \mathrm{Re}\,\frac{w(z)+a}{w(a)-a} = \mathrm{Re}\,\Phi_0(w(z))$ and

$$|\Phi_0(w(z))| \geq |c|\frac{m}{|z-A|^{\frac{1}{\nu}}}.$$

Since $\nu < p(A)$, we have $\Phi_0 \notin K^{p(\cdot)}(D)$ and therefore problem (1) has only a zero solution for $f = 0$. As to the function $\Phi(w)$, we should choose constants $B = B_f$ and $D = D_f$ such that $\Psi^+|z'|^{\frac{1}{r}}$ would belong to $L^{r(\cdot)}(\gamma)$. Taking $D_f = 0$, we have $2\Psi^+(\tau)|z'(\tau)|^{\frac{1}{r(\tau)}} = \varphi(\tau) + (T_1\varphi)(\tau)$, where $\varphi(\zeta) = g(\zeta)|z'(\zeta)|^{\frac{1}{r(\zeta)}-\frac{1}{\mu}}$,

$$(T_1\varphi)(\tau) = \frac{|z'(\tau)|^{\frac{1}{r(\tau)}}}{(\tau-a)\pi i}\int_\gamma \frac{\varphi(\zeta)(\zeta-a)}{|z'(\zeta)|^{\frac{1}{r(\zeta)}}}\frac{d\zeta}{\zeta-\tau} + \frac{B_f|z'(\tau)|^{\frac{1}{r(\tau)}}}{\tau-a}.$$

Let us write this function in the form

$$\begin{aligned}
T_1\varphi &= \frac{|z'(\tau)|^{\frac{1}{r(\tau)}}}{(\tau-a)\pi i}\int_\gamma \frac{\varphi(\zeta)(\zeta-\tau+\tau-a)}{|z'(\zeta)|^{\frac{1}{r(\zeta)}}}\frac{d\zeta}{\zeta-\tau} + \frac{B_f|z'(\tau)|^{\frac{1}{r(\tau)}}}{\tau-a} \\
&= \frac{|z'(\tau)|^{\frac{1}{r(\tau)}}}{(\tau-a)\pi i}\int_\gamma \frac{\varphi(\zeta)\,d\zeta}{|z'(\zeta)|^{\frac{1}{r(\zeta)}}} + \frac{|z'(\tau)|^{\frac{1}{r(\tau)}}}{\pi i}\int_\gamma \frac{\varphi(\zeta)}{|z'(\zeta)|^{\frac{1}{r(\zeta)}}}\frac{d\zeta}{\zeta-\tau} \\
&+ \frac{B_f|z'(\tau)|^{\frac{1}{r(\tau)}}}{\tau-a} = \frac{|z'(\tau)|^{\frac{1}{r(\tau)}}}{\tau-a}\left(\frac{1}{\pi i}\int_\gamma \frac{\varphi(\zeta)\,d\zeta}{|z'(\zeta)|^{\frac{1}{r(\zeta)}}} + B_f\right) + N\varphi,
\end{aligned} \tag{16}$$

where $N\varphi$ is define by (9).

As has been shown in considering case I_i, the function $N\varphi$ belongs to $L^{r(\cdot)}(\gamma)$ provided only that $\nu < r(a) = p(A)$. This condition is fulfilled by virtue of assumption I_{ii}. If now in (16) we take

$$B_f = -\frac{1}{\pi i}\int\limits_{\gamma} \frac{\varphi(\zeta)\,d\zeta}{|z'(\zeta)|^{\frac{1}{r(\zeta)}}} = -\frac{1}{\pi i}\int\limits_{\gamma} f(z(\zeta))\,d\zeta, \tag{17}$$

then $T_1\varphi \in L^{r(\cdot)}(\gamma)$ and thus (12) gives a unique solution of problem (1) when $u_0 \equiv 0$, while $u_f(w)$ is given by equality (4), where Φ is the function defined by (14), where $D = 0$ and the constant $B = B_f$ is given by equality (17).

Thus in both cases (I_i and I_{ii}) problem (1) has a unique solution.

II. $\nu > p(A)$.

Since in the considered case $\nu > \mu$, problem (2) has a solution depending on one parameter (see [40, pp. 158–160]) and given by equality (12), where u_0 is defined according to (13), while u_f according to (4), where this time

$$\Phi(w) = \frac{\sqrt[\mu]{z'(w)}}{w-a}\,\frac{1}{2\pi i}\int\limits_{\gamma} \frac{g(\zeta)(\zeta-a)}{\sqrt[\mu]{z'(w)}}\,\frac{d\zeta}{\zeta - w} + (\alpha w + \beta)\,\frac{\sqrt[\mu]{z'(w)}}{w-a} \tag{18}$$

$g(\zeta)$ given by (6), (see [40, p.159]). Applying the equality (15) we conclude that $u_0 = c\,\mathrm{Re}\,\Phi_0$, $c \in \mathbb{R}$, and

$$|\Phi_0(w(z))| \le |c|\,\frac{M}{|z-A|^{\frac{1}{\nu}}}\,.$$

Since $p(A) < \nu$, we have $\Phi_0(w(z)) \in E^{p(\cdot)}(D)$ and therefore (13) gives a general solution of problem (1) for $f = 0$. As to the function Φ defined by equality (18), we see that $|z'(\tau)|^{\frac{1}{r(\tau)}}|\Phi^+(\tau)|$ will belong to $L^{r(\cdot)}(\gamma)$ if the function $T\psi$ given by equality (10), where $\rho(\tau) = |\tau - a|^{\frac{\nu-1}{r(\tau)}-1}$, belongs to $L^{r(\cdot)}(\gamma)$ and $\alpha = \beta = 0$. For this we must have

$$-\frac{1}{r(a)} < \frac{\nu-1}{r(a)} - 1 < \frac{r(a)-1}{r(a)}\,,$$

i.e., $r(a) < \nu < 2r(a)$ or, which is the same,

$$p(A) < \nu < 2p(A). \tag{19}$$

This condition is fulfilled according to assumption II and thus the function given by equalities (12)–(13), (4) and (18) (for $\alpha = \beta = 0$ and an arbitrary real c in (13)) yields a solution of problem (1).

III. $\nu = p(A) \ge \mu$.

In this case the homogeneous problem (1) has only a zero solution.

Indeed, we have $|\Phi_0\,(w(z))\,| \ge \frac{|c|m}{|z-A|^{\frac{1}{\nu}}}$ and if $c \ne 0$, then $|u_0| \ge \frac{const}{|\tau-a|^{\frac{1}{p(A)}}}$, while the function $|\tau - a|^{-\frac{1}{p(A)}}$ does not belong to $L^{r(\cdot)}(\gamma)$.

The function u_f given by equality (4), where Φ is defined by (18), is a solution of problem (1) if and only if f is such that the inclusion

$$\frac{|z'(\tau)|^{\frac{1}{r(\tau)}}}{\tau-a}\int\limits_{\gamma}\frac{f(z(\zeta))\sqrt[\mu]{z'(\zeta)}}{|z'(\zeta)|^{\frac{1}{\mu}}}\frac{\zeta-a}{\zeta-\tau}d\zeta\in L^{r(\cdot)}(\gamma) \tag{20}$$

is fulfilled, i.e., assuming $f(z(\zeta))=\varphi(\zeta)|z'(\zeta)|^{-\frac{1}{r(\zeta)}}$, we have

$$|z_0(\tau)|^{\frac{1}{r(\tau)}}|\tau-a|^{\frac{\nu-1}{r(\tau)}-1}\int\limits_{\gamma}\frac{\varphi(\zeta)[z_0(\zeta)]^{-\frac{1}{r(\zeta)}}}{|\zeta-a|^{\frac{\nu-1}{r(\zeta)}-1}}\frac{d\zeta}{\zeta-\tau}\in L^{r(\cdot)}(\gamma). \tag{21}$$

Since $0<m\le|z_0(\zeta)|\le M<\infty$ and $r(\tau)\in\mathcal{P}(\gamma)$, this inclusion is equivalent to

$$T_2\varphi=|\tau-a|^{\frac{\nu-1}{r(a)}-1}\int\limits_{\gamma}\frac{\varphi(\zeta)}{|\zeta-a|^{\frac{\nu-1}{r(a)}-1}}\frac{d\zeta}{\zeta-\tau}\in L^{r(\cdot)}(\gamma),\quad \varphi\in L^{r(\cdot)}(\gamma). \tag{22}$$

It is obvious that when f runs through all $L^{p(\cdot)}(\Gamma)$, φ runs through all $L^{r(\cdot)}(\gamma)$. If problem (1) were solvable for an arbitrary function f, then (22) would hold for all $\varphi\in L^{r(\cdot)}(\gamma)$. But then the operator T_2, mapping $L^{r(\cdot)}(\gamma)$ onto $L^{r(\cdot)}(\gamma)$, would be continuous in $L^{r(\cdot)}(\gamma)$. (see Chapter 4, Theorem 2). Then we must have $\nu>r(a)=p(A)$ but in our case $\nu=p(A)$. Therefore inclusion (21) is fulfilled not for all φ, i.e., problem (1) is not solvable if $f(\zeta)=\varphi(\zeta)|z'(\zeta)|^{\frac{1}{r(\zeta)}}$, where $T_2\varphi\notin L^{r(\cdot)}(\gamma)$.

IV. $\nu=0$. In this case it is additionally assumed that $p(t)$ is such that

$$r(\tau)=p(z(\tau))\in\mathcal{P}(\gamma) \tag{23}$$

Instead of Warschawski's theorem we make use of the following result: If $\Gamma\subset C^1(A;0)$, $z=z(w)$ is a conformal mapping of the circle onto D and $z(a)=A$, then

$$z'(w)\sim(w-a)^{-1}z_0(w),\quad z_0(w)=\exp\frac{1}{2\pi}\int\limits_{\gamma}\frac{\beta(\zeta)\,d\zeta}{\zeta-w}, \tag{24}$$

where β is continuous real function on γ (see 0.25).

As above, applying the corresponding result from [40, p.164], we come to a conclusion that problem (1) is solvable only for those f's for which

$$Mf=|z_0(\tau)|^{\frac{1}{r(\tau)}}|\tau-a|^{-\frac{1}{r(a)}}\int\limits_{\gamma}\frac{f(z(\zeta))[z'(\zeta)]^{\frac{1}{r(\zeta)}}}{[z_0(\zeta)]^{\frac{1}{r(\zeta)}}(\zeta-a)^{-\frac{1}{r(a)}}}\frac{d\zeta}{\zeta-\tau}\in L^{r(\cdot)}(\gamma). \tag{25}$$

When this condition is fulfilled, the problem has a unique solution.

Note that condition (25) is fulfilled, not for all $f\in L^{p(\cdot)}(\Gamma)$. For instance, if $p(t)=p_0=const>1$ and if, for any $f\in L^{p_0}(\Gamma)$, $Mf\in L^{p_0}(\gamma)$, then the function

$\omega(\zeta) = (\zeta - a)^{-\frac{1}{p_0}}[z_0(\zeta)]^{\frac{1}{p_0}}$ would be a weight function for a operator S_γ in $L^{p_0}(\gamma)$. But then $\omega(\zeta)$ must belong to $L^{p_0+\varepsilon}(\gamma)$ for small ε ([121]), which is impossible because by virtue of the continuity of the function β we have $[z_0(\zeta)]^{\pm 1} \in \bigcap_{q>0} L^q(\gamma)$. Therefore for $\nu = 0$ the Dirichlet problem is solvable not for all $f \in L^{p_0}(\Gamma)$.

The reasoning of the cases I-IV implies that the following statement is true.

Theorem 1. *Let $\Gamma \in C^{1,L}(A;\nu)$ and $p(t) \in \mathcal{P}(\Gamma)$. In that case,*

(i) *if $0 < \nu < p(A)$, then problem* (1) *is uniquely solvable and its solutions are given by equalities* (12)–(13), (4), (14), *where $c = 0$, $D = 0$, and B is defined according to* (17);

(ii) *if $\nu > p(A)$, then problem* (1) *has a solution depending on one real parameter. All solutions are given by equalities* (12)–(13) *and* (18), *where $c \in \mathbb{R}$, $\alpha = \beta = 0$;*

(iii) *if $\nu = p(A)$, then problem* (1) *is, generally speaking, unsolvable. It has a solution only for those f's for which*

$$|z_0(\tau)|^{\frac{1}{r(\tau)}}|\tau - a|^{-\frac{1}{p(A)}} \int_\gamma \frac{f(z(\zeta))[z'(\zeta)]^{\frac{1}{r(\zeta)}} z_0^{-\frac{1}{r(\zeta)}}(\zeta)}{|\zeta - a|^{-\frac{1}{p(A)}}} \frac{d\zeta}{\zeta - \tau} \in L^{r(\cdot)}(\gamma) \tag{26}$$

and then a solution is given like in the case (ii);

(iv) *if $\nu = 0$ and $p(t)$ is such that $r(\tau) = p(z(\tau)) \in \mathcal{P}(\gamma)$, then problem* (1) *is solvable only for those f's for which* (25) *is valid. Moreover, the solution is unique.*

Finally we note that if $\Gamma \in C^{1,L}(A_1, A_2, \ldots, A_n; \nu_1, \nu_2, \ldots, \nu_n)$, $0 < \nu_k \le 2$ bounded the domain D then according to Theorem 15 of Chapter 3 $K^{p(\cdot)}(D) = E^{p(\cdot)}(D)$. Therefore in this case the statements i)–iii) of Theorem 1 about the Dirichlet problem in the class $e^{p(\cdot)}(D)$ are valid.

6.2 Solution of the Dirichlet problem in domain with arbitrary piecewise Lyapunov boundary

Theorem 2. *Let $\Gamma \in C^{1,L}(A_1, A_2, \ldots, A_n; \nu_1, \nu_2, \ldots, \nu_n)$, $0 \le \nu_k \le 2$, $p(t) \in \mathcal{P}(\Gamma)$ and $r(\tau) = p(z(\tau))$. If for some k we have $p(A_k) = \nu_k$ or $\nu_k = 0$ (in the latter case it assumed that $p(z(\tau)) \in \mathcal{P}(\gamma)$), then problem* (1) *is, generally speaking, unsolvable. If $p(A_{k_i}) = \nu_{k_i}$ or $\nu_{k_i} = 0$, $i = 1, \ldots, m$, then problem* (1) *is solvable only for those $f \in L^{p(\cdot)}(\Gamma)$ for which the inclusion*

$$|z_0(\tau)|^{\frac{1}{r(\tau)}}|\tau - a_{k_i}|^{-\frac{1}{p(A_{k_i})}} \int_\gamma \frac{f(z(\zeta))[z'(\zeta)]^{\frac{1}{r(\zeta)}}[z_0(\zeta)]^{-\frac{1}{r(\zeta)}}}{|\zeta - a_{k_i}|^{-\frac{1}{p(A_{k_i})}}} \chi_{a_{k_i}}(\zeta) \frac{d\zeta}{\zeta - \tau} \in L^{r(\cdot)}(\gamma)$$

holds, where $z(a_{k_i}) = A_{k_i}$ and $\chi_{a_{k_i}}(\zeta)$ is a characteristic function of such an arc on γ that of the points a_k contains only the point a_{k_i}.

If however $\{k : \nu_k = p(A_k) \text{ or } \nu_k = 0\} = \emptyset$, *then let* $p(A_{k_i}) < \nu_{k_i}$, $i = 1, \ldots, m$, $m \le n$ *and* $p(A_{k_i}) > \nu_{k_i}$, $k = m+1, \ldots, n$. *Then problem* (1) *is solvable and a general solution of the homogeneous problem is given by the equality*

$$u_0(z) = \sum_{i=1}^{m} c_{k_i} \operatorname{Re} \frac{w(z) + A_{k_i}}{w(z) - A_{k_i}}, \qquad c_{k_i} \in \mathbb{R}. \tag{27}$$

A particular solution of the nonhomogeneous problem can be constructed using the formulas indicated in section 6.1.

Part II

The Dirichlet problem in simply connected domains with piecewise smooth boundary

6.3 Reduction of the Dirichlet problem in $h^{p(\cdot)}(\omega)$ to the Riemann problem

Assume that (i) $p \in \mathcal{P}(\gamma)$ and (ii) ω is an analytic function in U and $\omega^{\pm 1} \in \bigcup\limits_{\delta>1} H^{\delta}$. Consider the Dirichlet problem formulated as follows: Find a function u if

$$\begin{cases} u \in h^{p(\cdot)}(\omega), \\ u^+(t) = f(t), \quad t \in \gamma, \quad f \in L^{p(\cdot)}(\gamma; \omega^+). \end{cases} \tag{28}$$

If u is a solution of the problem (28), then $u = \operatorname{Re} \Phi$ where $\Phi \in H^{p(\cdot)}(\omega)$. Therefore the boundary condition (28) can be rewritten in the form

$$\operatorname{Re}[\Phi^+(t)] = f(t), \quad \Phi \in H^{p(\cdot)}(\omega). \tag{29}$$

Assume

$$\Psi(w) = \Phi(w)\omega(w). \tag{30}$$

Then (29) takes the form

$$\operatorname{Re}\left[\frac{\Psi^+(t)}{\omega^+(t)}\right] = f(t), \quad \Psi \in H^{p(\cdot)},$$

or, which is the same,

$$\frac{\Psi^+(t)}{\omega^+(t)} + \overline{\left[\frac{\Psi^+(t)}{\omega^+(t)}\right]} = 2f(t).$$

Hence we obtain

$$\Psi^+(t) = -\frac{\omega^+(t)}{\overline{\omega^+(t)}}\,\overline{\Psi^+(t)} + g(t), \quad g(t) = 2f(t)\omega^+(t), \quad g \in L^{p(\cdot)}(\gamma). \tag{31}$$

Consider the function

$$\Omega(w)=\begin{cases}\Psi(w), & |w|<1,\\ \overline{\Psi\left(\frac{1}{\overline{w}}\right)}, & |w|>1.\end{cases} \tag{32}$$

Then the condition (31) takes the form

$$\Omega^+(t)=-\frac{\omega^+(t)}{\overline{\omega^+(t)}}\,\Omega^-(t)+g(t). \tag{33}$$

Note that any function of the form (32) has the property

$$\Omega(w)=\Omega_*(w), \tag{34}$$

where

$$\Omega_*(w)=\overline{\Omega\left(\frac{1}{\overline{w}}\right)},\quad |w|\neq 1. \tag{35}$$

Thus we have to seek for a solution Ω of the problem (33) that satisfies the condition (34) where Ω_* is defined by the equality (35).

Since $p\in\mathcal{P}(\gamma)$, we have $\underline{p}>1$ and therefore $H^{p(\cdot)}\subset H^1$. Then, by virtue of the well known property of functions from H^1, we have

$$\Psi(w)=\frac{1}{2\pi i}\int\limits_{\gamma}\frac{\Psi^+(t)\,dt}{t-w},\quad w\in U,\quad \Psi^+\in L^{p(\cdot)}(\gamma).$$

Lemma 3. *A function $\Omega_1(w)=\Omega(w)-\Omega(\infty)=\Omega(w)-\overline{\Psi(0)}$, $|w|\neq 1$, is representable by the Cauchy type integral with a density from $L^{p(\cdot)}(\gamma)$.*

Proof. Since $p\in\mathcal{P}(\gamma)$ and $\Omega_1(\infty)=0$, the restrictions of the function Ω_1 on U and $C\overline{U}$ belong to $H^{p(\cdot)}\subset H^1$ and $H^1(C\overline{U})$, respectively. Hence they are representable by Cauchy integrals in these domains and therefore

$$\Omega_1(w)=K_\gamma(\Omega_1^+-\Omega_1^-)(w),\quad (\Omega_1^+-\Omega_1^-)\in L^{p(\cdot)}(\gamma). \qquad \square$$

By Lemma 3 we conclude that $\Omega\in\widetilde{K}_0^{p(\cdot)}(\gamma)$.

We have established that any solution Φ of the problem (29) obtained by means of the equalities (30) and (32) generates a solution of the problem (31) belonging to $\widetilde{K}_0^{p(\cdot)}(\gamma)$. Though, the restriction of the solution $\Omega\in\widetilde{K}_0^{p(\cdot)}(\gamma)$ on U belongs to $H^{p(\cdot)}$. This fact, speaking in general, does not give the desired solution of the problem (31). For this it is necessary that the condition (34) be fulfilled. It is not difficult to verify that the restriction Ψ of any solution of such kind on U gives the desired solution of the problem (31) and, in that case, the function $u(w)=\mathrm{Re}[\Psi(w)(\omega(w))^{-1}]$ will be a solution of the problem (28).

If Ω is a solution of the problem (33) of the class $\widetilde{K}_0^{p(\cdot)}(\gamma)$, then the function $\Omega_*(w)$, too, will be such a solution. But then the function $\frac{1}{2}(\Omega(w)+\Omega_*(w))$, too, which satisfies the condition (34), will be a solution of the problem (33).

Thus the following theorem is valid.

Theorem 4. *If $p \in \mathcal{P}(\gamma)$ and $[\omega(w)]^{\pm 1} \in \bigcup\limits_{\delta>0} H^{\delta}$, then any solution of the problem (28) by means of the equalities (30) and (32) generates a solution Ω of the problem (33) that belongs to $\widetilde{K}_0^{p(\cdot)}(\gamma)$ and satisfies the condition (34).*

Conversely, if $\Omega \in \widetilde{K}_0^{p(\cdot)}(\gamma)$ is a solution of the problem (33) that satisfies the condition (34) and Ψ is the restriction of Ω on U, then the function $u(w) = \mathrm{Re}[\Psi(w)(\omega(w))^{-1}]$, $w \in U$, is a solution of the problem (28).

All solutions of the problem (28) are given by the equalities

$$u(w) = \frac{1}{2}\,\mathrm{Re}\,\frac{\Omega(w) + \Omega_*(w)}{\omega(w)}, \qquad w \in U, \tag{36}$$

where Ω is an arbitrary solution of the problem (33) of the class $\widetilde{K}_0^{p(\cdot)}(\gamma)$.

6.4 Solution of the Dirichlet problem in the class $h^{p(\cdot)}(\omega)$ with special ω

Let $\omega(w) = (w-a)^{\alpha} \exp \int\limits_{\gamma} \frac{\psi(\tau)d\tau}{\tau - w}$, $\alpha \in \mathbb{R}$, $a \in \gamma$, $\mathrm{Im}\,\psi = 0$, $\psi \in C(\gamma)$. Keeping in mind further applications, we investigate the problem (33) under the following assumptions as to α:

I. $-\dfrac{1}{p(a)} < \alpha < \dfrac{1}{p'(a)}$; II. $\dfrac{1}{p'(a)} < \alpha < \dfrac{1}{p'(a)} + 1$;

III. $\alpha = \dfrac{1}{p'(a)}$; IV. $\alpha = -\dfrac{1}{p(a)}$.

I. Let $-\frac{1}{p(a)} < \alpha < \frac{1}{p'(a)}$. Assume

$$X(w) = \begin{cases} -\omega(w), & |w| < 1, \\ \omega\left(\frac{1}{\overline{w}}\right), & |w| > 1. \end{cases} \tag{37}$$

It can be easily verified that $[X(w)]^{-1} \in \widetilde{K}_0^{p'(\cdot)}(\gamma) \subset \widetilde{K}_0^{1}(\gamma)$. Therefore only the functions

$$\Omega(w) = \frac{X(w)}{2\pi i} \int\limits_{\gamma} \frac{g(t)}{X^+(t)} \frac{dt}{t-w} + CX(w), \tag{38}$$

where C is an arbitrary (complex) constant, can be a solution of the problem (33). Since, by assumption I $\omega^+ \in W^{p(\cdot)}$, we have $\Omega \in \widetilde{K}_0^{p(\cdot)}(\gamma)$. Thus the function

$$\Omega(w) + \Omega_*(w) = \frac{X(w)}{2\pi i} \int\limits_{\gamma} \frac{g(t)}{X^+(t)} \frac{dt}{t-w} + \overline{\left(\frac{X\left(\frac{1}{\overline{w}}\right)}{2\pi i}\right)} \overline{\int\limits_{\gamma} \frac{g(t)}{X^+(t)} \frac{dt}{t - \frac{1}{\overline{w}}}}$$

$$+ [CX(w) + (CX(w))_*] \tag{39}$$

will be a solution satisfying the condition (34).

The condition $[CX(w)]_* = CX(w)$ implies that $\operatorname{Re} C = 0$. Now, from (36) and (39) we obtain

$$u(w) = \operatorname{Re} \frac{1}{2\pi i} \int_\gamma \frac{f(t)}{t} \frac{t+w}{t-w} \, dt. \tag{40}$$

II. $\frac{1}{p'(a)} < \alpha < \frac{1}{p'(a)} + 1$. If

$$X_1(w) = X(w)(w-a)^{-1}, \tag{41}$$

where $X(w)$ is given by the equality (37), then the function $\Omega(w)[X_1(w)]^{-1}$ may have a pole of first order at infinity. Therefore all solutions of the problem (33) lie in the set of functions given by the equality

$$\Omega(w) = \frac{X_1(w)}{2\pi i} \int_\gamma \frac{g(t)}{X_1^+(t)} \frac{dt}{t-w} + X_1(w)P(w), \quad g(t) = 2f(t)\omega^+(t), \tag{42}$$

where $P(w)$ is an arbitrary polynomial of first order. Since by virtue of condition II $(\alpha - 1) \in \left(-\frac{1}{p(a)}, \frac{1}{p'(a)}\right)$, we have $X_1^+ \in W^{p(\cdot)}(\gamma)$ and therefore the functions Ω given by the equality (42) are solutions of the problem (33) of the class $\widetilde{K}_0^{p(\cdot)}(\gamma)$.

Solutions of the homogeneous problem corresponding to (33) will be given by the function

$$\Omega_0(w) = X_1(w)(\delta w + \beta), \tag{43}$$

where δ, β are arbitrary complex numbers. From the condition (34) we establish that the number β in (43) can be taken arbitrarily and $\delta = \overline{\beta} a$ (see [40, p. 159]). Then

$$u_0(w) = M \operatorname{Re} \left[\frac{1}{\omega(w)} \left(\omega(w) \frac{w\overline{\beta}a + \beta}{w-a} \right) \right] = M \operatorname{Re} \frac{w\overline{\beta}a + \beta}{w-a},$$

where M is an arbitrary real constant. Hence it follows that (see [40, pp. 159–160])

$$u_0(w) = M \operatorname{Re} \frac{a+w}{a-w}. \tag{44}$$

According to (39), the function

$$u_f(w) = \operatorname{Re} \left[\frac{1}{w-a} \left(\frac{1}{2\pi i} \int_\gamma \frac{f(t)(t-a)}{t-w} \, dt - \frac{aw^2}{2\pi i} \int_\gamma \frac{f(t)}{t} \frac{\overline{t} - \overline{a}}{t-w} \, dt \right) \right] \tag{45}$$

will be a particular solution of the nonhomogeneous problem (28).

III. $\alpha = \frac{1}{p'(a)}$. First we consider the homogeneous problem. If X is the function defined by the equality (37), then the function $F(w) = \Omega(w)[X(w)]^{-1}$ in the circle U belongs to the class H^η for some $\eta > 0$, whereas in $C\overline{U}$ we have $[F(w) - F(\infty)] \in H^\eta(C\overline{U})$. Moreover, $F^+(t) = F^-(t)$, $t \in \gamma$. By the same reasoning as in [40, pp. 161–162] we find that F is an analytic function of the form $F(w) = \delta + \beta(w-a)^{-1}$, where δ, β are arbitrary constants. But then all solutions of the homogeneous problem corresponding to (33) lie in the set of functions

$$\Omega_0(w) = \delta X(w) + \beta(w-a)^{-1}X(w) = \delta X(w) + \beta X_1(w).$$

Let us consider two cases: a) $X_1 \notin H^{p(\cdot)}$ and b) $X_1 \in H^{p(\cdot)}$.

a) In this case Ω_0 may belong to $\widetilde{K}_0^{p(\cdot)}(\gamma)$ only when $\beta = 0$. Furthermore, the condition (34) gives the equality $\bar{\delta}X(w) = -\delta X(w)$, i.e. $\operatorname{Re}\delta = 0$.

Thus, for the condition a) a general solution of the homogeneous problem (33) that satisfies the condition (34) is given by the equality

$$\Omega_0(w) = \delta X(w), \quad \operatorname{Re}\delta = 0.$$

Therefore (36) implies that $u_0 = 0$.

b) In this case $\Omega_0(w) = X_1(w)P(w)$, where $P(w)$ is an arbitrary polynomial of first order, and a solution of the homogeneous Dirichlet problem is given by the equality (44).

Let us return to the nonhomogeneous problem.

Since $[X_1(w)]^{-1} \in \widetilde{K}_0^{p'(\cdot)}(\gamma)$, all possible solutions lie in the set of functions given by the equality (39) where $X(w)$ is replaced by the function $X_1(w)$. Using the reasoning from [40, pp. 162–163] we see that the desired solution can be constructed by the formula (39) only for those g for which the function

$$(Tg)(\zeta_0) = \frac{X_1^+(\zeta_0)}{2\pi i}\int\limits_\gamma \frac{g(\zeta)}{X_1^+(\zeta)}\frac{d\zeta}{\zeta-\zeta_0}, \qquad \zeta_0 \in \gamma,$$

belongs to $L^{p(\cdot)}(\gamma)$. Since $X_1^+(\zeta) = (\zeta-a)^{-1}\omega^+(\zeta)$, we have therefore proved the following Proposition.

Proposition 5. *If* $\omega(w) = (w-a)^{\frac{1}{p'(a)}}\exp\left(\int_\gamma \frac{\psi(\tau)\,d\tau}{\tau - w}\right)$ *and a function* f *is such that*

$$\frac{(\zeta_0-a)^{-1}\omega^+(\zeta_0)}{\pi i}\int\limits_\gamma \frac{g(\zeta)}{\omega^+(\zeta)(\zeta-a)^{-1}}\frac{d\zeta}{\zeta-\zeta_0} \in L^{p(\cdot)}(\gamma), \quad g = 2f\omega^+, \tag{46}$$

then the problem (28) *is solvable and its solution is given by the equality*

$$u(w) = \operatorname{Re}\frac{1}{2\pi i}\int\limits_\gamma \frac{f(t)}{t}\frac{t+w}{t-w}\,dt + u_0(w),$$

where $u_0(w) = M(p)\,\mathrm{Re}\,\frac{a+w}{a-w}$ *and*

$$M(p) = \begin{cases} 0, & \textit{if } X_1(w) = (w-a)^{-\frac{1}{p(a)}} \exp\left(\int_\gamma \frac{\psi(\tau)\,d\tau}{\tau - w}\right) \notin H^{p(\cdot)}, \\ M & \textit{is an arbitrary real constant if } X_1 \in H^{p(\cdot)}. \end{cases}$$

IV. Let $\alpha = -\frac{1}{p(a)}$. Then $[X(w)]^{-1} \in \widetilde{K}_0^{p'(\cdot)}(\gamma)$ and thus we find that $\Omega_0 = \delta X$, $\mathrm{Re}\,\delta = 0$. Therefore $u_0 = 0$. The nonhomogeneous problem is solvable if and only if

$$\frac{\omega^+(\zeta_0)}{\pi i} \int_\gamma \frac{g(\zeta)}{\omega^+(\zeta)} \frac{d\zeta}{\zeta - \zeta_0} \in L^{p(\cdot)}(\gamma), \quad g = 2f\omega^+ \tag{47}$$

Let us summarize the obtained results.

Theorem 6. *Let* $p \in \mathcal{P}(\gamma)$ *and*

$$\omega(w) = (w-a)^\alpha \exp\left(\int_\gamma \frac{\psi(\tau)\,d\tau}{\tau - w}\right), \quad a \in \gamma, \quad \mathrm{Im}\,\psi = 0, \quad \psi \in C(\gamma).$$

Consider the Dirichlet problem (28).
Then:

I. *For* $-\frac{1}{p(a)} < \alpha < \frac{1}{p'(a)}$, *the problem is solvable uniquely and its solution is given by the equality* (40);

II. *For* $\frac{1}{p'(a)} < \alpha < \frac{1}{p'(a)} + 1$ *the problem is solvable and its general solution contains an arbitrary real constant* M,

$$u(w) = u_f(w) + M\,\mathrm{Re}\,\frac{a+w}{a-w},$$

where u_f *is the function defined by the equality* (45).

III. *For* $\alpha = \frac{1}{p'(a)}$ *the problem is solvable only if the condition* (46) *is fulfilled and in that case the solution is defined by the equality*

$$u(w) = u_f(w) + M(p)\,\mathrm{Re}\,\frac{a+w}{a-w},$$

where $u_f(w)$ *is given by the formula* (40) *and if* X_1 *given by* (41)

$$M(p) = \begin{cases} 0, & \textit{if } X_1(w) = (w-a)^{-\frac{1}{p(a)}} \exp\left(\int_\gamma \frac{\varphi(\tau)\,d\tau}{\tau - w}\right) \notin H^{p(\cdot)}, \\ M & \textit{is an arbitrary real constant if } X_1 \in H^{p(\cdot)}. \end{cases}$$

IV. *For* $\alpha = -\frac{1}{p(a)}$ *the problem is solvable if the condition* (47) *is fulfilled and in that case it is solvable uniquely. The solution is given by the equality* (40).

6.5 The Dirichlet problem in the class $h^{p(\cdot)}(\omega)$ for the general weight functions

Let $\omega(\tau) = \prod_{k=1}^{n} (\tau - a_k)^{\alpha_k} \exp\left(\int_\gamma \frac{\psi(\tau)d\tau}{\tau - w}\right)$, $\operatorname{Im}\psi = 0$, $\psi \in C(\gamma)$, $a_k \in \gamma$, $\alpha_k \in \mathbb{R}$. Using the results of section 6.4 we come to a conclusion that the following statement is true.

Theorem 7. *Let $p \in \mathcal{P}(\gamma)$ and*

$$-\frac{1}{p(a_k)} < \alpha_k < \frac{1}{p'(a_k)}, \quad k = \overline{1, m};$$

$$\frac{1}{p'(a_k)} < \alpha_k < \frac{1}{p'(a_k)} + 1, \quad k = \overline{m+1, m+j};$$

$$\alpha_k = \frac{1}{p'(a_k)}, \quad k = \overline{m+j+1, m+j+s};$$

$$\alpha_k = -\frac{1}{p(a_k)}, \quad k = \overline{m+j+s+1, n},$$

then for the Dirichlet problem to be solvable in the class $h^{p(\cdot)}(\omega)$, it is necessary and sufficient that the condition

$$\frac{\omega_1^+(\zeta)}{\pi i} \int_\gamma \frac{g(\tau)}{\omega_1^+(\zeta)} \frac{d\tau}{\tau - \zeta} \in L^{p(\cdot)}(\gamma), \quad g = 2f\omega^+, \tag{48}$$

be fulfilled, where

$$\omega_1(w) = \frac{\omega(w)}{\rho(w)}, \quad \rho(w) = \begin{cases} \prod_{k=m+1}^{m+j+s} (w - a_k) & \text{if } j+s \geq 1, \\ 1 & \text{if } j+s = 0. \end{cases}$$

If (48) *holds, then a general solution is given by the equality*

$$u(w) = u_0(w) + u_f(w),$$

where

$$u_0(w) = \sum_{k=m+1}^{m+j+s} M_k(p) \operatorname{Re} \frac{a_k + w}{a_k - w}, \tag{49}$$

$$M_k(p) = \begin{cases} 0, \text{ for } k > m+j \text{ and } X_k \notin H^{p(\cdot)}, \\ \quad X_k(w) = (w - a_k)^{-\frac{1}{p(a_k)}} \exp \int_\gamma \frac{\psi(\tau)\,d\tau}{\tau - w}, \\ M_k \text{ is an arbitrary real constant for } m+1 \leq k \leq m+j \text{ and} \\ \quad m+j < k \leq m+j+s \text{ and } X_k \in H^{p(\cdot)}, \end{cases} \tag{50}$$

$$u_f(w) = \operatorname{Re}\left[\rho(w)\left(\frac{1}{2\pi i}\int\limits_{\gamma}\frac{f(\zeta)\rho(\zeta)}{\zeta - w}\,d\zeta\right.\right.$$

$$\left.\left. - \frac{(-1)^{j+1}w^{j+1}}{2\pi i}\prod_{k=m+1}^{m+j} a_k \int\limits_{\gamma}\frac{f(\zeta)}{\zeta}\,\overline{\frac{d\zeta}{(\zeta - w)\rho(\zeta)}}\right)\right], \tag{51}$$

where $\prod\limits_{k=m+1}^{m+j} a_k = 1$ *if* $j = 0$.

6.6 Class $H^{p(\cdot)}(C\overline{U};\widetilde{\omega})$ and Dirichlet problem in the class $h^{p(\cdot)}(C\overline{U};\widetilde{\omega})$

Let $C\overline{U} = \mathbb{C}\setminus\overline{U}$ and

$$\widetilde{\omega}(w) = \left(\frac{w-a}{w-w_0}\right)^{\alpha}\exp\int\limits_{\gamma}\frac{\psi(\tau)\,d\tau}{\tau - w}, w\in C\overline{U},\ w_0\in U,\ \operatorname{Im}\psi = 0,\ \psi\in C(\gamma). \tag{52}$$

Let $p = p(\theta)$ be periodical function on T with period equal 2π and $p\in\mathcal{P}(T)$. We will say that an analytic in $C\overline{U}$ function Φ belongs to the class $H^{p(\cdot)}(C\overline{U};\widetilde{\omega})$ if

$$\sup_{r>1}\int\limits_{|\zeta|=r}|\Phi(\zeta)\widetilde{\omega}(\zeta)|^{p(\theta)}|d\zeta| = \sup_{r>1}\int\limits_{0}^{2\pi}|\Phi(re^{i\vartheta})\widetilde{\omega}(re^{i\vartheta})|^{p(\vartheta)}r\,d\vartheta < \infty,\quad \zeta = re^{i\theta}. \tag{53}$$

Lemma 8. *Let* $\Phi\in H^{p(\cdot)}(C\overline{U};\widetilde{\omega})$ *and* $\varphi(w) = \overline{\Phi\left(\frac{1}{\overline{w}}\right)}$, $|w|<1$. *Then* $\varphi\in H^{p(\cdot)}\left(\overline{\widetilde{\omega}\left(\frac{1}{\overline{w}}\right)}\right)$ *and* $\varphi(0) = 0$. *Conversely, if* $\varphi\in H^{p(\cdot)}\left(\overline{\widetilde{\omega}\left(\frac{1}{\overline{w}}\right)}\right)$ *and* $\varphi(0)=0$, *then* $\Phi(w) = \overline{\varphi\left(\frac{1}{\overline{w}}\right)}$, $|w|>1$, *belongs to the class* $H^{p(\cdot)}(C\overline{U};\widetilde{\omega})$.

Proof. Since $\widetilde{\omega}(\infty)\neq 0$, from (53) it follows that $\Phi(\infty) = 0$ and therefore $\varphi(0) = 0$. Let us further consider the function $\varphi(w)\overline{\widetilde{\omega}\left(\frac{1}{\overline{w}}\right)} = \overline{\Phi\left(\frac{1}{\overline{w}}\right)\widetilde{\omega}\left(\frac{1}{\overline{w}}\right)}$; it is analytic in U and

$$\sup_{0<r<1}\int\limits_0^{2\pi}\left|\varphi(w)\overline{\widetilde{\omega}\left(\frac{1}{\overline{w}}\right)}\right|^{p(\vartheta)}d\vartheta = \sup_{0<r<1}\int\limits_0^{2\pi}\left|\Phi\left(\frac{1}{\overline{w}}\right)\widetilde{\omega}\left(\frac{1}{\overline{w}}\right)\right|^{p(\vartheta)}d\vartheta$$

$$= \sup_{0<r<1}\int\limits_0^{2\pi}\left|\Phi\left(\frac{e^{i\vartheta}}{r}\right)\widetilde{\omega}\left(\frac{e^{i\vartheta}}{r}\right)\right|^{p(\vartheta)}d\vartheta = \sup_{|\zeta|=\frac{1}{r}}\int\limits_0^{2\pi}|\Phi(\zeta)\widetilde{\omega}(\zeta)|^{p(\vartheta)}\frac{|d\zeta|}{|\zeta|} < \infty. \tag{54}$$

Therefore $\varphi\in H^{p(\cdot)}\left(\overline{\widetilde{\omega}\left(\frac{1}{\overline{w}}\right)}\right)$.

Conversely, let $\varphi\in H^{p(\cdot)}\left(\overline{C\overline{U};\widetilde{\omega}\left(\frac{1}{\overline{w}}\right)}\right)$ and $\varphi(0) = 0$. Then, since $\underline{p}>1$, from (54) it follows that $\Phi(w) = \overline{\varphi\left(\frac{1}{\overline{w}}\right)}\in H^{p(\cdot)}(C\overline{U};\widetilde{\omega}(w))$. □

Consider the problem: Define a harmonic function V in $C\overline{U}$ that satisfies the conditions

$$\begin{cases} \Delta V = 0, \quad V \in h^{p(\cdot)}(C\overline{U};\widetilde{\omega}) = \{V : V = \operatorname{Re}\Phi,\ \Phi \in H^{p(\cdot)}(C\overline{U};\widetilde{\omega})\}, \\ V^+(\tau) = f(\tau),\ \tau \in \gamma, \quad f \in L^{p(\cdot)}(\gamma;\widetilde{\omega}^+), \end{cases} \tag{55}$$

where

$$\widetilde{\omega}(w) = \prod_{k=1}^{n} \left(\frac{w-a_k}{w-w_0}\right)^{\alpha_k} \exp\left(\int_\gamma \frac{\psi(\tau)\,d\tau}{\tau - w}\right), \qquad \operatorname{Im}\psi = 0, \quad \psi \in C(\gamma).$$

We have $V = \operatorname{Re}\Phi$, $\Phi \in H^{p(\cdot)}(C\overline{U};\widetilde{\omega})$. Hence it follows that $u(w) = V\left(\frac{1}{\overline{w}}\right)$ is a harmonic function in U and $u = \operatorname{Re}\overline{\Phi\left(\frac{1}{\overline{w}}\right)}$. According to Lemma 8, we have $u \in h^{p(\cdot)}\left(\widetilde{\omega}\left(\frac{1}{\overline{w}}\right)\right)$. It is easy to verify that $h^{p(\cdot)}\left(\widetilde{\omega}\left(\frac{1}{\overline{w}}\right)\right) = h^{p(\cdot)}(w)$, where

$$\omega(w) = \prod_{k=1}^{n} (w-a_k)^{\alpha_k} \exp\left(\int_\gamma \frac{\psi(\tau)\,d\tau}{\tau - w}\right).$$

Thus the function $u(w) = V\left(\frac{1}{\overline{w}}\right)$ satisfies the conditions

$$\begin{cases} \Delta u = 0, \quad u \in h^{p(\cdot)}(\omega), \quad u(0) = 0, \\ u\big|_\gamma = f\left(\dfrac{1}{\overline{\tau}}\right) = f(\tau), \quad \tau \in \gamma. \end{cases} \tag{56}$$

If u is representable by a Poisson integral, then

$$u(0) = \frac{1}{2\pi}\int_0^{2\pi} u^+\,d\vartheta = \frac{1}{2\pi}\int_0^{2\pi} f(e^{i\vartheta})\,d\vartheta = 0.$$

Therefore in this case, to solve the problem (56) and thereby the problem (55), it is necessary that the condition

$$\int_0^{2\pi} f(e^{i\vartheta})\,d\vartheta = 0 \tag{57}$$

be fulfilled.

If this condition is fulfilled, then, under the assumptions of Theorem 7 as to the weight function ω, a solution u of the problem (56) is given by (49)–(51) provided that (48) is fulfilled, whereas the solution of the problem (55) will be the function $V_f(w) = u\left(\frac{1}{\overline{w}}\right)$. In particular,

$$V_0(w) = u_0\left(\frac{1}{\overline{w}}\right) = \sum_{k=1}^{n} M_k(p)\operatorname{Re}\frac{a_k + \frac{1}{\overline{w}}}{a_k - \frac{1}{\overline{w}}} = \sum_{k=1}^{n} M_k(p)\operatorname{Re}\frac{a_k\overline{w}+1}{a_k\overline{w}-1}$$

$$= \sum_{k=1}^{n} M_k(p) \operatorname{Re} \frac{a_k(\overline{w}_k + \overline{a}_k)}{a_k(\overline{w}_k - \overline{a}_k)} = \sum_{k=1}^{n} M_k(p) \operatorname{Re} \frac{w + a_k}{w - a_k},$$

where $\sum\limits_{k=1}^{n} M_k(p) = 0$, $M_k(p) = 0$ when $k \in \overline{0,m}$ or $k > m+j+s$, and $M_k(p)$ is defined according to (50) when $k = \overline{m+1, m+j+s}$.

Let write solutions of the homogeneous problem (55) in the class $h^{p(\cdot)}(C\overline{U}_\rho; \widetilde{\omega})$, where $U_\rho = \{w : |w| > \rho\}$. Then $V_0(w) = u\left(\frac{\rho}{\overline{w}}\right)$ and therefore for $a_k = \rho e^{i\vartheta_k}$ we obtain

$$\begin{aligned} V_0(w) &= \sum_{k=1}^{n} M_k(p) \operatorname{Re} \frac{\frac{\rho}{a_k} + \frac{\rho}{\overline{w}}}{\frac{\rho}{a_k} - \frac{\rho}{\overline{w}}} = \sum_{k=1}^{n} M_k(p) \operatorname{Re} \frac{e^{i\vartheta_k} + \frac{\rho}{\overline{w}}}{e^{i\vartheta_k} - \frac{\rho}{\overline{w}}} \\ &= \sum_{k=1}^{n} M_k(p) \operatorname{Re} \frac{\overline{w} + \rho e^{-i\vartheta_k}}{\overline{w} - \rho e^{-i\vartheta_k}} = \sum_{k=1}^{n} M_k(p) \operatorname{Re} \frac{\overline{w} + \overline{a}_k}{\overline{w} - \overline{a}_k} \\ &= \sum_{k=1}^{n} M_k(p) \operatorname{Re} \frac{w + a_k}{w - a_k}, \quad \sum_{k=1}^{n} M_k(p) = 0. \end{aligned}$$

Part III

6.7 The Dirichlet problem in the class $e^{p(\cdot)}(D)$ for a domain D with an arbitrary piecewise smooth boundary

In this section we employ the class $e^p(D)$ of harmonic functions. For the definitions of this class and other related classes see Section 3.1.

Let $\Gamma \in C^1(t_1, \ldots, t_n; \nu_1, \ldots, \nu_n)$, $0 \le \nu_k \le 2$, and $p \in \widetilde{\mathcal{P}}(\Gamma)$. Consider the problem: Define a harmonic function $\mathcal{U}(z)$ in D that satisfies the conditions

$$\begin{cases} \mathcal{U}(z) \in e^{p(\cdot)}(D), \\ \mathcal{U}^+(t) = f(t), \quad t \in \Gamma, \quad f \in L^{p(\cdot)}(\Gamma). \end{cases} \tag{58}$$

From the definition of the classes $E^{p(\cdot)}(D)$ and $H^{p(\cdot)}(\omega)$ it follows that: if $\mathcal{U}(z) \in e^{p(\cdot)}(D)$, then the function $u(w) = \mathcal{U}(z(w))$ belongs to the class $h^{p(z(\tau))}(\mathfrak{m})$ where

$$\mathfrak{m}(w) = |z'(re^{i\vartheta})|^{\frac{1}{p(z(\tau))}}, \quad w = re^{i\vartheta}, \quad \tau = e^{i\vartheta}, \tag{59}$$

and, conversely, if $u \in h^{p(z(\tau))}(\mathfrak{m})$, where $\mathfrak{m}$ is the function defined by (59), then $\mathcal{U}(z) = u(w(z)) \in e^{p(t)}(D)$. Therefore the problem (58) is equivalent to the following problem: Find a function u if

$$\begin{cases} u \in h^{\ell(\cdot)}(\mathfrak{m}), & \ell(\tau) = p(z(\tau)), \\ u^+(\tau) = g(\tau), & g(\tau) = f(z(\tau)) \in L^{\ell(\cdot)}(\gamma; \mathfrak{m}^+). \end{cases} \tag{60}$$

In the sequel, to investigate the problem (60) we need that ℓ would belong to $\widetilde{\mathcal{P}}(\gamma)$. Having this in mind let us introduce the class of functions $Q(\Gamma)$ as follows.

Definition 9. Let Γ be a simple closed curve bounding the domain D and $z = z(w)$ be the conformal mapping of the circle U onto the domain D. We will say that a function p given on Γ belongs to the class $Q(\Gamma)$ if $p \in \widetilde{\mathcal{P}}(\Gamma)$ and $\ell \in \widetilde{\mathcal{P}}(\gamma)$, where $\ell(\tau) = p(z(\tau))$, $\tau \in \gamma$.

As follows from Theorem 31 of Chapter 3, if Γ is a piecewise smooth curve without internal zero angles, then the class $Q(\Gamma)$ coincides with $\widetilde{\mathcal{P}}(\Gamma)$. Note that if Γ is an arbitrary piecewise smooth curve and p is a function from $\widetilde{\mathcal{P}}(\Gamma)$ such that p is constant in the neighborhood of the points with zero internal angles, then $p \in Q(\Gamma)$.

So, we assume that $p \in Q(\Gamma)$.

Thus, we have to investigate the problem (60) under the assumption: $\ell \in \widetilde{\mathcal{P}}(\gamma)$.

Let $t_k = z(a_k)$, then for z' we have

$$z'(w) \sim \prod_{k=1}^{n} (w - a_k)^{\nu_k - 1} \exp \int_\gamma \frac{\psi(\tau)\, d\tau}{\tau - w}, \quad w \in U, \quad \operatorname{Im} \psi = 0, \quad \psi \in C(\gamma). \tag{61}$$

Together with the weight $\mathfrak{m}$ given by the equality (59), consider the weight

$$\mathfrak{m}_0(w) = \prod_{k=1}^{n} (w - a_k)^{\frac{\nu_k - 1}{\ell(a_k)}} \exp \left(\frac{1}{\ell(\tau)} \int_\gamma \frac{\psi(\xi)\, d\xi}{\xi - w} \right), \quad w = re^{i\vartheta}, \quad \tau = e^{i\vartheta}. \tag{62}$$

Lemma 10. *If $\mathfrak{m}$ and $\mathfrak{m}_0$ are the functions given by the equalities* (59) *and* (62), *respectively, then $\mathfrak{m} \sim \mathfrak{m}_0$, and also $\mathfrak{m}_0^+ \in L^{\ell(\cdot)}(\gamma)$.*

Proof. We have

$$F(w) = \frac{\mathfrak{m}(w)}{\mathfrak{m}_0(w)} = \prod_{k=1}^{n} (w - a_k)^{\left[\frac{1}{\ell(\tau)} - \frac{1}{\ell(a_k)}\right](\nu_k - 1)}.$$

In the proof of Proposition 14 in Chapter 3 it is established that for $\ell \in \widetilde{\mathcal{P}}(\gamma)$, the functions $[F(w)]^{\pm 1}$ are bounded in U. Hence it follows that the relation $\mathfrak{m} \sim \mathfrak{m}_0$ is valid. Let us show that $\mathfrak{m}_0^+ \in L^{\ell(\cdot)}(\gamma)$.

Suppose $I = |\mathfrak{m}_0^+(e^{i\vartheta})|^{\ell(e^{i\vartheta})} = |\mathfrak{m}_0^+(\tau)|^{\ell(\tau)}$, $\tau = e^{i\vartheta}$. Then

$$I \leq \exp(2\pi \max \psi(\tau)) \left| \prod_{k=1}^{n} (\tau - a_k)^{\frac{\ell(\tau)(\nu_k - 1)}{\ell(a_k)}} \exp \int_\gamma \frac{\psi(\zeta)\, d\zeta}{\zeta - \tau} \right|$$

$$\leq M \left| \prod_{k=1}^{n} \exp\left(\frac{\ell(\tau)(\nu_k - 1)}{\ell(a_k)} \ln(\tau - a_k) \right) \exp \int_{\gamma} \frac{\psi(\zeta)\, d\zeta}{\zeta - \tau} \Bigg) \right|$$

$$= M \prod_{k=1}^{n} \Bigg| \exp(\nu_k - 1) \ln(\tau - a_k) \exp \frac{\ell(\tau) - \ell(a_k)}{\ell(a_k)} (\nu_k - 1)$$

$$\times \ln(\tau - a_k) \exp \int_{\gamma} \frac{\psi(\zeta)\, d\zeta}{\zeta - \tau} \Bigg) \Bigg|$$

$$= M \prod_{k=1}^{n} \left| (\tau - a_k)^{\nu_k - 1} \exp \int_{\gamma} \frac{\psi(\zeta)\, d\zeta}{\zeta - \tau} \right| \exp \left| \frac{\ell(\tau) - \ell(a_k)}{\ell(a_k)} (\nu_k - 1) \ln(\tau - a_k) \right|.$$

Since $\ell \in \widetilde{\mathcal{P}}(\gamma)$, we obtain

$$I \sim \left| \prod_{k=1}^{n} (\tau - a_k)^{\nu_k - 1} \exp \int_{\gamma} \frac{\psi(\zeta)\, d\zeta}{\zeta - \tau} \right| \sim |z'(\tau)|.$$

But $z' \in H^1$. Therefore $\int\limits_0^{2\pi} |\mathfrak{m}_0^+(e^{i\vartheta})|^{\ell(e^{i\vartheta})} d\vartheta < \infty$ and thus $\mathfrak{m}_0^+ \in L^{\ell(\cdot)}(\gamma)$. $\square$

From the previous discussion we conclude

Lemma 11. *If the domain D is bounded by a simple closed curve $\Gamma \in C^1(t_1, \dots, t_n; \nu_1, \dots, \nu_n)$ and $p \in Q(\Gamma)$, then all solutions of the problem (58) are given by the equality $\mathcal{U}(z) = u(w(z))$, where u is a solution of the following problem*

$$\begin{cases} \Delta u = 0, \quad u \in h^{\ell(\cdot)}(\mathfrak{m}_0), \quad \ell(\tau) = p(z(\tau)), \\ u^+(\tau) = g(\tau), \quad g(\tau) = f(z(\tau)) \in L^{\ell(\cdot)}(\gamma; \mathfrak{m}_0^+), \end{cases} \tag{63}$$

where $\ell \in \widetilde{\mathcal{P}}(\gamma)$ and $\mathfrak{m}_0$ is the function given by the equality (62).

Lemma 12. *Let*

$$\omega_0(w) = \prod_{k=1}^{n} (w - a_k)^{\frac{\nu_k - 1}{\ell(a_k)}} \exp \int_{\gamma} \frac{\varphi(\zeta)\, d\zeta}{\zeta - w}, \quad \varphi(\zeta) = \frac{\psi(\zeta)}{\ell(\zeta)}, \quad w \in U,$$

and $\ell \in \widetilde{\mathcal{P}}(\gamma)$. Then $\mathfrak{m}_0(w) \sim \omega_0(w)$, $w \in U$.

Proof. We have

$$\exp\left(\frac{1}{\ell(\tau)} \exp \int_{\gamma} \frac{\psi(\zeta)\, d\zeta}{\zeta - w} \right) = \exp \int_{\gamma} \frac{\varphi(\zeta)}{\ell(\zeta)} \frac{d\zeta}{\zeta - w} \exp \int_{\gamma} \frac{\ell(\zeta) - \ell(\tau)}{\ell(\zeta)\ell(\tau)} \frac{\varphi(\zeta)\, d\zeta}{\zeta - w}. \tag{64}$$

In the book [98, §15, pp. 50–52], it is proved that if the function $\ell(\zeta)$ satisfies Hölder's condition, then the function $\int\limits_{\gamma} \frac{\ell(\zeta)-\ell(\tau)}{\zeta-w}\,d\zeta$ uniformly tends to the limit $\int\limits_{\gamma} \frac{\ell(\zeta)-\ell(\tau)}{\zeta-\tau}\,d\tau$ as $w \to \tau$. Due to this proof we easily establish that for $\ell \in \widetilde{\mathcal{P}}(\gamma)$,

$$\int\limits_{\gamma} \frac{\ell(\zeta)-\ell(\tau)}{\zeta - w}\varphi(\zeta)\,d\zeta \to \int\limits_{\gamma} \frac{\ell(\zeta)-\ell(\tau)}{\zeta-\tau}\varphi(\zeta)\,d\zeta, \quad \varphi(\zeta) = \frac{\psi(\zeta)}{\ell(\zeta)},$$

uniformly with respect to $\tau \in \gamma$ as $w \to \tau$. By virtue of this fact, from (64) we conclude that $(\mathfrak{m}_0\omega_0^{-1})^{\pm 1}$ are bounded functions in U. Therefore $\mathfrak{m}_0 \sim \omega_0$ in U. □

Since by Lemma 12, in the problem (63) we can take ω_0 instead of $\mathfrak{m}_0$ and ω_0 has the form considered in section 6.5, Theorem 7 is applicable to the problem (63). If we assume in it that $\alpha_k = \frac{\nu_k - 1}{p(t_k)} \left(= \frac{\nu_k-1}{\ell(a_k)}\right)$, then the following statement is valid.

Theorem 13. *Let* i) D *be a simply connected bounded domain with a boundary* $\Gamma \in C^1(t_1,\dots,t_n;\nu_1,\dots,\nu_n)$; ii) $w = w(z)$ *be the conformal mapping of the domain* D *onto the unit circle* U; iii) $p \in Q(\Gamma)$; iv) ψ *be a real continuous function from the relation* (61) *for* z' *and* $\varphi(\tau) = \frac{\psi(\tau)}{\ell(\tau)}$, $\ell(\tau) = p(z(\tau))$, $\tau \in \gamma$.

Assume

$$\begin{gathered} 0 < \nu_1 < p(t_1), \dots, 0 < \nu_m < p(t_m); \\ \nu_{m+1} > p(t_{m+1}), \dots, \nu_{m+j} > p(t_{m+j}); \\ \nu_{m+j+1} = p(t_{m+j+1}), \dots, \nu_{m+j+s} = p(t_{m+j+s}); \\ \nu_{m+j+s+1} = 0, \dots, \nu_n = 0. \end{gathered}$$

Then

I) *If among the points* t_k *there are no points such that* $\nu_k = p(t_k)$ *or* $\nu_k = 0$, *then the problem* (63) *and therefore the problem*

$$\begin{cases} u(z) \in e^{p(\cdot)}(D), \\ u^+(z) = f(t), \quad t \in \Gamma, \quad f \in L^{p(\cdot)}(\Gamma), \end{cases}$$

is solvable and its solution contains j *arbitrary real constants (i.e. as many as there are points at which* $\nu_k > p(t_k)$*). In that case, a solution of the problem is given by the equality*

$$u(z) = u_f(z) + u_0(z), \tag{65}$$

where

$$u_0(z) = \sum_{k=m+1}^{m+j} M_k \operatorname{Re} \frac{w(t_k)+w(z)}{w(t_k)-w(z)}, \tag{66}$$

$$u_f(z) = \operatorname{Re}\Bigg[\frac{1}{\rho(w(z))}\bigg(\frac{1}{2\pi i}\int\limits_{\gamma}\frac{f(z(\tau))\rho(\tau)}{\tau - w(z)}\,d\tau$$

$$-\frac{(-1)^{j+1}}{2\pi i}\,w^{j+1}(z)\prod_{k=m+1}^{m+j} a_k \int\limits_{\gamma}\frac{f(z(\tau))\overline{\rho(\tau)}d\tau}{\tau(\tau - w(z))}\bigg)\Bigg], \tag{67}$$

$\rho(w) = \prod\limits_{k=m+1}^{m+j}(w - a_k)$ for $j \geq 1$; if $\{k : \nu_k > p(t_k)\} = \varnothing$, then $\rho(w) = 1$, and $\prod\limits_{k=m+1}^{m+j} a_k = 1$ for $j = 0$; M_k are arbitrary real constants.

In particular, if Γ is a smooth curve or at all points t_k we have $0 < \nu_k < p(t_k)$, then the problem is uniquely solvable and the solution is given by the equality

$$u(z) = \operatorname{Re}\frac{1}{2\pi i}\int\limits_{\gamma}\frac{f(z(\tau))}{\tau}\,\frac{\tau + w(z)}{\tau - w(z)}\,d\tau. \tag{68}$$

II) *If among the points t_k there are such points that $\nu_k = p(t_k)$ or $\nu_k = 0$, then, speaking in general, the problem is not solvable. For it to be solvable, it is necessary and sufficient that the function $f(z(\tau))$ satisfy the condition*

$$\frac{\omega^+(\zeta)}{2\pi i}\int\limits_{\gamma}\frac{f(z(\tau))\rho(\tau)\,d\tau}{\tau - \zeta} \in L^{\ell(\cdot)}(\gamma), \tag{69}$$

where

$$\omega(w) = \prod_{k\in T}(w - w(t_k))^{-\frac{1}{p(t_k)}}\exp\int\limits_{\gamma}\frac{\varphi(\tau)\,d\tau}{\tau - w}, \quad T = \{k : \nu_k = p(t_k) \ \text{ or } \ \nu_k = 0\}.$$

If the condition (69) is fulfilled, then the problem is solvable and its general solution is given by the equality (65) where u_f is defined by the equality (67), whereas this time $u_0(z)$ is given by the equality

$$u_0(z) = \sum_{k=m+1}^{m+j+s} M_k(p)\operatorname{Re}\frac{w(t_k) + w(z)}{w(t_k) - w(z)},$$

where $M_{m+1}, \dots, M_{m+j}$ are arbitrary constants, and for $m + j < k \leq m + j + s$

$$M_k(p) = \begin{cases} 0 \ \text{ if } X_k \notin H^{\ell(\cdot)}, \ X_k(w) = (w - a_k)^{-\frac{1}{\ell(a_k)}}\exp\displaystyle\int\limits_{\gamma}\frac{\varphi(\tau)\,d\tau}{\tau - w}, \\ M_k \ \text{ is an arbitrary constant if } X_k \in H^{\ell(\cdot)}. \end{cases}$$

III) *If under all other assumptions of Theorem 13 it is assumed that $\Gamma \in C^{1,L}(t_1, \dots, t_n; \nu_1, \dots, \nu_n)$, then for $\nu_k = p(t_k)$ we have $X_k \notin H^{\ell(\cdot)}$ and therefore in that case $u_0(z)$ is given by the equality (66).*

In formulating statement III), we took into account that $X_k \notin H^{\ell(\cdot)}$. This follows from Warschawski's theorem.

Part III

The Dirichlet problem in double connected domains

6.8 The Dirichlet problem in circular ring

In this Section we consider the Dirichlet problem in the class $h^{p_1(\cdot),p_2(\cdot)}(K;\omega)$. Remind that (see Section 3.9)

$$h^{p_1(\cdot),p_2(\cdot)}(K;\omega) = \Big\{u:\ u = \operatorname{Re}\Phi,\quad \Phi \in H^{p_1(\cdot),p_2(\cdot)}(K;\omega)\Big\}.$$

Lemma 14. *When $p_i \in \mathcal{P}(\gamma_i)$ and $\omega \in W^{p_1(\cdot),p_2(\cdot)}(K)$, every function $u(z)$ of the class $h^{p_1(\cdot),p_2(\cdot)}(K;\omega)$ is representable as*

$$u(re^{i\vartheta}) = \frac{1}{2\pi}\int_0^{2\pi} \mu_1(\zeta)\,\frac{1-r^2}{1+r^2-2r\cos(\zeta-\vartheta)}\,d\zeta + \frac{1}{2\pi}\int_0^{2\pi} \mu_2(\zeta)\,\frac{\rho^2-r^2}{\rho^2+r^2-2\rho r\cos(\zeta-\vartheta)}\,d\zeta, \tag{70}$$

where

$$\mu_i \in L^{p_i(\cdot)}(\gamma_i;\omega_i),\quad \omega_1 = \omega(e^{i\theta}),\quad \omega_2 = \omega(\rho e^{i\theta}) \text{ and } \int_0^{2\pi} \mu_2(\zeta)\,d\zeta = 0. \tag{71}$$

Proof. By virtue of the assumptions we have

$$u(z) = \operatorname{Re}(\Phi_1(z) + \Phi_2(z)), \tag{72}$$

where

$$\Phi_i(z) = \frac{1}{2\pi i}\int_{\gamma_i} \frac{\varphi_i(t)\,dt}{t-z},\qquad \varphi_i \in L^{p_i(\cdot)}(\gamma_i;\omega_i)$$

(see Chapter 3 Lemmas 23 and 24).

Since $\omega_i \in W^{p_i(\cdot)}(\gamma_i)$ and $\varphi_i \in L^{p(\cdot)}(\gamma_i;\omega_i)$, there exists $\varepsilon > 0$ such that $\varphi_i \in L^{1+\varepsilon}(\gamma_i)$. Therefore $\Phi_i \in E^1(K_i)$. The functions of the class $E^1(K_i)$ are representable by the Schwartz integral (see Chapter 1 Lemma 2)

$$\Phi_1(z) = \frac{1}{2\pi}\int_0^{2\pi} \operatorname{Re}\Phi_1^+(e^{i\zeta})\,\frac{e^{i\zeta}+z}{e^{i\zeta}-z}\,d\zeta + \operatorname{Im}\Phi_1(0),\quad z \in K_1,$$

$$\Phi_2(z) = \frac{1}{2\pi}\int\limits_0^{2\pi} \operatorname{Re}\Phi_2^+(\rho e^{i\zeta})\,\frac{\rho e^{i\zeta}+z}{\rho e^{i\zeta}-z}\,d\zeta, \quad \Phi_2(\infty)=0, \quad z\in K_2.$$

Assume $\mu_1(\zeta) = \operatorname{Re}\Phi_1^+(e^{i\zeta})$, $\mu_2(\zeta) = \operatorname{Re}\Phi_2^+(\rho e^{i\zeta})$. Then $\mu_i \in L^{p_i(\cdot)}(\gamma_i;\omega_i)$ and

$$\int\limits_0^{2\pi} \mu_2(\zeta)\,d\zeta = \int\limits_0^{2\pi} \operatorname{Re}\Phi_2^+(\rho e^{i\zeta})\,d\zeta = 0.$$

By virtue of (72) this proves the validity of the equalities (70)–(71). □

Let us consider the Dirichlet problem formulated as follows: define a harmonic function $u(z)$ in K that satisfies the conditions

$$\begin{cases} u\in h^{p_1(\cdot),p_2(\cdot)}(K;\omega), \\ u|_{\gamma_i} = f_i, \quad f_i\in L^{p_i(\cdot)}(\gamma_i;\omega_i), \quad \omega_1^+ = \omega(e^{i\vartheta}), \quad \omega_2^+ = \omega(\rho e^{i\vartheta}). \end{cases} \tag{73}$$

By virtue of Lemma 14 the function u is representable by the formula (70) and the condition (71) is fulfilled. Taking the boundary conditions from (73) into account, we obtain

$$\begin{cases} \mu_1(\zeta) - \dfrac{1}{2\pi}\displaystyle\int\limits_0^{2\pi} \mu_2(\alpha)\,\frac{1-\rho^2}{1+\rho^2-2\rho\cos(\alpha-\zeta)}\,d\alpha = f_1(\zeta), \\ -\mu_2(\zeta) + \dfrac{1}{2\pi}\displaystyle\int\limits_0^{2\pi} \mu_1(\alpha)\,\frac{1-\rho^2}{1+\rho^2-2\rho\cos(\alpha-\zeta)}\,d\alpha = f_2(\zeta) \end{cases} \tag{74}$$

with the additional condition

$$\int\limits_0^{2\pi} \mu_2(\zeta)\,d\zeta = 0. \tag{75}$$

A solution (μ_1,μ_2) of the system (74) must be defined under the assumption $\mu_i\in L^{p_i(\cdot)}(\gamma_i;\omega_i)$. It is obvious that any such solution of the system (74) satisfying the condition (75) gives by virtue of (70) a solution of the Dirichlet problem (73).

Theorem 15. *If $f_i\in L^{p_i(\cdot)}(\gamma_i;\omega_i)$, $\omega_i\in W^{p_i}(\gamma_i)$ and a solution of the system (74) is sought for under the assumption $\mu_i\in L^{p_i(\cdot)}(\gamma_i;\omega_i)$, then this system is solvable if and only if*

$$\int\limits_0^{2\pi} f_1(\zeta)\,d\zeta = \int\limits_0^{2\pi} f_2(\zeta)\,d\zeta, \quad f_1(\zeta)\equiv f_1(e^{i\zeta}), \quad f_2(\zeta)\equiv f_2(\rho e^{i\zeta}). \tag{76}$$

If this condition is fulfilled, then the problem (73) *has a unique solution given by the equality* (70) *where*

$$\mu_1 = \widetilde{\mu}_1 + C_0, \tag{77}$$

$$\mu_2(\zeta) = \frac{1}{2\pi} \int_0^{2\pi} \mu_1(\alpha) \frac{1-\rho^2}{1+\rho^2 - 2\rho\cos(\alpha-\zeta)}\, d\alpha + C_0 - f_2(\zeta), \tag{78}$$

$\widetilde{\mu}_1$ *is a particular solution from the class* $L^{p_1(\cdot)}(\gamma_1;\omega_1)$ *of the equation*

$$\mu_1(\zeta) + \frac{1}{2\pi} \int_0^{2\pi} \mu_1(\beta) K(\beta,\zeta)\, d\beta = \nu(\zeta), \tag{79}$$

$$K(\beta,\zeta) = -\frac{1}{2\pi} \int_0^{2\pi} \frac{1-\rho^2}{1+\rho^2-2\rho\cos(\beta-\alpha)} \, \frac{1-\rho^2}{1+\rho^2-2\rho\cos(\alpha-\zeta)}\, d\alpha, \tag{80}$$

$$\nu(\zeta) = f_1(\zeta) - \frac{1}{2\pi} \int_0^{2\pi} f_2(\alpha) \frac{1-\rho^2}{1+\rho^2-2\rho\cos(\alpha-\zeta)}\, d\alpha, \tag{81}$$

$$C_0 = -\frac{1}{2\pi} \int_0^{2\pi} \left[\frac{\widetilde{\mu}_1(\zeta)}{2\pi} - f_2(\zeta) \right] d\zeta. \tag{82}$$

Proof. First we note that the function $\nu(\zeta)$ defined by the equality (81) belongs to the class $L^{p_1(\cdot)}(\gamma_1;\omega_1)$. Further we take into account the fact that the equation (79) and its adjoint equation

$$h(\zeta) + \int_0^{2\pi} h(\beta) K(\zeta,\beta)\, d\beta = m(\zeta) \tag{83}$$

considered in the class $L^{p_1'(\cdot)}\left(\gamma; \frac{1}{\omega_1}\right)$ when $K(\beta,\zeta) = K(\zeta,\beta)$ is a continuous function, the Fredholm theorems are valid [68]. All the reasoning carried out in [39, pp. 46–52], for constant p_1 applies to the case $p = p_1(\vartheta) \in \mathcal{P}(\gamma)$, too. As a result we come to a conclusion that the equations (79) and (83) when $\nu = 0$, $m = 0$ (homogeneous equations) have only constant functions as solutions.

Returning to the nonhomogeneous equation (79), we find that for this equation to be solvable it is necessary and sufficient that the equality

$$\int_0^{2\pi} C\nu(\zeta)\, d\zeta = 0,$$

would hold for any constant C, i.e.,

$$\int_0^{2\pi}\left[f_1(\zeta)-\frac{1}{2\pi}\int_0^{2\pi}f_2(\alpha)\frac{1-\rho^2}{1+\rho^2-2\rho\cos(\alpha-\zeta)}\,d\alpha\right]d\zeta=0.$$

Hence the condition (76) follows.

If the condition (76) is fulfilled, then the equation (79) has a solution μ_1 depending on one arbitrary constant C:

$$\mu_1=\widetilde{\mu}_1+C,\tag{84}$$

where $\widetilde{\mu}_1$ is a particular solution of (79) of the class $L^{p_1(\cdot)}(\gamma_1;\omega_1)$. Substituting this value μ_1 into the second equation of the system (74) we define the function μ_2,

$$\mu_2(\zeta)=\frac{1}{2\pi}\int_0^{2\pi}\mu_1(\alpha)\frac{1-\rho^2}{1+\rho^2-2\rho\cos(\alpha-\zeta)}\,d\alpha+C-f_2(\zeta),\tag{85}$$

which belongs to $L^{p_2(\cdot)}(\gamma_2;\omega_2)$. The condition (75) gives

$$0=\frac{1}{2\pi}\int_0^{2\pi}\mu_2(\zeta)\,d\zeta=\frac{1}{(2\pi)^2}\int_0^{2\pi}\widetilde{\mu}_1(\alpha)\,d\alpha+C-\frac{1}{2\pi}\int_0^{2\pi}f_1(\zeta)\,d\zeta,$$

i.e., C_0 is number given by (82).

Therefore, replacing in the equalities (84) and (85) C by C_0, we obtain the solution of the system (74) that satisfies the condition (75). Having functions μ_i, we construct the solution of the problem (73) (by the formula (70)).

It is easy to show that for $f_i=0$, $i=1,2$, we have $u=0$. Indeed, in that case, the system (74) has only constant functions as a solution, whereas the condition (75) gives $\mu_2=0$, but then (74) implies that $\mu_1=0$, too. □

6.9 On the conformal mapping of a circular ring onto a doubly-connected domain with a piecewise smooth boundary

In [39] is proved the following

Statement 16. *If D is a doubly-connected domain with the boundary $\Gamma=\Gamma_1\cup\Gamma_2$, $\Gamma\in C^1(t_1,\dots,t_n;\nu_1,\dots,\nu_n)$, $0\le\nu_k\le 2$, $k=\overline{1,n}$ and $z=z(w)$ is conformal mapping of the ring $K=\{w:\ \rho<|w|<1\}$ onto D, such that $z(a_k)=t_k$, $z(\gamma_i)=\Gamma_i$, $\gamma_1=\{t:\ |t|=1\}$, $\gamma_2=\{t:\ |t|=\rho\}$, then*

$$z'(w)\sim\prod_{k=1}^{n}(w-a_k)^{\nu_k-1}z_0(w),$$

where

$$[z_0(w)]^{\pm 1} \in \bigcap_{\delta>1} E^\delta(K),$$
$$[z_0(e^{i\vartheta})]^{\pm 1} \in \bigcap_{\delta>1} W^\delta(\gamma_1), \quad [z_0(\rho e^{i\vartheta})]^{\pm 1} \in \bigcap_{\delta>1} W^\delta(\gamma_2).$$

In addition to this statement we first prove that the statement below is valid.

Statement 17. *In the assumptions of Statement 16, for every point t_k, $k = \overline{1,n}$, there exists the real continuous on γ function φ_k, such that in some subdomain $G_k \subset K$ containing the point $a_k = w(t_k)$ and with the boundary having with boundary $\gamma_1 \cup \gamma_2$ of the ring K a common arc l_k, $a_k \in l_k$ we have*

$$z'(w) \sim (w - a_k)^{\nu_k - 1} \chi_{\varphi_k}(\zeta_k(w)),$$

where $\zeta = \zeta_k(w)$ is conformal mapping of the circle U onto G_k and $\chi_{\varphi_k}(\zeta) = \exp{(S_\gamma \varphi_k)(\zeta)}$.

Proof. Let G_k be such a subdomain of the ring K which is bounded by a closed Lyapounov curve Λ_k having one arc l_k, common with the boundary of K, where $a_k \in l_k$, and l_k does not contain points from the set $\{a_1, \dots, a_n\}$, except the point a_k. Consider a restriction of the function $z(w)$ on G_k and denote it by $z_k(w)$. Then z_k is conformal mapping of G_k onto some simply connected domain D_k, the subdomain of D, containing the point t_k, but not containing another angular points Γ. Let $w = w_k(\zeta)$ be conformal mapping of U onto G_k, and $w_k(b_k) = a_k$; then $z_k(w_k(\zeta))$ is conformal mapping of U onto D_k. Since the boundary of the domain D_k is a piecewise smooth curve with one angular point t_k, therefore $z_k'(w_k(\zeta)) \sim (\zeta - b_k)^{\nu_k - 1} \widetilde{z}_0(\zeta)$, where

$$\widetilde{z}_0(\zeta) \sim \chi_{\varphi_k}(\zeta), \ \ \operatorname{Im} \varphi_k = 0, \ \ \varphi_k \in C(\gamma),$$

(see Section 3.8). Thus we find that

$$z_k'(w) \sim \big(\zeta_k(w) - \zeta_k(a_k)\big) \chi_{\varphi_k}(\zeta_k(w)).$$

Since $w_k(\zeta)$ is conformal mapping of the circle U onto the domain G_k with Lyapounov boundary, therefore $\zeta_k(w) - \zeta_k(a_n) \sim (w - a_k)\zeta_{k_0}(w)$, where ζ_{k_0} belongs to the Hölder class, and $\zeta_{k_0}(a_k) \neq 0$ (see, for e.g., [40, p. 146]). Consequently,

$$z_k'(w) \sim (w - a_k)^{\nu_k - 1} \chi_{\varphi_k}(\zeta_k(w)). \qquad \square$$

Theorem 18. *Let D be the doubly-connected domain with the boundary $\Gamma = \Gamma_1 \cup \Gamma_2$, $\Gamma \in C_D^1(t_1, \dots, t_n; \nu_1, \dots, \nu_n)$, $z = z(w)$ conformally maps the ring $K = \{w : \rho <$*

$|w|<1\}$ with the boundary $\gamma_1\cup\gamma_2$, $\gamma_1=\{\tau:\ |\tau|=1\}$, $\gamma_2=\{\tau:\ |\tau|=\rho\}$ onto the domain D, and $z(a_k)=t_k$. Then

$$z'(w)\sim\prod_{k=1}^{n}(w-a_k)^{\nu_k-1}\exp\int\limits_{\gamma_1}\frac{\psi_1(\tau)}{\tau-w}\,d\tau\exp\int\limits_{\gamma_2}\frac{\psi_2(\tau)}{\tau-w}\,d\tau\,, \tag{86}$$

where ψ_i is the continuous real function on γ_i, $i=1,2$.

Proof. Consider first only the points t_k lying on Γ_1. Let G_k be subdomains of the ring K with Lyapounov boundaries constructed when proving Statement 17. If $\zeta_k=\zeta_k(w)$ maps conformally the circle U onto G_k, $\varphi_k(t)=\varphi_k(\zeta_k(\tau))$, then

$$\int\limits_{\gamma}\frac{\varphi_k(\tau)}{\tau-\zeta_k(w)}\,d\tau=\int\limits_{\Lambda_k}\frac{\varphi_k(t)\zeta_k'(t)}{\zeta_k(t)-\zeta_k(w)}\,dt$$

$$=\int\limits_{\Lambda_k}\varphi_k(t)\Big[\frac{\zeta_k'(t)}{\zeta_k(t)-\zeta_k(w)}-\frac{1}{t-w}\Big]\,dt+\int\limits_{\Lambda_k}\frac{\varphi_k(t)}{t-w}\,dt=I_{k1}(w)+I_{k2}(w).$$

Since ζ_k' is Hölder continuous in U and $|\zeta(t)-\zeta(w)|\geq m|t-w|$, it is not difficult to show that

$$\Big|\frac{\zeta_k'(t)}{\zeta_k(t)-\zeta_k(w)}-\frac{1}{t-w}\Big|\leq\frac{M}{|t-w|^{\lambda}},\quad\lambda<1,$$

(analogously to the reasoning in [42, p. 18], where is considered the difference $\frac{\zeta'(s)}{\zeta(s)-\zeta(s_0)}-\operatorname{ctg}\frac{s-s_0}{2}$); thus we have that $(\exp I_{k_1})^{\pm1}$ are the bounded functions. Consequently,

$$z'(w)\sim(w-a)^{\nu_k-1}\chi_{\varphi_k}(w),\quad w\in G_k.$$

The curve Λ_k contains the arc $l_k\in\gamma_1$, hence

$$\int\limits_{\Lambda_k}\frac{\varphi_k(t)}{t-w}\,dt=\int\limits_{l_k}\frac{\varphi_k(t)}{t-w}\,dt+\int\limits_{\Lambda_k-l_k}\frac{\varphi_k(t)}{t-w}\,dt.$$

The second summand here is the bounded in the subdomain $G_k'\subset G_k$ function adjoining to γ_1 along the curve $l_k'\subset l_k$, such that $\overline{l_k'}\subset l_k$. Therefore in G_k' we have

$$z_k'(w)\sim(w-a_k)^{\nu_k-1}\chi_{\varphi_k}(w),$$

where we recall that

$$\chi_{\varphi_k}(w)=\exp\int\limits_{\gamma}\frac{\varphi_k(t)}{t-w}\,dt.$$

The domains G_k and G_k' can be chosen in such a way that the following conditions are fulfilled:

(i) $\cup G'_k$ covers a one-sided neighborhood $G^{(1)}$ of the curve γ_1, i.e., a set of those $w \in K$ for which $1-\delta < |w| < 1$;

(ii) the curves l_k intersect only with the curves l_{k-1} and l_{k+1} ($l_0 = l_{n_1}, l_{n_1+1} = l_1$), where n_1 is a number of points a_k, lying on γ_1. Then

$$z'(w) \sim \prod_{a_k \in \gamma_1} (w-a_k)^{\nu_k - 1} \exp \int_{l_k} \frac{\varphi_k(t)}{t-w}\, dt, \quad w \in G^{(1)}.$$

Assume

$$\psi_1(t) = \begin{cases} \varphi_k(t), & t \in l_k \setminus \widetilde{l}_k, \ \widetilde{l}_k = l_k \setminus \left[(l_k \cap l_{k-1} \cup (l_k \cap l_{k+1})\right], \\ \varphi_k(t) + \varphi_{k-1}(t), & t_k \in l_k \setminus (l_k \cap l_{k-1}), \\ \varphi_k(t) + \varphi_{k+1}(t), & t_k \in l_k \setminus (l_k \cap l_{k+1}). \end{cases} \quad t \in \gamma_1.$$

Then for $n_1 > 0$, we have

$$\prod_{k=1}^{n_1} \exp \int_{l_k} \frac{\varphi_k(t)}{t-w}\, dt \sim \exp \int_{\gamma_1} \frac{\psi_1(t)}{t-w}\, dt, \quad w \in G^{(1)}.$$

Obviously, ψ_1 is real and continuous on γ_1. Thus

$$z'(w) \sim \prod_{a_k \in \gamma_1} (w-a_k)^{\nu_k - 1} \exp \int_{\gamma_1} \frac{\psi_1(t)}{t-w}\, dt, \quad w \in G^{(1)}. \tag{87}$$

Analogously, we prove the existence of the neighborhood $G^{(2)}$ of the curve γ_2 and of the function $\psi_2 \in C(\gamma_2)$, such that

$$z'(w) \sim \prod_{a_k \in \gamma_2} (w-a_k)^{\nu_k - 1} \exp \int_{\gamma_2} \frac{\psi_2(t)}{t-w}\, dt, \quad w \in G^{(2)}. \tag{88}$$

The validity of the theorem follows from the above-proven relations (87)–(88). □

6.10 The Dirichlet problem in the class $e^{p_1(\cdot),p_2(\cdot)}(D)$

Let D be the doubly-connected domain bounded by simple closed curves Γ_1 and Γ_2, where Γ_2 lies in the bounded domain with boundary Γ_1, and let $\Gamma = \Gamma_1 \cup \Gamma_2$.

We assume that $\Gamma \in C^1(t_1, \ldots, t_n; \nu_1, \ldots, \nu_n)$ and $p_i \in Q(\Gamma_i)$, $i = 1, 2$.

Consider the Dirichlet problem formulated as follows: Find a harmonic function $\mathcal{U}(z)$ satisfying the conditions

$$\begin{cases} \mathcal{U}(z) \in e^{p_1(\cdot),p_2(\cdot)}(D), \\ \mathcal{U}\big|_{\Gamma_i} = f_i, \ \ f_i \in L^{p_i(\cdot)}(\Gamma_i), \ \ i = 1, 2. \end{cases} \tag{89}$$

Reduction of the problem (89) to the problem for a ring

Lemma 19. *If $\mathcal{U}(z) \in e^{p_1(\cdot),p_2(\cdot)}(D)$ and $p_i \in Q(\Gamma_i)$, then the function $u(w) = \mathcal{U}(z(w))$, where $z = z(w)$ is conformal mapping of the ring K onto D, belongs to the class $h^{l_1(\cdot),l_2(\cdot)}(K;\omega)$, where $l_i(\zeta)=p_i(z(\zeta))$, $\zeta\in \gamma_i$,*

$$\omega(w)=\prod_{k=1}^{n}(w-a_k)^{\frac{\nu_k-1}{p(t_k)}} \exp\int\limits_{\gamma_1}\frac{\psi_1(\zeta)}{l_1(\zeta)}\,\frac{d\zeta}{\zeta-w} \exp\int\limits_{\gamma_2}\frac{\psi_2(\zeta)}{l_2(\zeta)}\,\frac{d\zeta}{\zeta-w}, \tag{90}$$

ψ_i, $i = 1, 2$, are the functions defined by means of z' appearing in (86), $a_k = w(t_k)$, $p(t_k) = p_i(t_k)$, $t_k \in \Gamma_i$.

Proof. Let $\mathcal{U}(z) = \operatorname{Re}\Phi(z)$, $\Phi \in E^{p_1(\cdot),p_2(\cdot)}(D)$. This implies that

$$I = \bigg(\sup_{r_0<r<1} \int\limits_0^{2\pi} \big|\Phi(z(re^{i\vartheta}))\big|^{p_1(\vartheta)} \big|z'(re^{i\vartheta})\big|\,d\vartheta \\ + \sup_{\rho<r\le r_0} \int\limits_0^{2\pi} \big|\Phi(z(re^{i\vartheta}))\big|^{p_2(\vartheta)} \big|z'(re^{i\vartheta})\big|\,d\vartheta \bigg) < \infty,$$

where

$$p_1(\vartheta) = p_1(z(e^{i\vartheta}))\ \big(= l_1(\zeta),\ \ \zeta\in\gamma_1\big),\ p_2(\vartheta)\ = p_2(z(\rho e^{i\vartheta}))\ \big(= l_2(\zeta),\ \ \zeta\in\gamma_2\big).$$

From Lemma 10 it directly follows that the functions

$$\big(|z'(w)|^{\frac{1}{p_1(\vartheta)}}(\omega(w))^{-1}\big)^{\pm 1},\quad w = re^{i\vartheta},$$

are bounded in K. We now obtain

$$I = \bigg(\sup_{r_0<r<1} \int\limits_0^{2\pi} \big|\Phi(z(re^{i\vartheta}))\omega(z(re^{i\vartheta}))\big|^{p_1(\vartheta)}\,d\vartheta \\ + \sup_{\rho<r\le r_0} \int\limits_0^{2\pi} \big|\Phi(z(re^{i\vartheta}))\omega(z(re^{i\vartheta}))\big|^{p_2(\vartheta)}\,d\vartheta \bigg) < \infty.$$

This means that $\Phi(z(re^{i\vartheta})) = \Phi(z(w)) \in H^{l_1(\cdot),l_2(\cdot)}(K;\omega)$ and since $u(w) = \mathcal{U}(z(w)) = \operatorname{Re}\Phi(z(w))$, therefore $u(w) \in h^{l_1(\cdot),l_2(\cdot)}(K;\omega)$. □

Corollary 20. *Every solution* $\mathcal{U}(z)$ *of the problem* (89) *generates a solution* $u(w) = \mathcal{U}(z(w))$ *of the problem*

$$\begin{cases} \Delta u = 0, & u \in h^{l_1(\cdot),l_2(\cdot)}(K;\omega), \ \ l_i(\tau) = p_i(z(\tau)), \\ u\big|_{\gamma_i} = g_i, & g_i(\tau) = f_i(z(\tau)), \ \ g_i \in L^{\ell(\cdot)(\gamma_i;\omega_i^+)}, \end{cases} \tag{91}$$

where ω *is the function given by equality* (90) *and* $\omega_i^+(\tau) = \lim\limits_{w \widehat{\to} \tau} \omega(w)$, $\tau \in \gamma_i$. *Conversely, if* $u(w)$ *is a solution of the problem* (91), *then* $\mathcal{U}(z) = u(w(z))$ *is a solution of the problem* (89).

Remark 21. The condition $p_i \in Q(\Gamma_i)$ allows us to conclude that $l_i(\zeta) = p_i(z(\zeta))$, $\zeta \in \gamma_i$ belongs to $\widetilde{P}(\gamma_i)$, $i = 1, 2$.

Now we give a solution of the problem (91). First of all we mentioned some properties of a solution of (91). In the assumptions adopted regarding Γ and the function p_i, every function $u \in h^{l_1(\cdot),l_2(\cdot)}(K;\omega)$ is representable in the form $u = u_1 + u_2$, where $u_i \in h^{l_i(\cdot)}(K_i;\omega_i)$, $u_2(\infty) = 0$, and

$$\omega_1(w) = \prod_{a_k \in \gamma_1} (w - a_k)^{\frac{\nu_k - 1}{l_1(a_k)}} \exp \int_{\gamma_1} \frac{\psi_1(\zeta)}{l_1(\zeta)} \frac{d\zeta}{\zeta - w},$$

$$w \in K_1, \ \ l_1(a_k) = p_1(t_k), \ \ a_k = w(t_k),$$

$$\omega_2(w) = \prod_{a_k \in \gamma_2} \left(\frac{w - a_k}{w - w_0}\right)^{\frac{\nu_k - 1}{l_2(a_k)}} \exp \int_{\gamma_2} \frac{\psi_2(\zeta)}{l_2(\zeta)} \frac{d\zeta}{\zeta - w},$$

$$w \in K_2, \ \ |w_0| < \rho, \ \ l_2(a_k) = p_2(t_k).$$

Note that

$$\omega_i^+(t) \sim \prod_{a_k \in \gamma_i} (t - a_k)^{\frac{\nu_k - 1}{\ell(a_k)}} \exp \int_{\gamma_i} \frac{\psi_i(\zeta)}{\ell_i(\zeta)} \frac{d\zeta}{\zeta - t}, \quad t \in \gamma_i. \tag{90_1}$$

According to Lemmas 10 and 12, we have $\omega_i \in H^{l_i(\cdot)}(K_i)$. Moreover, if u_2 is representable by the Poisson integral, then

$$\int_0^{2\pi} u_2(\rho e^{i\vartheta})\, d\vartheta = 0.$$

If now $u = u_1 + u_2$ is a solution of the problem (91), then we can conclude that

$$\begin{cases} \Delta u_1 = 0, & u_1 \in h^{l_1(\cdot)}(K_1;\omega_1), \\ u_1^+(e^{i\vartheta}) = g_1(e^{i\vartheta}) - u_2(e^{i\vartheta}), & g_1 \in L^{l_1(\cdot)}(\gamma_1;\omega_1^+), \end{cases} \tag{92}$$

$$\begin{cases} \Delta u_2 = 0, & u_2 \in h^{l_2()}(K_2;\omega_2), \\ u_2^+(\rho e^{i\vartheta}) = g_2(\rho e^{i\vartheta}) - u_1(\rho e^{i\vartheta}), & g_2 \in L^{l_2(\cdot)}(\gamma_2;\omega_2^+). \end{cases} \tag{93}$$

6.11 The Dirichlet problem when Γ has points at which $\nu_k = 0$ or $\nu_k = p(t_k)$

Lemma 22. *If Γ has points at which $\nu_k = 0$, then the problem (91), generally speaking, unsolvable.*

Proof. Let, for example, $t_0 \in \Gamma_1$, $\nu(t_0) = 0$ and $z(t_0) = a$, then $\frac{\nu(t_0)-1}{l(a)} = -\frac{1}{l(a)}$, and

$$\omega_1(w) = (w-a)^{-\frac{1}{l(a)}} \chi_{\varphi_1}(w), \quad \varphi_1 = \psi_1/l_1.$$

Assume now that the statement of the lemma is invalid. Then for any fixed $g_2 \in L^{l_2(\cdot)}(\gamma_2; \omega_2^+)$ and any $g_1 \in L^{l_1(\cdot)}(\gamma_1; \omega_1^+)$ there exists a solution of the problem (91), $u = u_1 + u_2$, where $u_1 = \operatorname{Re} \Phi_1$, $\Phi_1 \in H^{l_1(\cdot)}(K_1; \omega_1) = H^{l_1(\cdot)}(\omega_1)$, and for u_1 the condition (92) is fulfilled. Thus $\Phi_1 = \omega_1^{-1} F_1$, $F_1 \in H^{l_1(\cdot)}$. Since $\omega_1^{-1} \in H^{l_1'(\cdot)}$ we have $\Phi_1 \in H^1$. But every function Φ of H^1 is representable by the Schwartz integral with density $\operatorname{Re}\Phi$ (see, Chapter 1, Lemma 2). This easily results in

$$\Phi(w) = \frac{1}{\pi i} \int_{\gamma_i} \frac{\operatorname{Re}\Phi}{t-w}\, dt - \overline{\Phi(0)}. \tag{94}$$

Therefore

$$\Phi_1(w) = \frac{1}{\pi i} \int_{\gamma_1} \frac{\operatorname{Re}\Phi_1(t)}{t-w}\, dt + \text{const} = \frac{1}{\pi i} \int_{\gamma_1} \frac{g_1(t) - u_2(t)}{t-w}\, dt + \text{const}\,.$$

Since u_2 is differentiable on γ_1, the Cauchy type integral $K_{\gamma_1} u_2$ is the bounded function, and hence, it belongs to $H^{l_1(\cdot)}(\omega_1)$ because $\omega_1 \in H^{l_1(\cdot)}$. The last statement follows directly from Lemmas 10 and 12. Thus since $\Phi_1 = K_{\gamma_1} g_1 - K_{\gamma_1} u_2 + \text{const}$ and $\Phi_1 \in H^{l_1(\cdot)}(\omega_1)$, it follows that $K_{\gamma_1} g_1 \in H^{l_1(\cdot)}(\omega_1)$ for any $g_1 \in L^{l_1(\cdot)}(\gamma_1; \omega_1^+)$. By virtue of Plemelj–Sokhotskiĭ formulas, we conclude that $S_{\gamma_1} g_1 \in L^{l_1(\cdot)}(\gamma_1; \omega_1^+)$ for any $g_1 \in L^{l_1(\cdot)}(\gamma_1; \omega_1^+)$. This implies that $\omega_1^+ \in W^{l_1(\cdot)}(\gamma_1)$ (see Chapter 4, Theorem 2). Moreover, $(\omega_1^+)^{-1} \sim (\tau - a)^{\frac{1}{l_1(a)}} (\chi_{\varphi_1}(\tau)^{-1}$, and hence $(\omega_1^+)^{-1} \in L^{l_1'(\cdot)+\varepsilon}(\gamma_1)$. Therefore according to Corollary 11 from Chapter 4, we conclude that $\omega_1^+(w)\chi_\psi(w)$ belongs to $W^{l_1(\cdot)}(\gamma_1)$ for any real, continuous on γ_1 functions ψ. Assuming $\psi = -\varphi_1$, we obtain $(\tau - a)^{-\frac{1}{l(a)}} \in W^{l_1(\cdot)}(\gamma_1)$, but this is impossible because $(\tau - a)^\alpha$ belongs to $W^{l_1(\cdot)}(\gamma_1)$ if and only if $\alpha \in \big(-\frac{1}{l(a)}, \frac{1}{l'(a)}\big)$.

Similar contradiction is obtained under the assumption that $t_0 \in \gamma_2$. □

Lemma 23. *If Γ has points t_k at which $\nu_k = p(t_k)$, then the problem (91) is, generally speaking, unsolvable.*

Proof. For the sake of simplicity, let there is only one point $t_{k_0} \in \Gamma_i$, such that $\nu(t_{k_0}) = p_i(t_{k_0})$ and $w(t_{k_0}) = a$, $a \in \gamma_i$. Then according to definition of functions

$\omega_i(w)$ from Section 6.10 we have

$$\omega_i(w) = (w-a)^{\frac{1}{l_i'(a)}} \chi_{\varphi_i}(w), \quad \varphi_i = \psi_i/l_i.$$

Assume that the problem is solvable for any given $g_i \in L^{l_i(\cdot)}(\gamma_i; \omega_i^+)$, $i = 1, 2$. For its solution $u = u_1 + u_2$ we have $u_i = \operatorname{Re} \Phi_i$, $\Phi_i \in H^{l_i(\cdot)}(K_i; \omega_i)$. Consider the functions $\Psi_i(w) = \Phi_i(w)(w-a)^{-1}$. Let $\widetilde{\omega}(w) = \omega(w)(w-a)^{-1}$ then $\Psi_i \in H^{l_i(\cdot)}(\widetilde{\omega}_i)$, where

$$\widetilde{\omega}_i(w) = \omega_i(w)(w-a)^{-1} \sim (w-a)^{-\frac{1}{l_i(a)}} \chi_{\varphi_i}(w), \quad i = 1, 2. \tag{95}$$

Let $\widetilde{g}_i(\tau) = g_i(\tau)(\tau - a)^{-1}$, then $\widetilde{g}_i \in L^{l_i(\cdot)}(\gamma_i; \widetilde{\omega}_i^+)$, and when g_i runs through $L^{l_i(\cdot)}(\gamma_i; \omega_i^+)$, then $\widetilde{g}_i$ runs through $L^{l_i(\cdot)}(\gamma_i; \widetilde{\omega}_i^+)$. The functions $\widetilde{u}_i = \operatorname{Re} \Psi_i$ belong to $h^{l_1(\cdot)}(\widetilde{\omega}_i)$, and $\widetilde{u} = \widetilde{u}_1 + \widetilde{u}_2$ is a solution of the problem

$$\begin{cases} \Delta \widetilde{u} = 0, & \widetilde{u} \in h^{l_1(\cdot), l_2(\cdot)}(\widetilde{\omega}), \ \widetilde{\omega}(w) = \omega(w)(w-a)^{-1}, \\ \widetilde{u}\big|_{\gamma_i} = \mu_i, & \mu_i \in L^{l_i(\cdot)}(\gamma_i; \widetilde{\omega}_i^+), \end{cases}$$

where $\widetilde{\omega}(w) = \omega(w)(w-a)^{-1}$ and $\mu_i \sim \widetilde{g}_i$.

According to our assumption, this problem turns out to be solvable for any μ_i from $L^{l_i(\cdot)}(\gamma_i;\widetilde{\omega}_i^+)$. But this is impossible on account of Lemma 22. □

From Lemmas 22 and 23 follows

If Γ has angular points t_k for which $\nu_k = p(t_k)$, or $\nu_k = 0$ ($p(t_k) = p_i(t_k)$, $t_k \in \Gamma_i$), then the problem (91) (and hence the Problem (89)) is, generally speaking, unsolvable.

Let us give one necessary condition for the solvability of problem (91). Change in equality (48) the function g with the function $2f_i(z(\tau))\omega_i^+(\tau)$ from the Theorem 13 and formulas (90_1) it follows that for the conditions (92) and (93) to be fulfilled, it is necessary that the functions $g_i(\tau) = f_i(z(\tau))$ satisfy the conditions

$$\frac{\omega_i^+(\tau)}{\rho_i(\tau)} \int_{\gamma_i} \frac{\omega_i^+(\zeta) g_i(\zeta) \rho_i(\zeta)}{\omega_i^+(\zeta)} \frac{d\zeta}{\zeta - \tau} \in L^{l_i(\cdot)}(\gamma_i), \quad i = 1, 2, \tag{96}$$

where

$$\omega_i(w) = \prod_{a_k \in \gamma_i} (w - a_k)^{\frac{\nu_k - 1}{l_i(a_k)}} \chi_{\varphi_i}(w), \quad \varphi_i = \psi_i/l_i, \quad \rho_i(w) = \prod_{k \in T_i} (w - a_k)$$

for $T_i = \{k : \ \nu_k \geq p_i(t_k), \ t_k \in \Gamma_i\} \neq \varnothing$, and $\rho_i(w) = 1$ for $T_i = \varnothing$.

Remark 24. If $0 < \nu_k < p(t_k)$ then $\omega_i^+(\tau) \in W^{\ell_i(\cdot)}(\gamma_i)$ and hence the conditions (96) are fulfilled.

Lemma 25. *If the conditions* (96) *are fulfilled, then* $g_i \in L^{\underline{l_i}(\cdot) - \varepsilon}(\gamma_i)$, *for some* $\varepsilon > 0$.

Proof. Since $g_i(\tau) = f_i(z(\tau))$ and $f_i \in L^{p_i(\cdot)}(\Gamma_i)$, in view of the fact that $l_i \in \widetilde{P}(\gamma_i)$, we obtain $g_i \in L^{l_i(\cdot)}(\gamma_i; \omega_i^+)$, i.e., $g_i = q_i(\omega_i^+)^{-1}$, where $q_i \in L^{l_i(\cdot)}(\gamma)$. Since in the absence of the points from the set $\{t_k : \nu_k = p(t_k) \text{ or } \nu_k = 0,\ t_k \in \Gamma_i\}$ the functions $\omega_i^+ q_i$ belong to $W^{l_i(\cdot)}(\gamma_i)$, the conditions (96) are equivalent to the conditions

$$\lambda_i(\tau) = \frac{\omega_i^+(\tau)}{\pi i} \int\limits_i \frac{q_i(\zeta)}{\omega_i^+(\zeta)} \frac{d\zeta}{\zeta - \tau} \in L^{l_i(\cdot)}(\gamma_i),$$

where this time

$$\omega_i(w) = \prod_{k \in T_i'} (w - a_k)^{-\frac{1}{l_i(a_k)}} \exp \int\limits_{\gamma_i} \frac{\psi_i(\zeta)}{l_i(\zeta)} \frac{d\zeta}{\zeta - w},$$

$$T_i' = \{k : \nu_k = p(t_k), \text{ or } \nu_k = 0,\ t_k \in \Gamma_i\},$$

whence

$$\frac{1}{\pi i} \int\limits_{\gamma_i} \frac{q_i(\zeta)}{\omega_i^+(\zeta)} \frac{d\zeta}{\zeta - \tau} = \frac{\lambda_i(\tau)}{\omega_i^+(\tau)}, \quad \text{i.e.,} \quad S_{\gamma_i} \frac{q_i}{\omega_i^+} = \frac{\lambda_i}{\omega_i^+}.$$

The above equality yields $S_{\gamma_i}\left(S_{\gamma_i} \frac{q_i}{\omega_i^+}\right) = S_{\gamma_i} \frac{\lambda_i}{\omega_i^+}$. Since $\frac{q_i}{\omega_i^+} \in L^{l_i(\cdot)}(\gamma_i) \subset L^{\underline{l_i}(\gamma_i)}$ and $\underline{l_i} > 1$, we have $S_{\gamma_i}\left(S_{\gamma_i} \frac{q_i}{\omega_i^+}\right) = \frac{\lambda_i}{\omega_i^+}$ (see, for e.g., [42, p. 35–36]). Finally, we obtain $q_i = \omega_i^+ S_{\gamma_i} \frac{\lambda_i}{\omega_i^+}$. But then $g_i = q_i(\omega_i^+)^{-1} = S_{\gamma_i} \frac{\lambda_i}{\omega_i^+} \in L^{\underline{l_i}(\cdot)-\varepsilon}(\gamma_i)$ (because $\lambda_i \in L^{l_i(\cdot)}(\gamma_i)$, and $(\omega_i^+)^{-1} \in \bigcap\limits_{\delta>1} L^{\delta}(\gamma_i)$). □

In the sequel we need the existence of a particular solution $u = u_1 + u_2$ of the problem (91) when u_i are representable by the Poisson integral in the domains K_i.

Relying on Lemma 25, we will show that under the conditions (96) there exists a particular solution of the problem (91) admitting the representation $u = u_1 + u_2$, where u_i, $i = 1, 2$, are representable by the Poisson integrals in the domains K_i with density from $L^{\underline{l_i}-\varepsilon}(\gamma_i)$, if and only if

$$\int\limits_0^{2\pi i} g_1(e^{i\vartheta})\, d\vartheta = \int\limits_0^{2\pi i} g_2(\rho e^{i\vartheta})\, d\vartheta. \tag{97}$$

Lemma 26. *If $u = u_1 + u_2$ is a solution of the problem* (91), *where g_i, $i = 1, 2$, satisfy the condition* (96), *then the functions u_i are representable by the Poisson integrals in domains K_i with density from $L^{\underline{l_i}-\varepsilon}(\gamma_i)$, $\varepsilon > 0$.*

Proof. Let us prove that $u_1 \in h^{\underline{l_i}-\varepsilon}(K_1)$. u_1 satisfies the condition (92). Without restriction of generality, we may assume that $\{t_k : \nu_k > p(t_k),\ t_k \in \gamma_1\} = \varnothing$. (Indeed, in the presence of such points we would be able to reduce the problem (91) to an analogous problem of the class $h^{l_1(\cdot),l_2(\cdot)}(\omega_*)$, $\omega_*(t) = \omega(t) \prod\limits_{\{k:\, \nu_k > p(t_k)\}} (t - t_k)^{-1})$.

Now, according to Theorem 7, u_1 is given by the equality

$$u_1(w) = \operatorname{Re}\left\{\frac{\omega_1(w)}{2\pi i}\int\limits_{\gamma_1}\frac{g_1(e^{i\vartheta}) - u_2(e^{i\vartheta})}{\omega_1^+(e^{i\vartheta})}\frac{\zeta + w}{\zeta - w}\,d\zeta\right\}, \quad \zeta = e^{i\vartheta}.$$

Since $u_2(e^{i\vartheta}) = u_2(\zeta)$ is differentiable on γ_1, and $\omega_1 \in H^{l_1(\cdot)-\varepsilon}$, we can easily verify that

$$\frac{\omega_1(w)}{2\pi i}\int\limits_{\gamma_1}\frac{u_2(\zeta)}{\omega_1^+(\zeta)}\frac{\zeta + w}{\zeta - w}\,d\zeta \in H^{l_1(\cdot)-\varepsilon}.$$

Therefore, our lemma will be completed if we show that

$$(G_1(g_1))(w) = \omega_1(w)\int\limits_{\gamma_1}\frac{g_1(\zeta)}{\omega_1^+(\zeta)}\frac{d\zeta}{\zeta - w} \in H^{l_1(\cdot)-\varepsilon}.$$

As $\omega_1 \in H^{\eta}$, for some $\eta > 0$, and the Cauchy type integral with density $\frac{g_1}{\omega_1^+}$ belongs to $\bigcap\limits_{\delta<1} H^{\delta}$ (see, for e.g., [19, p. 33]), we have $G_1(g_1) \in H^{\eta_0}$, for some $\eta_0 > 0$. Show that $G_1(g_1) \in H^{\underline{l_i}-\varepsilon}$. Towards this end, owing to the Smirnov's theorem, it suffices to prove that $(G_1 g_1)^+ \in L^{\underline{l_i}-\varepsilon}(\gamma_1)$, and to this end, by virtue of the Plemelj–Sokhotskiĭ formulas and Lemma 25 it suffices to establish that $\omega_1^+ S_{\gamma_1}\frac{g_1}{\omega_1^+} \in L^{\underline{l}_1-\varepsilon}$. But this function in view of (96) belongs to $L^{l_1(\cdot)}(\gamma_1) \subset L^{\underline{l}_1-\varepsilon}(\gamma_1)$. Analogously, we can prove that $u_2 \in h^{\underline{l}_2-\varepsilon}(K_2)$. □

6.12 Integral equation with respect to the function $u_2(\rho e^{i\vartheta})$

To solve the problem (91), we refer to the condition (93). This problem will be considered as the Dirichlet problem in the class $h^{l_2(\cdot)}(K_2;\omega_2^+)$. To make use of the results from Section 6.6, we have to fulfil the condition (57).

Lemma 27. *If the conditions* (96) *and* (97) *are fulfilled, and* $u = u_1 + u_2$ *is a solution of the problem* (91), *then*

$$I = \int\limits_0^{2\pi}\left[g_2(\rho e^{i\vartheta}) - u_1(\rho e^{i\vartheta})\right]d\vartheta = 0. \tag{98}$$

Proof. Since

$$u_1(\rho e^{i\vartheta}) = \frac{1}{2\pi}\int\limits_0^{2\pi} u_1(e^{i\alpha})\,P(\rho, \alpha - \theta)d\alpha,$$

then

$$I=\int_0^{2\pi}\left[g_2(\rho e^{i\vartheta})-\frac{1}{2\pi}\int_0^{2\pi}u_1(e^{i\alpha})\,\frac{1-\rho^2}{1+\rho^2-2\rho\cos(\alpha-\vartheta)}\,d\alpha\right]d\vartheta=\int_0^{2\pi}g_2(\rho e^{i\vartheta})\,d\vartheta$$
$$-\int_0^{2\pi}u_1(e^{i\alpha})\,\frac{1}{2\pi}\int_0^{2\pi}\frac{1-\rho^2}{1+\rho^2-2\rho\cos(\alpha-\vartheta)}\,d\vartheta\,d\alpha=\int_0^{2\pi}g_2(\rho e^{i\vartheta})\,d\vartheta-\int_0^{2\pi}u_1(e^{i\alpha})\,d\alpha.$$

Moreover, since $u_2\in h^1(K_2)$, therefore $\int\limits_0^{2\pi}u_2(e^{i\alpha})\,d\alpha=0$. As far as $u=u_1+u_2$ is the solution of the problem (91), we have $u_1(e^{i\alpha})+u_2(e^{i\alpha})=g_1(e^{i\alpha})$. Consequently, $\int\limits_0^{2\pi}u_1(e^{i\alpha})\,d\alpha=\int\limits_0^{2\pi}g_1(e^{i\alpha})\,d\alpha$, by virtue of the condition (97), we conclude that $I=0$. □

Thus assuming that the conditions (96) and (97) are fulfilled, we have equality (98), and hence we are able to apply Theorem 7 according to which the problem (93) is solvable, and any its solution lias in the set of functions $u_0(w)+u_2(w)$, where u_2 is some particular solution and u_0 is given by (49). Thus the restriction of the function $u_2(w)$ on γ_1 is contained in a set of functions

$$u_2(e^{i\vartheta})+\sum_{k\in T_2}M_k(p)\,\mathrm{Re}\,\frac{e^{i\vartheta}+a_k}{e^{i\vartheta}-a_k}=u_2(e^{i\vartheta})+u_{2,0}(\vartheta),$$
$$T_2=\big\{k:\ \nu_k\geq p_2(t_k),\ t_k\in\Gamma_2\big\},$$

where

$$u_{2,0}(e^{i\vartheta})=\sum_{k\in T_2}M_k(p)\,\mathrm{Re}\left[(e^{i\vartheta}+a_k)(e^{i\vartheta}-a_k)^{-1}\right],\ \ \sum_{k\in T_2}M_k(p)=0,$$

and the real constants $M_k(p)$ are defined by virtue of (50).

u_1 now is contained in a set of functions satisfying the condition

$$\begin{cases}u_1\in h^{l_1(\cdot)}(K_1;\omega_1^+),\ \ K_1=U,\\ u_1(\tau)=g_1(\tau)-u_2(\tau)-\sum\limits_{k\in T_2}M_k(p)\,\mathrm{Re}\,\dfrac{\tau+a_k}{\tau-a_k},\ \ \tau\in\gamma_1.\end{cases}\tag{99}$$

In view of Theorem 7, if for g_1 the condition (96) is fulfilled, then the problem (99) is solvable, and

$$u_1(w)=\sum_{k\in T_1}M_k(p)\,\mathrm{Re}\,\frac{w+a_k}{w-a_k}+u_{1\widetilde{g}}(w),\ \ w\in U,\tag{100}$$

where $u_{1\widetilde{g}}$ is a particular solution of the problem (99) representable by the Poisson integral. Hence

$$
\begin{aligned}
u_{1\widetilde{g}}(re^{i\vartheta}) &= \frac{1}{2\pi}\int_0^{2\pi} g_1(e^{i\alpha})\,\frac{1-r^2}{1+r^2-2r\cos(\alpha-\vartheta)}\,d\alpha \\
&\quad -\frac{1}{2\pi}\int_0^{2\pi} u_2(e^{i\alpha})\,\frac{1-r^2}{1+r^2-2r\cos(\alpha-\vartheta)}\,d\alpha \\
&\quad -\sum_{k\in T_2}\frac{1}{2\pi}\int_0^{2\pi} M_k(p)\,\mathrm{Re}\,\frac{e^{i\alpha}+a_k}{e^{i\alpha}-a_k}\,\frac{1-r^2}{1+r^2-2r\cos(\alpha-\vartheta)}\,d\alpha \\
&= (Pg_1)(r,\vartheta)-(Pu_2)(r,\vartheta)-(Pu_{2,0})(r,\vartheta),\quad re^{i\theta}\in K_1,
\end{aligned} \tag{101}
$$

where

$$
(Pu_{2,0})(r,\vartheta)=\frac{1}{2\pi}\int_0^{2\pi}\sum_{k\in T_2} M_k(p)\,\mathrm{Re}\,\frac{e^{i\alpha}+a_k}{e^{i\alpha}-a_k}\,\frac{1-r^2}{1+r^2-2r\cos(\alpha-\vartheta)}\,d\alpha.
$$

This implies that in the ring K,

$$
u(w)=\sum_{k\in T_1} M_k(p)\,\mathrm{Re}\,\frac{w+a_k}{w-a_k}-(Pu_2)(w)+(Pg_1)(w)-(Pu_{2,0})(w)+u_2(w).
$$

Since u is a solution of the problem (91), therefore $u(\rho e^{i\vartheta})=g_2(\rho e^{i\vartheta})$, and the last equality results in

$$
\begin{aligned}
u_2(\rho e^{i\vartheta}) &- \frac{1}{2\pi}\int_0^{2\pi} u_2(e^{i\alpha})\,\frac{1-\rho^2}{1+\rho^2-2\rho\cos(\alpha-\vartheta)}\,d\alpha+(Pg_1)(\rho,\vartheta) \\
&-(Pu_{2,0})(\rho,\vartheta)+\sum_{k\in T_1} M_k(p)\,\mathrm{Re}\,\frac{\rho e^{i\vartheta}+a_k}{\rho e^{i\vartheta}-a_k}=g_2(\rho e^{i\vartheta}).
\end{aligned} \tag{102}
$$

As far as $u_2\in h^1(K_2)$,

$$
u_2(re^{i\vartheta})=\frac{1}{2\pi}\int_0^{2\pi} u_2(\rho e^{i\beta})\,\frac{r^2-\rho^2}{r^2+\rho^2-2r\rho\cos(\beta-\vartheta)}\,d\beta,
$$

hence

$$
u_2(e^{i\alpha})=\frac{1}{2\pi}\int_0^{2\pi} u_2(\rho e^{i\beta})\,\frac{1-\rho^2}{1+\rho^2-2\rho\cos(\beta-\alpha)}\,d\beta.
$$

Substituting this value into (102), we obtain

$$u_2(\rho e^{i\vartheta}) - \frac{1}{2\pi}\int_0^{2\pi}\left[\frac{1}{2\pi}\int_0^{2\pi} u_2(\rho e^{i\beta})\,\frac{1-\rho^2}{1+\rho^2-2\rho\cos(\alpha-\beta)}\,d\beta\right] \times \frac{1-\rho^2}{1+\rho^2-2\rho\cos(\alpha-\vartheta)}\,d\alpha + (Pg_1)(\rho,\vartheta) - (Pu_{2,0})(\rho,\vartheta) + \sum_{k\in T_1} M_k(p)\,\mathrm{Re}\,\frac{\rho e^{i\vartheta}+a_k}{\rho e^{i\vartheta}-a_k} = g_2(\rho e^{i\vartheta}), \tag{103}$$

that is,

$$u_2(\rho e^{i\vartheta}) + (Nu_2)(\rho,\vartheta) = \widetilde{g}_2(\rho,\vartheta),$$

where

$$(Nu_2)(\rho,\vartheta) = \frac{1}{2\pi}\int_0^{2\pi}\left[\frac{1}{2\pi}\int_0^{2\pi} u_2(\rho e^{i\beta})\,\frac{1-\rho^2}{1+\rho^2-2\rho\cos(\alpha-\beta)}\,d\beta\right] \times \frac{1-\rho^2}{1+\rho^2-2\rho\cos(\alpha-\vartheta)}\,d\alpha, \tag{104}$$

$$\widetilde{g}_2(\rho,\vartheta) = g_2(\rho e^{i\vartheta}) - (Pg_1)(\rho,\vartheta) + (Pu_{2,0})(\rho,\vartheta) - \sum_{k\in T_1} M_k(p)\,\mathrm{Re}\,\frac{\rho e^{i\vartheta}+a_k}{\rho e^{i\vartheta}-a_k}.$$

6.13 Solution of the Dirichlet problem in the weighted class $h^{\ell_1(\cdot),\ell_2(\cdot)}(K;\omega)$

Since it suffices to find only one solution of equation (103), we put $M_k(p)=0$, $k\in T_1$. Then $\widetilde{g}_2$ satisfies the condition (96), and hence by Lemma 25, $g_2\in L^{\underline{l}_2-\varepsilon}$.

We will now proceed to investigating equation (103). We fix $\varepsilon>0$ such that $\widetilde{l}_2=\underline{l}_2-\varepsilon>1$ and consider equation (103) in the space $L^{\widetilde{l}_2}(\gamma_2)$.

The kernel of the operator N is a continuous function, hence it is compact in $L^{\widetilde{l}_2}(\gamma)$. The equations $u+Nu=0$ and $v+N^*v=0$ in the capacity of solutions have only constant functions [39]. Taking this fact into account, equation (103) is solvable, if and only if

$$\int_0^{2\pi} \widetilde{g}_2(\rho,\vartheta)\,d\vartheta = 0,$$

i.e.,

$$\int_0^{2\pi} (Pg_1)(\rho,\vartheta)\,d\vartheta + \sum_{k\in T_1} M_k(p)\int_0^{2\pi} \mathrm{Re}\,\frac{\rho e^{i\vartheta}+a_k}{\rho e^{i\vartheta}-a_k}\,d\vartheta$$

$$-\int_0^{2\pi}(Pu_{2,0})(\rho,\vartheta)\,d\vartheta=\int_0^{2\pi}g_2(\rho e^{i\vartheta})\,d\vartheta. \quad (105)$$

It can be easily verified that

$$\int_0^{2\pi}(Pg_1)(\rho,\vartheta)\,d\vartheta=\int_0^{2\pi}g_1(e^{i\vartheta})\,d\vartheta,$$

$$\int_0^{2\pi}\frac{\rho e^{i\vartheta}+a_k}{\rho e^{i\vartheta}-a_k}\frac{de^{i\vartheta}}{ie^{i\vartheta}}=\int_{\gamma_2}\frac{t+a_k}{t-a_k}\frac{dt}{it}=1,\quad |a_k|=1.$$

Moreover,

$$\int_0^{2\pi}(Pu_{2,0})\,d\vartheta=\int_0^{2\pi}\frac{1}{2\pi}\int_0^{2\pi}\sum_{k\in T_2}M_k(p)\,\mathrm{Re}\,\frac{e^{i\vartheta}+a_k}{e^{i\vartheta}-a_k}\frac{1-\rho^2}{1+\rho^2-2\rho\cos(\alpha-\vartheta)}\,d\alpha\,d\vartheta$$

$$=\sum_{k\in T_2}M_k(p)\int_0^{2\pi}\mathrm{Re}\,\frac{e^{i\alpha}+a_k}{e^{i\alpha}-a_k}\,d\alpha=0.$$

With regard for the above equalities, the condition (105) takes the form

$$\int_0^{2\pi}g_1(e^{i\vartheta})\,d\vartheta-\sum_{k\in T_1}M_k(p)=\int_0^{2\pi}g_2(\rho e^{i\vartheta})\,d\vartheta.$$

It now suffices to take $M_k(p)=0$, $k\in T_1$, and the last equality coincides with the condition (97).

Thus if the conditions (96) and (97) are fulfilled, and $M_k(p)=0$, $k\in T_1$, then the condition (105) is likewise fulfilled, and hence equation (103) is solvable in $L^{\widetilde{\ell}_2}(\gamma_2)$. Since the homogeneous equation has in the capacity of solutions only constant functions, it follows that a solution satisfying the condition $\int_0^{2\pi}u_2(\rho e^{i\vartheta})\,d\vartheta=0$ is unique. Further, let $u_2=\mathrm{Re}\,\Phi_2$, $\Phi_2\in H^{\widetilde{l}_2}$. If we apply formula (94) to the function Φ_2 and take into account that $u_2(\rho e^{i\vartheta})=g_2(\rho e^{i\vartheta})-u_1(\rho e^{i\vartheta})$, then making use of the condition (96) and the fact that $\omega_2\in h^{l_2(\cdot)}$, we can conclude that $u_2\in h^{l_2(\cdot)}(C\overline{U}_\rho;\omega_2)$.

Having the function $u_2(\rho e^{i\vartheta})$ at hand, by means of the Poisson integral we find the function $u_2(w)$, $|w|>\rho$, and hence the function $u_2(e^{i\alpha})$, as well. Next, equality (102) allows us to find the function $u_{1\widetilde{g}}$. In view of the condition (96) and taking into account the fact that $\omega_1\in h^{l_1(\cdot)}$ (see proof of Lemma 22), from equality (101) we conclude that $u_{1\widetilde{g}} \in h^{l_1(\cdot)}(\omega_1)$. Having $u_{1\widetilde{g}}$, by means of equality (100) we find the function $u_1(w)$. It is not difficult to verify that if $u(w) = u_1(w) + u_2(w)$, where $u_i \in h^{l_i(\cdot)}(K_i;\omega_i)$, and ω_i are given by equalities from Section 6.10, then $u \in h^{l(\cdot)}(K;\omega_1\omega_2) = h^{l(\cdot)}(K;\omega)$, in which ω is the function defined by equality (90). Thus we have proved that u is a solution of the problem (91) of the class $h^{\ell_1(\cdot),\ell_2(\cdot)}(K;\omega)$.

6.14 The basic result referring to the Dirichlet problem in the class $e^{p_1(\cdot),p_2(\cdot)}(D)$

Having a picture of solvability of the Dirichlet problem (91) in the ring K, we can, relying on corollary 20, get a picture of solvability of the Dirichlet problem in the class $e^{p_1(\cdot),p_2(\cdot)}(D)$. From the results of Sections 6.10–6.12 follows

Theorem 28. *Let*

(1) *the doubly-connected domain D be bounded by simple closed curves Γ_1 and Γ_2, where Γ_2 lies inside of Γ_1, while $\Gamma = \Gamma_1\cup\Gamma_2$ belongs to $C^1_D(t_1,\dots,t_n;\nu_1,\dots,\nu_n)$, $0 \le \nu_k \le 2$;*

(2) *$w = w(z)$ be conformal mapping of the domain D onto the ring $K = \{w : \rho < |w| < 1\}$; $z = z(w)$ be inverse mapping, $w(t_k) = a_k$, $z(\gamma_i) = \Gamma_i$, where $\gamma_1 = \{\tau : |\tau| = 1\}$, $\gamma_2 = \{\tau : |\tau| = \rho\}$;*

(3) *$p_i \in Q(\Gamma_i)$, and $\ell_i(\tau) = p_i(z(\tau))$, $i = 1,2$.*

Then for the Dirichlet problem

$$\begin{cases} \Delta\mathcal{U} = 0, & \mathcal{U} \in e^{p_1(\cdot),p_2(\cdot)}(D), \\ \mathcal{U}\big|_{\Gamma_i} = f_i, & f_i \in L^{p_i}(\Gamma_i), \end{cases}$$

the following statements are valid.

I. *The homogeneous problem is solvable and its general solution is given by the equality*

$$\mathcal{U}_0(z) = \sum_{k\in T_1\cup T_2} M_k(p)\,\mathrm{Re}\,\frac{w(z)+a_k}{w(z)-a_k},$$
$$T_i = \big\{k : \nu_k \ge p(t_k),\ t_k \in \Gamma_i\big\},\ \sum_{k\in T_2} M_k(p) = 0, \tag{106}$$

$$M_k(p) = \begin{cases} 0, \ \textit{if}\ 0 \le \nu_k < p(t_k),\ \textit{or}\ \nu = p(t_k) \\ \qquad \textit{for}\ a_k \in \gamma_i,\ X_k \overline{\in} H^{l_i(\cdot)}, \\ \qquad X_k(w) = (w-a_k)^{-\frac{1}{l_i(a_k)}} \exp\displaystyle\int_{\gamma_i} \frac{\psi_i(\tau)}{l_i(\tau)}\,\frac{d\tau}{\tau - w}, \\ M_k\ \textit{is an arbitrary constant for}\ p(t_k) < \nu_k \le 2\ \textit{or} \\ \qquad \nu_k = p(t_k),\ X_k \in H^{l_i(\cdot)}, \end{cases}$$

where ψ_i are the functions defined by means of z' appearing in (86).

II. *If among the points t_k there are such that $\nu_k = p(t_k)$ or $\nu_k = 0$, then the Dirichlet problem is, generally speaking, unsolvable.*

III. *The problem is solvable if and only if for the functions $g_i(\tau) = f_i(z(\tau))$, $\tau \in \gamma_i$, the following conditions are fulfilled:*

(a)

$$\frac{\omega_i^+(\zeta)}{\rho_i(\zeta)} \int\limits_{\gamma_i} \frac{\rho_i(\tau) g_i(\tau)}{\omega_i^+(\tau)} \frac{d\tau}{\tau - \zeta} \in L^{l_i(\cdot)}(\gamma_i), \tag{107}$$

where

$$\omega_i(w) = \prod_{a_k \in \gamma_i} (w - a_k)^{\frac{\nu_k - 1}{\ell_i(a_k)}} \exp \int\limits_{\gamma_i} \frac{\psi_i(\tau)}{l_i(\tau)} \frac{d\tau}{\tau - w},$$

$$\rho_i(\tau) = \prod_{k \in T_i'} (\tau - a_k),$$

where

$$T_i' = \{k : \nu_k = p(t_k) \text{ or } \nu_k = 0, \ t_k \in \Gamma_i\} \quad \text{and} \quad \rho_i(\tau) = 1 \quad \text{if} \quad T_i' = \emptyset.$$

(b)

$$\int\limits_0^{2\pi} g_1(e^{i\vartheta})\, d\vartheta = \int\limits_0^{2\pi} g_2(\rho e^{i\vartheta})\, d\vartheta. \tag{108}$$

In particular, if $0 < \nu_k < p(t_k)$, $k = \overline{1,n}$, then the condition (107) *is automatically fulfilled since the function $\omega_i^+ \in W^p(\Gamma_i)$. Hence the Dirichlet problem is solvable if and only if the condition* (108) *is satisfied.*

IV. *If the conditions* (107) *and* (108) *are fulfilled, then the Dirichlet problem is solvable, and its general solution is given by the equality*

$$\mathcal{U}(z) = \mathcal{U}_0(z) + u^*(w(z)),$$

where $\mathcal{U}_0(z)$ is given by equality (106)*, and $u^*(w) = u_1(w) + u_2(w)$, where $u_1(w)$ is the function given by equalities* (100) *and* (101)*, and*

$$u_2(w) = u_2(re^{i\vartheta}) = \int\limits_0^{2\pi} u_2(\rho e^{i\alpha}) \frac{r^2 - \rho^2}{r^2 + \rho^2 - 2r\rho\cos(\alpha - \vartheta)}\, d\alpha,$$

where $u_2(\rho e^{i\vartheta})$ is a solution of equation (103).

Remark 29. For $p(t) = \text{const}$, the work [40, Chapter 4] presents easily verifiable condition referring to the function $f(t)$ which guarantees the existence of a solution.

It consists in that the function

$$f(t)\prod_{\{\nu(t_k)=0 \text{ or } \nu(t_k)=p\}} \ln|w(t)-w(t_k)|$$

to belongs to $L^p(\Gamma)$.

Part IV

6.15 The Dirichlet problem with data from the grand Lebesgue spaces

In this section it is assumed that D is the internal domain bounded by a simple closed Lyapunov curve Γ. Put

$$K^{p),\theta}(D;\omega)=\left\{\Phi:\Phi(z)=\frac{1}{2\pi i}\int_\Gamma \frac{\varphi(t)\,dt}{t-z},\ \ z\in D,\ \ \varphi\in L^{p),\theta}(\Gamma;\omega)\right\},$$

$$e^{p),\theta}(D;\omega)=\left\{u:u=\operatorname{Re}\Phi,\ \ \Phi\in K^{p),\theta}(\Gamma;\omega)\right\}.$$

Assume that the weight function

$$\omega(t)=\prod_{k=1}^{n}(t-t_k)^{\alpha_k},\quad t_k\in\Gamma,\quad \alpha_k\in\left(\frac{1}{p},\frac{1}{p'}\right).$$

Here we will solve the Dirichlet problem formulated as follows

$$\begin{cases}\Delta\mathcal{U}=0, & \mathcal{U}\in e^{p),\theta}(D;\omega),\\ \mathcal{U}^+(t)=f(t), & f\in L^{p),\theta}(\Gamma;\omega).\end{cases}\quad p>1,\ \ \theta>0, \tag{109}$$

Let $z=z(w)$ denotes the conformal mapping of U onto the domain D, and $w=w(z)$ is its inverse mapping. Consider the function

$$\rho(\tau)=\prod_{k=1}^{n}|\tau-\tau_k|^{\alpha_k},\quad \tau_k=w(t_k). \tag{110}$$

Since Γ is a Lyapunov curve, by virtue of the Kellog's theorem (see [27, p.411]) we have

$$|z(\tau)-z(\tau_k)|\sim|\tau-\tau_k| \text{ and } \inf_{t\in\gamma} z'(\tau)>0,\ \ \sup_{\tau\in\gamma}|z'(\tau)|<\infty. \tag{111}$$

Hence it readily follows that

$$|\omega(z(\tau))|\sim|\rho(\tau)| \quad\text{and}\quad |\rho(w(t))|\sim|\omega(t)|. \tag{112}$$

Along with the function $\mathcal{U}(z)$, we will consider the harmonic in U function $u(w)=\mathcal{U}(z(w))$.

Lemma 30. *If* $\mathcal{U} \in e^{p),\theta}(D;\omega)$ *and* $u(w) = \mathcal{U}(z(w))$, *then* $u \in e^{p),\theta}(\mathrm{U};\rho)$. *Conversely, if* $u \in e^{p),\theta}(\mathrm{U},\rho)$, *then* $\mathcal{U}(z) = u(w(z)) \in e^{p),\theta}(D;\omega)$.

Proof. As $u(w) = \operatorname{Re}\Phi(z(w))$ it suffices to establish that if $\mathcal{U}(z) = \operatorname{Re}\Phi(z)$, $\Phi \in K^{p),\theta}(D;\omega)$, then the function $\Psi(w) = \Phi(z(w))$ belongs to the class $K^{p),\theta}(\mathrm{U};\rho)$. Since $\Phi \in K^{p),\theta}(D;\omega)$, we have $\Phi(z) = (K_\Gamma\varphi)(z)$, $\varphi \in L^{p),\theta}(\Gamma;\omega)$. But then $\varphi\omega \in \bigcap\limits_{0<\varepsilon<p} L^{p-\varepsilon}(\Gamma)$ and since $\frac{1}{\omega} \in L^{p'+\delta}(\Gamma)$, $\delta > 0$, there exists $\eta > 0$ such that $\varphi \in L^{1+\eta}(\Gamma)$. Therefore $\Phi \in E^{1+\eta}(D)$ (see [40, p. 29]). Thus $\Phi(z(w))\ \sqrt[1+\eta]{z'(w)}$ belongs to the class $H^{1+\eta}$. Furthermore, by (111) we conclude that $\Psi(w) = \Phi(z(w)) \in H^1$ and therefore $\Psi(w) = (K_\gamma\Psi^+)(w)$.

For the function Ψ^+ according (112) we have

$$\int\limits_\gamma |\Psi^+(\tau)\rho(\tau)|^{p-\varepsilon}|d\tau| = \int\limits_\Gamma \left|\Phi^+(z(\tau))\omega(z(\tau))\frac{\rho(\tau)}{\omega(z(\tau))}\right|^{p-\varepsilon}|d\tau|$$
$$\leq M\int\limits_\gamma |\Phi^+(z(\tau))\omega(z(\tau))|^{p-\varepsilon}|d\tau| = M\int\limits_\Gamma |\Phi^+(t)\omega(t)|^{p-\varepsilon}\frac{|dt|}{|z'(t)|}$$
$$\leq M_1\int\limits_\Gamma |\Phi^+(t)\omega(t)|^{p-\varepsilon}ds.$$

Since a Cauchy singular integral operator is continuous in the space $L^{p),\theta}(\Gamma;\omega)$, the inclusion $K_\Gamma\varphi \in K^{p),\theta}(D;\omega)$ implies that $\Phi^+ \in L^{p),\theta}(\Gamma;\omega)$. Now from the above estimates we establish that $\Psi^+ \in L^{p),\theta}(\gamma;\rho)$, whereas from the equality $\Psi(w) = (K_\gamma\Psi^+)(w)$ we obtain $\Psi \in K^{p),\theta}(\mathrm{U};\rho)$.

The converse statement is proved analogously. □

By virtue of this lemma we have that problem (109) is equivalent to the problem

$$\begin{cases} \Delta u = 0, \quad u \in e^{p),\theta}(\mathrm{U};\rho), \\ u^+(t) = f(z(\tau)), \quad f(z(\tau)) = g(\tau) \in L^{p),\theta}(\gamma;\rho). \end{cases} \tag{113}$$

As easily seen, $g \in L^{1+\eta}(\gamma)$, for some $\eta > 0$, whereas from the proof of the lemma 30 it follows that $e^{p),\theta}(\mathrm{U};\rho) \subset \mathrm{h}^{1+\eta}$. Therefore if the solution $u(w)$ of problem (113) exists, it belongs to $h^{1+\eta}$, too. But the function

$$u(w) = \operatorname{Re}\frac{1}{2\pi}\int\limits_\gamma \frac{f(z(\tau))}{\tau}\,\frac{\tau+w}{\tau-w}\,d\tau \tag{114}$$

is the only solution of such kind (see Chapter 1, Statement 24). Furthermore,

$$\Psi(w) = \frac{1}{2\pi}\int\limits_\gamma \frac{f(z(\tau))}{\tau}\,\frac{\tau+w}{\tau-w}\,d\tau = \frac{1}{2\pi}\int\limits_\gamma \frac{f(z(\tau))}{\tau-w}\,d\tau + \frac{w}{2\pi}\int\limits_\gamma \frac{f(z(\tau))}{\tau}\,\frac{d\tau}{\tau-w}.$$

The function $\Psi(w)$ belongs to $K^{p),\theta}(\gamma;\rho)$ since both summands belong to the Hardy class H^1, and, by the continuity of the operator S_γ in the space $L^{p),\theta}(\gamma;\rho)$, their boundary values belong to the class $L^{p),\theta}(\gamma;\rho)$. Hence it follows that the function defined by equality (114) belongs to $e^{p),\theta}(\mathrm{U};\rho)$.

Thus we have proved

Theorem 31. *If D is the internal domain bounded by a closed Lyapunov curve, then the Dirichlet problem* (109) *is uniquely solvable and its solution is given by the equality*

$$\mathcal{U}(z) = \mathrm{Re}\,\frac{1}{2\pi}\int\limits_{\gamma} \frac{f(z(\tau))}{\tau}\,\frac{\tau + w(z)}{\tau - w(z)}\,d\tau,$$

where $w = w(z)$ is the conformal mapping of D onto U.

6.16 Notes and comments on Chapter 6

The Dirichlet problem for harmonic functions being the real parts of Cauchy type integrals with densities from classical weighted Lebesgue spaces has been studied in [42], [39], [40]. The similar problem in the frame of variable exponent analysis was explored in [56], [59], and [60].

The Dirichlet problem for real parts of weighted variable exponents Smirnov classes both in simply and doubly-connected domains with piecewise smooth boundaries is solved in the papers [61], [41].

The material of the Chapter 6 is taken from the latter two papers.

Chapter 7

The Riemann–Hilbert–Poincaré Problem

In many boundary value problems of function theory the boundary conditions contain not only the sought function, but also its derivatives up to certain order. Therefore it is useful to have formulas giving an integral representation of such analytic function. One form of such representations, quite convenient for applications, was proposed by I. Vekua ([128], [129]). N. Muskhelishvili expounded them in his book, where they are called I. Vekua's integral representations (see [98, pp. 224–232]). I. Vekua's result reads as follows:

Theorem (I. Vekua). *Let D^+ be a finite domain bounded by a simple closed Lyapunov curve Γ and $\Phi(z)$ be a analytic function in D^+, whose derivative of order m is continuous in $\overline{D^+}$ and the boundary belong to the Hölder's class $H(\Gamma)$. Then, assuming that the origin is in D^+, the function Φ is representable for $m = 0$ as*

$$\Phi(z) = \int\limits_{\Gamma} \frac{\varphi(t)\, dt}{1 - \frac{z}{t}} + id \tag{1}$$

and for $m \geq 1$ as

$$\Phi(z) = \int\limits_{\Gamma} \varphi(t) \left(1 - \frac{z}{t}\right)^{m-1} \ln\left(1 - \frac{z}{t}\right) ds + \int\limits_{\Gamma} \varphi(t)\, ds + id, \tag{2}$$

where $\varphi(t)$ is a real function from the class $H(\Gamma)$, and d is a real constant; $\varphi(t)$ and d are defined uniquely with respect to $\Phi(z)$.

Subsequently, B. Khvedelidze [42] gave a generalization of this theorem to the case where a derivative of order m of the function $\Phi(z)$ is representable in D^+ by a Cauchy type integral with a density from $L^p(\Gamma; \omega)$, where $p > 1$ and

$$\omega(t) = \prod_{k=1}^{n} |t - t_k|^{\alpha_k}, \ \ t_k \in \Gamma, \ \ -\frac{1}{p} < \alpha_k < \frac{1}{p'}, \ \ p' = \frac{p}{p-1}.$$

In that case, φ belongs to $L^p(\Gamma;\omega)$. In [3] was generalized this result to the case $\omega(t)=1$ for a wide subclass of piecewise smooth curves.

I. Vekua used these representations for investigating quite a general boundary value problem, namely, the Riemann–Hilbert–Poincaré problem. An analogous investigation was carried out in [42] and [3].

In this Chapter, we consider the question whether integrals of form (1)–(2) can be used to present analytic functions whose m-th derivative is representable by a Cauchy type integral with a density from $L^{p(\cdot)}(\Gamma;\omega)$, where Γ is a piecewise smooth curve. Relations are found between the values of the function $p(t)$ at the angular points of the curve Γ, the values of these angles and the power exponents of weight functions, which provide the required representation. This has been achieved thanks to the well studied Riemann–Hilbert problem in the considered classes. For this we had to extend the results obtained in Chapter 5 for bounded domains to the case of unbounded domains. Also note that in the mentioned works only the existence of φ and d such that (1), (2) holds is proved. Our approach enables us to indicate a more concrete construction of the function φ by means of the prescribed function $\Phi(z)$.

The representations obtained are used in investigating the Riemann–Hilbert–Poincaré problem under more general assumptions than in mentioned works.

The statements we make in connection with this problem are seemingly analogous to the above-mentioned results, but, in fact, the relations between the values essentially depend on the geometry of the boundary Γ, multipliers $|t-t_k|^{\alpha_k}$ from a weight function $\omega(t)$, discontinuity points of the principal coefficient from the boundary condition, as well as on the values of a function $p(t)$ at the angular points of Γ and at points t_k.

Some definitions and auxiliary statements

If D^+ and D^- are simply connected domains with the common boundary Γ and Γ is a curve of the class $C^1_{D^+}(A_1,\dots,A_i;\nu_1,\dots,\nu_i)$, then Γ is a curve of the class $C^1_{D^-}(A_1,\dots,A_i;\lambda_1,\dots,\lambda_i)$, where $\nu_k+\lambda_k=2$, $k=\overline{1,i}$.

We assume

$$\omega(t)=\prod_{k=1}^{n}|t-t_k|^{\alpha_k},\qquad -\frac{1}{p(t_k)}<\alpha_k<\frac{1}{p'(t_k)}. \tag{3}$$

Definition 1. If $m\ge 0$ is an integer number, then we denote by $K^{p(\cdot)}_{D,m}(\Gamma;\omega)$ the set of functions Φ analytic in D for which $\Phi^{(m)}(z)\in K^{p(\cdot)}(\Gamma;\omega)$, $(=K^{p(\cdot)}(D;\omega))$. It is assumed that $\Phi^{(0)}(z)=\Phi(z)$ and thus $K^{p(\cdot)}_{D,0}(\Gamma;\omega)=K^{p(\cdot)}(\Gamma;\omega)$, (see Sections 5.1–5.2).

Some time we write $K^{p(\cdot)}_m(\Gamma;\omega)$ instead of $K^{p(\cdot)}_{D,m}(\Gamma;\omega)$.

The problem of representation of analytic functions in terms of new assumptions reduces to the investigation of the Riemann–Hilbert problem in the class $K^{p(\cdot)}_{D^-}(\Gamma;\omega)$. In Chapter 5, this problem is solved in the class $K^{p(\cdot)}_{D}(\Gamma;\omega)$ when D is a bounded domain with the boundary Γ. The case of an unbounded domain can be easily investigated by reducing it to the considered one.

Let: I. Γ be a curve of the class $C^1_{D^+}(A_1,\dots,A_i;\nu_1,\dots,\nu_i)$, $0<\nu_k\le 2$, $k=\overline{1,i}$, $p\in\widetilde{\mathcal{P}}(\Gamma)$; a, b, c be real functions, a and b being piecewise continuous and $c\in L^{p(\cdot)}(\Gamma;\omega)$, where ω is the weight function given by equality (3) or II. Γ be a curve of the class $C^{1,L}_{D^+}(A_1,\dots,A_i;\nu_1,\dots,\nu_i)$, $0<\nu_k\le 2$, $k=\overline{1,i}$, $p\in\mathcal{P}(\Gamma)$; a and b be piecewise Hölder functions on Γ and $c\in L^{p(\cdot)}(\Gamma;\omega)$.

Let us consider the Riemann–Hilbert problem – Find a function $\Phi\in K^{p(\cdot)}_{D^+}(\Gamma;\omega)$ for which the equality

$$\operatorname{Re}\big[(a(t)+ib(t))\Phi^+(t)\big]=c(t) \tag{4}$$

is valid a.e. on Γ.

We denote by B_k, $k=\overline{1,\lambda}$, the discontinuity points of the function $G(t)=-[a(t)-ib(t)][a(t)+ib(t)]^{-1}$ and assume that $\inf|G(t)|>0$ and $G(B_k^-)/G(B_k^+)=\exp(2\pi i u_k)$.

Let $z=z(w)$ be the conformal mapping of the circle U on D^+ and $w=w(z)$ be the inverse mapping. We assume

$$T=\{\tau_k:\ \tau_k=w(t_k)\},\quad A=\{a_k:\ a_k=w(A_k)\},\quad B=\{b_k:\ b_k=w(B_k)\}$$

and let the numbers ω_k, δ_k are defined by formulas (72) and (73) from Chapter 5.

Assume that

$$\{\delta_k\}\ne\frac{1}{\ell'(w_k)},\quad k=\overline{1,j},\quad j=\mu+r+q+p+M+n+s, \tag{5}$$

and let

$$\gamma_k=\begin{cases}[\delta_k] & \text{if } \{\delta_k\}<\frac{1}{\ell'(w_k)},\\ [\delta_k]+1 & \text{if } \{\delta_k\}>\frac{1}{\ell'(w_k)},\end{cases}\qquad Q(w)=\prod_{k=1}^{j}(w-w_k)^{\gamma_k}. \tag{6}$$

Furthermore, assuming that the point $z=0$ belongs to D^+, we write (see [98, p. 147])

$$r_k(w)=\begin{cases}(w-b_k)^{u_k}, & |w|<1,\\ \left(\frac{1}{\overline{w}}-b_k\right), & |w|>1,\end{cases}\qquad r(w)=\prod_{k=1}^{\lambda}r_k(w), \tag{7}$$

$$R_k(\tau)=r_k^+(\tau)[r_k^-(\tau)]^{-1}=\exp(2i\arg(\tau-b_k)), \tag{8}$$

$$G_1(\tau)=G(z(\tau))\prod_{k=1}^{\lambda}R_k(\tau),\quad \varkappa_1=\operatorname{ind}G_1(\tau),\quad \widetilde{G}_1(\tau)=G_1(\tau)\tau^{-\varkappa_1}, \tag{9}$$

$$X_1(w) = \begin{cases} C\exp\left\{\frac{1}{2\pi i}\int_\gamma \frac{\ln \widetilde{G}_1(\tau)d\tau}{\tau - w}\right\}, & |w| < 1, \\ Cw^{-\varkappa_1}\exp\left\{\frac{1}{2\pi i}\int_\gamma \frac{\ln \widetilde{G}_1(\tau)d\tau}{\tau - w}\right\}, & |w| > 1, \end{cases} \tag{10}$$

$$X(w) = X_1(w)r(w), \quad C = \exp\left(-\frac{i\alpha}{2}\right), \quad \alpha = \frac{1}{2\pi}\int_0^{2\pi} \arg\left[-\tau^{-\varkappa_1}G_1(\tau)\right] d\tau. \tag{11}$$

We denote by $\varkappa_0$ the order of the rational function $Q(z)$ at infinity and assume that $\varkappa = \varkappa(\mu, G, \omega, D^+) = \varkappa_0 + \varkappa_1$.

Then, by assumptions I (or II) and (5) Theorem 27 (16) from Chapter 5 is valid.

Remark 2. If the point $z = 0$ lies in D^+ and we search for a solution $\Phi(z)$ of problem (4) with the condition $\Phi(0) = 0$, then, as readily follows by the reasoning of Chapter 5 the Theorem 16 remains valid when $\varkappa$ in it is replaced by $\varkappa - 1$. Thus for $\varkappa < 0$ the solvability conditions have the form

$$\int_\Gamma \frac{c(t)Q(w(t))w^k(t)w'(t)}{X^+(w(t))[a(t) + ib(t)]}\,dt = 0, \quad k = \overline{0, |\varkappa| - 1}, \tag{12}$$

and their fulfillment implies that there exists a unique solution $\Phi(z) = \widetilde{\Omega}(w(z))$.

When $\varkappa \geq 0$, the problem is solvable and all solutions are given by the equality

$$\Phi(z) = \widetilde{\Omega}(w(z)) + X(w(z))[Q(w(z))]^{-1}P_{\varkappa-1}(w(z)). \tag{13}$$

Remark 3. The investigation of problem (4) reduces to that of the Riemann problem

$$\Psi^+(\tau) = G(\tau)\Psi^-(\tau) + \frac{2c(z(\tau))}{X^+(\tau)}, \quad G(\tau) = -\frac{\widetilde{a}(\tau) - i\widetilde{b}(\tau)}{\widetilde{a}(\tau) + i\widetilde{b}(\tau)},$$
$$\widetilde{a}(\tau) = a(z(\tau)), \ \widetilde{b}(\tau) = b(z(\tau)), \ \tau \in \gamma, \tag{14}$$

in the class $\widetilde{K}_0^{\ell(\tau)}(\gamma;\omega)$, $\ell(\tau) = p(z(\tau))$ with the additional condition $\Psi(w) = \Psi_*(w)$, where $\Psi_*(w) = \overline{\Psi\left(\frac{1}{\overline{w}}\right)}$, $|w| \neq 1$.

The function $Y(w) = X(w)Q(w)$ constructed in Chapter 5 is a canonical function for $G(\tau)$. The order $\varkappa$ of the function $Y(w)$ at infinity is the index of the Riemann problem (14) (or, which is the same, the index of the singular operator $K\psi = \psi - \frac{1}{2}(G-1)(-\psi + S_\gamma\psi)$ in $L^{p(\cdot)}(\gamma;\omega)$). Problem (4) is solvable if and only if problem (14), too, is solvable. Note that when $\varkappa > 0$, the homogeneous problem (14) has a general solution depending on $\varkappa$ arbitrary complex constants, while the homogeneous problem (4) has a general solution depending on such the polinom $P(z) = h_0 + h_1 z + \cdots + h_\varkappa z^\varkappa$ for which

$$\overline{h_k} = Ah_{\varkappa-k}, \ k = \overline{0,\varkappa}, \ \ A = (-1)^{\varkappa_0}\prod_{k=1}^{j} w_k.$$

7.1 The Riemann-Hilbert Problem in Unbounded Domain

Let Γ be a curve of the class $C^1_{D^-}(A_1,\dots,A_i;\nu_1,\dots,\nu_i)$, $p\in\widetilde{\mathcal{P}}(\Gamma)$ and the Riemann–Hilbert problem be considered in the class $K^{p(\cdot)}_{D^-}(\Gamma;\omega)$. This time the boundary condition is written in the form

$$\operatorname{Re}\left[(a(t)+ib(t))\Phi^-(t)\right]=c(t). \tag{15}$$

We assume that the point $z=0$ does not lie in $\overline{D}^-$. By transforming $\zeta=\frac{1}{z}$ the domain D^- becomes a finite domain which we denote by D_1^+. The function $\Phi(z)$ transforms to the function $F(\zeta)=\Phi(\frac{1}{\zeta})$. Also $F(0)=0$. It is obvious that the boundary Γ_1 of the domain D_1^+ is a curve of the class $C^1_{D_1^+}(\frac{1}{A_1},\dots,\frac{1}{A_i};\lambda_1,\dots,\lambda_i)$, $\lambda_k=2-\nu_k$, $k=\overline{1,i}$.

It is easy to verify that $p^1(\eta)=p(\frac{1}{\eta})$ belongs to $\widetilde{\mathcal{P}}(\Gamma_1)$. The weight function $\omega(t)$ transforms to the function

$$\omega^1(\eta)=\prod_{k=1}^{n}\left|\frac{1}{\eta}-\frac{1}{\eta_k}\right|^{\alpha_k}=\prod_{k=1}^{n}\frac{1}{|\eta\eta_k|^{\alpha_k}}\prod_{k=1}^{n}|\eta-\eta_k|^{\alpha_k}\sim\prod_{k=1}^{n}|\eta-\eta_k|^{\alpha_k},\ \ \eta_k=\frac{1}{t_k}\in\Gamma_1\,,$$

and therefore conditions (3) are fulfilled for $\omega^1(\eta)$.

The boundary condition (15) transforms to the condition

$$\begin{gathered}\operatorname{Re}\left[F^+(\eta)(a^1(\eta)+ib^1(\eta))\right]=c_1(\eta),\\ a^1(\eta)=a\left(\frac{1}{\eta}\right),\quad b^1(\eta)=b\left(\frac{1}{\eta}\right),\quad c^1(\eta)=c\left(\frac{1}{\eta}\right).\end{gathered}\tag{16}$$

It can be easily verified that $F\in K^{p^1(\cdot)}_{D_1^+}(\Gamma_1;\omega^1)$. Thus we should solve problem (16) in the class $K^{p^1(\cdot)}_{D_1^+}(\Gamma_1;\omega^1)$ using the additional condition $F(0)=0$. If we require that $0<\lambda_k\le 2$ (which is equivalent to the requirement $0\le\nu_k<2$, $k=\overline{1,i}$), then all the conditions required of Γ_1, p^1, ω^1 in Section 5.5 will be fulfilled. If $F(\zeta)$ is a solution of this problem, then $\Phi(z)=F(\frac{1}{z})$ will be a solution of problem (15) of the class $K^{p(\cdot)}_{D^-}(\Gamma;\omega)$.

Thus we are to find a solution $F(\zeta)$ of problem (16) of the class $K^{p^1(\cdot)}_{D_1^+}(\Gamma_1;\omega^1)$, where Γ_1 is a curve of the class $C^1_{D_1^+}(\frac{1}{A_1},\dots,\frac{1}{A_i};\lambda_1,\dots,\lambda_i)$, $\lambda_k=2-\nu_k$, $k=\overline{1,i}$, with the condition $F(0)=0$.

Let $\zeta=\zeta(w)$ be a conformal mapping of the circle U onto D_1^+ with the condition $\zeta(0)=0$, and $w=\psi(\zeta)$ be its inverse mapping. Besides, we denote

$$G^1(\eta)=-\left[a^1(\eta)-ib^1(\eta)\right]\left[a^1(\eta)+ib^1(\eta)\right]^{-1},\quad \ell^1(\tau)=p^1(\zeta(\tau))=p\left(\frac{1}{\zeta(\tau)}\right).$$

For the domain D_1^+ we define the numbers ω_k^1, δ_k^1 and γ_k^1 by analogy with Section 5.4 (see (17), (19), (20)). The numbers δ_k^1 are obtained from formulas (19) of Chapter 5 in which ν_k is replaced by λ_k (i.e., by $2-\nu_k$), while when defining the numbers γ_k^1, $\ell(\tau)$ is understood as $p(\frac{1}{\zeta(\tau)})$. Define the function $X^1(w)$ and $Q^1(w)$ in accordance with the points w_k^1. Assume $\varkappa^- = \varkappa_0^1 + \varkappa_1^1$. Now, applying the Remark 2, we obtain

Proposition 4. *If Γ is a curve of the class $C_{D^-}^1(A_1,\dots,A_i;\nu_1,\dots,\nu_i)$, $0\le \nu_k<2$, $k=\overline{1,i}$, and problem (15) is considered in the class $K_{D^-}^{p(\cdot)}(\Gamma;\omega)$, then, assuming that $\varkappa^-<0$, for the problem to be solvable it is necessary and sufficient that the conditions*

$$\int\limits_{\Gamma_1} \frac{c^1(\eta)Q^1(\psi(\eta))\psi^k(\eta)\psi'(\eta)}{X^1(\psi(\eta))(a^1(\eta)+ib^1(\eta))}\,d\eta = 0, \quad k=\overline{0,|\varkappa^-|-1}, \tag{17}$$

is fulfilled, and when they are fulfilled, the problem has a unique solution

$$\Phi(z) = \widetilde{\Omega}\left(\psi\left(\frac{1}{z}\right)\right) = \frac{1}{2}\left(\Omega\left(\psi\left(\frac{1}{z}\right)\right) + \Omega_*\left(\psi\left(\frac{1}{z}\right)\right)\right), \tag{18}$$

where

$$\Omega(w) = X_1^1(w)[Q^1(w)]^{-1}\frac{1}{2\pi i}\int\limits_{\gamma} \frac{2c^1(\zeta(\tau))Q^1(\tau)}{X^1(\tau)[a^1(\zeta(\tau))+ib^1(\zeta(\tau))]}\,d\tau, \quad |w|\ne 1.$$

When $\varkappa^-\ge 0$, the problem is unconditionally solvable and all solutions are given by the equality

$$\Phi(z) = \widetilde{\Omega}\left(\psi\left(\frac{1}{z}\right)\right) + X^1\left(\psi\left(\frac{1}{z}\right)\right)\left[Q^1\left(\psi\left(\frac{1}{z}\right)\right)\right]^{-1} P_{\varkappa^- -1}\left(\psi\left(\frac{1}{z}\right)\right),$$

where for the polynomial $P_{\varkappa^- -1}(w) = h_0 + h_1 w + \cdots + h_{\varkappa^- -1}w^{\varkappa^- -1}$ we have

$$\overline{h}_k = A h_{\varkappa^- -k-1}, \quad k=\overline{0,\varkappa^- -1}, \quad A=(-1)^{\varkappa_0}\prod_{k=1}^{j} w_k^1. \tag{19}$$

If we reject the condition $\Phi(\infty)=0$, then, assuming that $\varkappa^- < -1$, for problem (15) to be solvable in the class $\widetilde{K}_0^{p(\cdot)}(\Gamma;\omega)$, where $\widetilde{K}_0^{p(\cdot)}(\Gamma;\omega) = \{\Phi : \Phi = \Phi_0 + const,\ \Phi_0 \in K_{D^-}^{p(\cdot)}(\Gamma;\omega)\}$, it is necessary and sufficient that conditions (17) are fulfilled, where $k=\overline{0,|\varkappa^-|-2}$, and when they are fulfilled, there exists a unique solution given by equality (18).

When $\varkappa^- \ge -1$, the problem is unconditionally solvable and the solution is given by the equality

$$\Phi(z) = \widetilde{\Omega}\left(\psi\left(\frac{1}{z}\right)\right) + X^1\left(\psi\left(\frac{1}{z}\right)\right)\left[Q^1\left(\psi\left(\frac{1}{z}\right)\right)\right]^{-1} P_{\varkappa^-}\left(\psi\left(\frac{1}{z}\right)\right),$$

where, we recall, $\psi=\psi(\zeta)$ is a conformal mapping of D_1^+ onto the unit circle U; $P_{-1}(w)=0$, while when $\varkappa^-\ge 0$, conditions (19) are fulfilled for $P_{\varkappa^-}(w)$.

7.2 I. Vekua's integral representation of analytic functions from the class $K^{p(\cdot)}_{D^+,m}(\Gamma;\omega)$

Theorem 5. *Let*

i) Γ *be a curve of the class* $C^1_{D^+}(A_1,\dots,A_i;\nu_1,\dots,\nu_i)$, $0\le\nu_k<2$, $k=\overline{1,i}$ *and* $p\in\widetilde{\mathcal{P}}(\Gamma)$ *or* $\Gamma\in C^{1,L}_{D^+}(A_1,\dots,A_i;\nu_1,\dots,\nu_i)$, $0\le\nu_k<2$, $k=\overline{1,i}$ *and* $p\in\mathcal{P}(\Gamma)$;

ii) ω *be a power function of form* (3);

iii) *the point* $z=0$ *lie in* D^+; $z=z(w)$ *be a conformal mapping of the circle* U *onto the domain* $D^-=C\overline{D^+}$; $z(0)=\infty$ *and* $w=w(z)$ *be its inverse mapping. Let* $a_k=w(A_k)$, $k=\overline{1,i}$, $\tau_k=w(t_k)$, $k=\overline{1,n}$, *and the points of the set* $\{a_1,\dots,a_i\}\cup\{\tau_1,\dots,\tau_n\}$ *be numbered so that*

$$\begin{aligned}&w_1=a_1=\tau_1,\dots,w_\mu=a_\mu=\tau_\mu,\\&w_{\mu+1}=a_{\mu+1},\dots,w_{\mu+p}=a_{\mu+p},\\&w_{\mu+p+1}=\tau_{\mu+1},\dots,w_{\mu+p+M}=\tau_{\mu+M}\end{aligned}\tag{20}$$

and

$$\delta_k=\begin{cases}\alpha_k\lambda_k+\frac{\nu_k-1}{\ell(w_k)}+\nu_k-1, & k=\overline{1,\mu},\ \ \lambda_k=2-\nu_k,\\ \frac{\nu_k-1}{\ell(w_k)}+\nu_k-1, & k=\overline{\mu+1,\mu+p},\\ \alpha_{k-p}, & k=\overline{\mu+p+1,\mu+p+M},\end{cases}\tag{21}$$

where $\ell(\tau)=p(z(\tau))$, $|\tau|=1$;

iv) $\Phi\in K^{p(\cdot)}_{D^+,m}(\Gamma;\omega)$.

If

$$\{\delta_k\}\ne\frac{1}{\ell'(w_k)},\quad k=\overline{1,j},\ \ j=n+i-\mu=\mu+p+M,\tag{22}$$

then there exist a real function $\varphi\in L^{p(\cdot)}(\Gamma;\omega)$ *and a real constant* d *such that*

$$\Phi(z)=\int_\Gamma\frac{\varphi(t)\,ds}{1-\frac{z}{t}}+id\tag{23}$$

for $m=0$,

$$\Phi(z)=\int_\Gamma\varphi(t)\left(1-\frac{z}{t}\right)^{m-1}\ln\left(1-\frac{z}{t}\right)ds+\int_\Gamma\varphi(t)\,ds+id\tag{24}$$

for $m\ge1$.

The function φ *and the constant* d *are defined in a unique manner.*

Before proceeding to the proof, we would like to note the particular cases of the theorem. Representations (23) and (24) are valid in the following cases:

I. a) Γ is a smooth curve and $p\in\widetilde{\mathcal{P}}(\Gamma)$ or Γ is a Lyapunov curve and $p\in\mathcal{P}(\Gamma)$;

b) ω is a weight function of form (3).

II. a) Γ is a curve of the class $C^1(A_1,\dots,A_i;\nu_1,\dots,\nu_i)$ and $p\in\widetilde{\mathcal{P}}(\Gamma)$ or Γ is a curve of the class $C^{1,L}(A_1,\dots,A_i;\nu_1,\dots,\nu_i)$ and $p\in\mathcal{P}(\Gamma)$ (in both cases $0<\nu_k<2$, $k=\overline{1,i}$); b) $\omega(t)=1$; c) $\left\{\frac{\nu_k-1}{p(A_k)}+\nu_k-1\right\}\neq\frac{1}{p'(A_k)}$.

Proof of the theorem. Since (23), (24) imply that $d=\operatorname{Im}\Phi(0)$, we can assume that $d=0$. Let us prove the validity of equalities (23) and (24). Having these representations for $d=0$, from the equality $\Phi(z)=[\Phi(z)-i\operatorname{Im}\Phi(0)]+i\operatorname{Im}\Phi(0)$ we will obtain the representations in the general case with $d=\operatorname{Im}\Phi(0)$.

Let $m=0$ and representation (23) hold. If $t=t(s)$ is the equation of the curve Γ with respect to the arc abscissa, then $ds=(t')^{-1}dt=\overline{t'}dt$ (see [98, p. 225]). Therefore (23) can be rewritten in the form

$$\Phi(z)=\frac{1}{2\pi i}\int\limits_{\Gamma}\frac{2\pi i t\,\overline{t'}\,\varphi(t)}{t-z}\,dt. \tag{25}$$

Since $\Phi\in K^{p(\cdot)}_{D^+}(\Gamma;\omega)$ and $\omega\in W^{p(\cdot)}(\Gamma)$, we have $\Phi\in E^1(D^+)$ (Chapter 4 Theorem 2). Therefore, $\Phi(z)=(K_\Gamma\Phi^+)(z)$, $\Phi^+\in L^{p(\cdot)}(\Gamma;\omega)$. Combined with (25), this implies that

$$\frac{1}{2\pi i}\int\limits_{\Gamma}\frac{\Phi^+(t)-2\pi i t\,\overline{t'}\varphi(t)}{t-z}\,dt=\begin{cases}0, & z\in D^+,\\ N(z), & z\in D^-.\end{cases}$$

Since Γ is a piecewise smooth curve, and Φ^+ and φ belong to $L^{p(\cdot)}(\Gamma;\omega)$, we have $N(z)\in E^1(D^-)\cap K^{p(\cdot)}_{D^-}(\Gamma;\omega)$. Therefore from the latter equality it follows that

$$\Phi^+(t)-2\pi i t\,\overline{t'}\varphi(t)=N^-(t). \tag{26}$$

Let us divide both parts of equality (26) by $t\,\overline{t'}$ and separate the real parts. We obtain

$$\operatorname{Re}\frac{N^-(t)}{t\,\overline{t'}}=\operatorname{Re}\frac{\Phi^+(t)}{t\,\overline{t'}},\quad t\in\Gamma. \tag{27}$$

We have come to the Riemann–Hilbert problem with regard to the function N which we must solve in the class $K^{p(\cdot)}_{D^-}(\Gamma;\omega)$.

For $m\geq 1$, from representation (24) we obtain

$$\Phi^{(m)}(z)=(-1)^m(m-1)!\int\limits_{\Gamma}\frac{\varphi(t)\,\overline{t'}\,dt}{t^{m-1}(t-z)}. \tag{28}$$

By the assumption of the theorem, $\Phi^{(m)}\in K^{p(\cdot)}_{D^+}(\Gamma;\omega)$. This function, as mentioned above, belongs to $E^1(D^+)$. By this fact and (28) we obtain

$$\frac{1}{2\pi i}\int\limits_{\Gamma}\frac{\Phi^{(m)}(t)-2\pi i(-1)^m(m-1)!t^{1-m}\,\overline{t'}\varphi(t)}{t-z}\,dt=\begin{cases}0, & z\in D^+,\\ N(z), & z\in D^-,\end{cases} \tag{29}$$

where $N \in K^{p(\cdot)}_{D^-}(\Gamma;\omega)$, $\Phi^{(m)}(t) = \left[\Phi^{(m)}\right]^+(t)$. From (28) we obtain

$$\Phi^{(m)}(t) - 2\pi i(-1)^m(m-1)!t^{1-m}\overline{t'}\varphi(t) = N^-(t). \tag{30}$$

Therefore for $m \geq 1$ we have

$$\operatorname{Re}\frac{N^-(t)}{t^{1-m}\,\overline{t'}} = \operatorname{Re}\frac{\Phi^{(m)}(t)}{t^{1-m}\,\overline{t'}}. \tag{31}$$

For $m = 0$ equality (31) coincides with (27).

If we solve problem (31) in the class $K^{p(\cdot)}_{D^-}(\Gamma;\omega)$, then, using equalities (26) and (30), we can define the function φ.

We will first show that representations (23) and (24) are unique.

Both cases are treated in the same manner.

Let, for instance,

$$0 = \int_\Gamma \frac{\varphi(t)}{1-\frac{z}{t}}\,ds + id, \quad z \in D^+.$$

For $z = 0$ we have $\int_\Gamma \varphi(t)\,ds + id = 0$. Hence it follows that $d = 0$. Let us prove that $\varphi(t) = 0$.

Expanding the integral $\int_\Gamma \frac{\varphi(t)}{1-\frac{z}{t}}\,ds$ into a series with respect to z in the neighborhood of zero, we obtain

$$\int_\Gamma \varphi(t)t^{-k}\,ds = 0, \quad k = 0, 1, \dots,$$

i.e.

$$\int_\Gamma \varphi(t)t^k\,\overline{t'}\,dt = 0, \quad k = 0, -1, -2\dots.$$

By the Golubev–Privalov theorem, for the external domain [108, p. 202] we obtain $\varphi(t)\overline{t'} = m^-(t)$, where $m(z) \in E^1(D^-)$. Further, absolutely in the same manner as in [3] we establish that $\lim\limits_{z\hat{\to}t,\, z\in D^-} \operatorname{Im} m_0(z) = 0$, where $m_0(z) = \int_{z_0}^{z} m^-(z)\,dz$, $z_0 \in D^-$, and therefore $m_0(z) = const$, i.e., $m^-(z) = m_0'(z) = 0$. Hence we conclude that $\varphi(t) = 0$.

Let us apply the results from of the preceding sections to problem (31).

In that case, $\Gamma_1 \in C^1_{D_1^+}\left(\frac{1}{A_1},\dots,\frac{1}{A_i};2-\nu_1,\dots,2-\nu_i\right)$, $a(t)+ib(t) = \frac{1}{t^{1-m}\overline{t'}}$, $c(t) = \operatorname{Re}\frac{\Phi^{(m)}(t)}{t^{1-m}\overline{t'}}$. Therefore

$$G(t) = -\frac{a(t)-ib(t)}{a(t)+ib(t)} = -\frac{|a(t)+ib(t)|^2}{(a(t)+ib(t))^2} = -\frac{(t^{1-m}\overline{t'})^2}{|(t^{1-m}\overline{t'})^2|}.$$

This function possesses discontinuity at the points A_k, $k = \overline{1,i}$. But then the function $G^1(\eta) = G\left(\frac{1}{\eta}\right)$ is discontinuous at the points $B_k^1 = \frac{1}{A_k} \in \Gamma_1$. It is not difficult to verify that $G^1(B_k^1-)/G^1(B_k^1+) = \exp[-2i(1-\nu_k)\pi] = \exp[2i(\nu_k - 1)\pi]$. Therefore $u_k = u_k^1 = \nu_k - 1$.

Furthermore, all discontinuity points of $G(t)$ coincide with the angular points of Γ_1. By assumption, on Γ there exist μ angular points coinciding with some points t_k from the weight ω; accordingly, on Γ_1 there exist μ angular points coinciding with some points $\tau_k = \frac{1}{t_k}$ from the weight ω^1. Furthermore, there are $p = i - \mu$ angular points not appearing among the singular points of the weight ω. Finally, in the weight ω (in the weight ω^1) there are M multipliers of the form $(t-t_k)^{\alpha_k}$ $((\tau - \tau_k)^{\alpha_k})$ in which the points t_k (τ_k) do not coincide with angular points. Let us assume that $\zeta = \zeta(w)$ is a conformal mapping of U onto D_1^+, $\zeta(0) = 0$ and $w = \psi(\zeta)$ is its inverse mapping. The function $z = \frac{1}{\zeta(w)} = z(w)$ gives the conformal mapping of U onto D^-; let $w = w(z)$ $(= \psi(\frac{1}{z}))$ be the inverse mapping.

We have

$$a_k = \zeta\left(\frac{1}{A_k}\right) = w(A_k), \quad \tau_k = \zeta\left(\frac{1}{t_k}\right) = w(t_k).$$

These points have been numbered according to (20). Then equalities (19) from Chapter 5 take form (21). Using these numbers, we define numbers γ_k by means of equalities (6). Then $\varkappa_0^1 = \sum\limits_{k=1}^{j} \gamma_k$, $j = \nu + i - \mu$ and the index $\varkappa^-$ of problem (15) is calculated by the equality $\varkappa^- = \varkappa(\Gamma; m; p; \omega) = \operatorname{ind} G_1 + \sum\limits_{k=1}^{j} \gamma_k$.

If $\varkappa^- < -1$, then the conditions of solvability of problem (31) in the class $K_{D^-}^{p(\cdot)}(\Gamma;\omega)$ (or, which is the same, of problem (16) in the class $K_{D_1^+}^{p^1(\cdot)}(\Gamma_1;\omega^1)$ with the condition $F(0) = 0$) take the form

$$I_k = \int\limits_{\Gamma_1} \frac{\operatorname{Re}[\Phi^{(m)}(\frac{1}{\eta})(a^1(\eta) + ib^1(\eta))]}{X^{1+}(\psi(\eta))(a^1(\eta) + ib^1(\eta))} Q^1(\psi(\eta))\psi^k(\eta)\psi'(\eta)\, d\eta = 0, \tag{32}$$

$$k = \overline{0, |\varkappa^-| - 1}.$$

Hence we obtain

$$\begin{aligned} I_k &= \frac{1}{2} \int\limits_{\Gamma_1} \frac{\Phi^{(m)}(\frac{1}{\eta})(a^1(\eta)+ib^1(\eta)) + \overline{\Phi^{(m)}(\frac{1}{\eta})}(a^1(\eta)-ib^1(\eta))}{X^{1+}(\psi(\eta))(a^1(\eta) + ib^1(\eta))} Q^1(\psi(\eta))\psi^k(\eta)\psi'(\eta)\, d\eta \\ &= I_{k,1} + I_{k,2}\,, \end{aligned}$$

where

$$I_{k,1} = \frac{1}{2} \int\limits_{\Gamma_1} \frac{\Phi^{(m)}(\frac{1}{\eta})Q^1(\psi(\eta))}{X^{1+}(\psi(\eta))} \psi^k(\eta)\psi'(\eta)\, d\eta,$$

$$I_{k,2} = \frac{1}{2}\int_{\Gamma_1} \frac{\overline{\Phi^{(m)}(\frac{1}{\eta})Q^1(\psi(\eta))}}{X^{1+}(\psi(\eta))} \frac{(a^1(\eta) - ib^1(\eta))}{a^1(\eta) + ib^1(\eta)} \psi^k(\eta)\psi'(\eta)\, d\eta.$$

The function $\Psi_m(\zeta) = \Phi^{(m)}(\frac{1}{\zeta})[X^1(\psi(\zeta))]^{-1}Q^1(\psi(\zeta))\psi^k(\zeta)\psi'(\zeta)$ is analytic in D_1^+ and belongs to $E^\delta(D_1^+)$ for some $\delta > 0$. Indeed, $\Phi^{(m)} \in E^1(D_1^+)$, $\psi' \in E^1(D_1^+)$, $|\psi^k(\zeta)| \leq 1$. Furthermore $X^1(w) = X_1^1(w)r(w)$ and $Q^1(w)$ is a rational function with zeros and poles on γ, $r(w)$ is a power function, while $[X_1^1]^{\pm 1} \in \bigcap\limits_{\beta>0} E^\beta(D_1^+)$. Hence it follows that $Q^1(\psi(\zeta))[X^1(\psi(\zeta))]^{-1}$ and thereby $\Psi_m(\zeta)$, too, belongs to $E^\delta(D_1^+)$, $\delta > 0$. Moreover, since Γ is a piecewise smooth curve without zero angles (since $0 < 2 - \nu_k \leq 2$), the construction of the functions X^1 and Q^1 implies that the function

$$\Psi_m^+(\eta) = \Phi^{(m)}\left(\frac{1}{\eta}\right) Q^1(\psi(\eta))[X^{1+}(\psi(\eta))]^{-1}\psi^k(\eta)\psi'(\eta)$$

belongs to $L(\Gamma_1)$ (see Section 5.5). Thus, according to Smirnov's theorem, $\Psi_m \in E^1(D_1^+)$ and therefore $I_{k,1} = 0$.

Let us write $I_{k,2}$ in the form

$$I_{k,2} = \frac{1}{2}\int_{\gamma} \frac{\overline{\Phi^{(m)}(\frac{1}{\zeta(\tau)})Q^1(\tau)}}{X^{1+}(\tau)} \frac{(a^1(\zeta(\tau)) - ib^1(\zeta(\tau)))}{a^1(\zeta(\tau)) + ib^1(\zeta(\tau))} \tau^k\, d\tau.$$

But

$$\frac{a^1(\zeta(\tau)) - ib^1(\zeta(\tau))}{a^1(\zeta(\tau)) + ib^1(\zeta(\tau))} = \frac{X^{1+}(\tau)}{X^{1-}(\tau)},$$

and therefore

$$I_{k,2} = \frac{1}{2}\int_{\gamma} \frac{\overline{\Phi^{(m)}(\frac{1}{\zeta(\tau)})Q^1(\tau)}}{X^{1-}(\tau)} Q^1(\tau)\tau^k\, d\tau.$$

The function $\Phi^{(m)}\left(\frac{1}{\zeta(w)}\right)$ is analytic in the unit circle U and therefore $\overline{\Phi^{(m)}\left(\frac{1}{\zeta(\tau)}\right)}$ is a boundary value, on γ, of a analytic function in $U^- = \mathbb{C} \setminus \overline{U}$ (see [98, p. 143]). Denote it by M_-. It is easy to verify that the functions $M^-(w)[X^1(w)]^{-1}Q^1(w)w^k$, $k = \overline{0, |\varkappa| - 1}$, belong to $E^1(U^-)$. By virtue of this fact we have

$$I_{k,2} = \frac{1}{2}\int_{\gamma} \frac{M^-(\tau)Q^1(\tau)}{X^{1-}(\tau)} \tau^k\, d\tau = 0, \quad k = \overline{0, |\varkappa| - 1}.$$

Thereby $I_k = 0$ and thus conditions (32) sufficient for problem (31) to be solvable are fulfilled. The solution is given by equality (18). When $\varkappa^- \geq 0$, problem (15) is solvable in $K_{D^-}^{p(\cdot)}(\Gamma; \omega)$ and the solution is given by the equality (18). □

Let us indicate how to define the function φ when the solution of problem (31) is known.

When the solution $N(z)$ is unique, φ is defined immediately from equalities (26) and (30). We will consider the case where $\varkappa^- > 0$.

By differentiation with respect to z, from (24) we find

$$(-1)^n \frac{n!}{2\pi i} \int_\Gamma \frac{2\pi i(-1)^m(m-1)!t^{1-m}\overline{t'}}{(t-z)^{n+1}} \varphi(t)\, dt = \Omega^{(n)}(z) + \sum_{k=0}^{\varkappa^- -1} h_k [w^k(z)]^{(n)}.$$

Setting here $z = 0$, $n = 0, 1, \dots, \varkappa^- - 1$, we obtain a system with respect to the unknowns $h_0, h_1 \dots, h_{\varkappa^- -1}, \varphi_0, \varphi_1, \dots, \varphi_{\varkappa^- -1}$, where

$$\varphi_k = \frac{(-1)^{k+1}k!}{2\pi i} \int_\Gamma \frac{2\pi i(-1)^m(m-1)!t^{1-m}\overline{t'}}{t^{k+1}} \varphi(t)\, dt. \tag{33}$$

Thus we have $2\varkappa^-$ complex unknowns and $\varkappa^-$ equations. To these equations we add $\varkappa^-$ equations more: $\overline{h_k} = Ah_{\varkappa^- -k-1}$, $k = \overline{0, \varkappa^- - 1}$ (see Section 7.1). As a result we obtain a linear system of $2\varkappa^-$ equations with respect to $2\varkappa^-$ unknowns. As has been proved above, there exists a unique desired solution φ. Therefore the obtained system is uniquely solvable. Substituting the numbers $h_0, \dots, h_{\varkappa^- -1}$ found from this solution into (33) we find $N(z)$ and then from (30) we define φ.

7.3 The Riemann–Hilbert–Poincaré problem in the class $K^{p(\cdot)}_{D^+,m}(\Gamma; \omega)$

I. Vekua applied the representations (1), (2) to the investigation of the Riemann–Hilbert–Poincaré problem

$$\operatorname{Re}\left[\sum_{k=0}^{m} \left(a_k(t_0)\Phi^{(k)}(t_0) + \int_\Gamma H_k(t_0,t)\Phi^{(k)}(t)\, ds\right)\right] = f(t_0), \tag{34}$$

where a_k, H_k, f are the given functions of Hölder's class, Γ is a Lyapunov curve bounding the finite domain D^+, and the sought function Φ has a continuous derivative of order m in $\overline{D^+}$ and with boundary values from H ([128], [129]). In [42] this problem is considered when $a_k(t)$ are continuous and $\Phi^{(m)}(z)$ is representable by a Cauchy type integral with a density from $L^p(\Gamma; \omega)$, where $p > 1$ and ω is a power function.

Here we assume that for p, Γ and ω the conditions of Theorem 5 are fulfilled. We want to solve problem (34) in the class $K^{p(\cdot)}_{D^+.m}(\Gamma; \omega)$, therefore it is assumed that $f \in L^{p(\cdot)}(\Gamma; \omega)$. Since $\Phi^{(m)} \in K^{p(\cdot)}_{D^+}(\Gamma; \omega) \subset E^1(D^+)$, the functions $\Phi^{(0)}(z) = \Phi(z)$, $\Phi'(z), \dots, \Phi^{(m-1)}(z)$ are continuous in $\overline{D^+}$ and absolutely continuous on Γ with respect to the arc abscissa (see, for instance, [108, p. 208]). Thus it is natural to assume that in condition (34) the coefficients $a_k(t)$, $k=\overline{0,m-1}$, belong to $L^{p(\cdot)}(\Gamma;\omega)$. As to the coefficient $a_m(t)$, we should assume that it is bounded. However this is not enough. Following [128], [129], [98], we reduce the problem to a singular integral equation in the class $L^{p(\cdot)}(\Gamma;\omega)$, which is investigated in various conditions depending on the assumptions made for p, Γ and ω ([36], [68], [66]). It is assumed for simplicity that $a_m(t)$ is piecewise continuous on Γ and $\inf|a_m(t)|>0$.

So, let Γ be a curve of the class $C^1_{D^+}(A_1,\dots,A_i;\nu_1,\dots,\nu_i)$, $0\le\nu_k<2$, ω be the power function (3), the coefficients $a_0,a_1,\dots,a_{m-1}$ belong to $L^{p(\cdot)}(\Gamma;\omega)$, $p\in\widetilde{\mathcal{P}}(\Gamma)$, $a_m\in C(\widetilde{B}_1,\dots,\widetilde{B}_\lambda)$ and the operators

$$\mathcal{H}_k\varphi=\int\limits_\Gamma H_k(t_0,t)\varphi(t)\,dt,\quad t_0\in\Gamma,$$

be compact in $L^{p(\cdot)}(\Gamma;\omega)$.

It is required to find a function $\Phi\in K^{p(\cdot)}_{D^+,m}(\Gamma;\omega)$ for which equality (34) holds a.e. on Γ. Note that the compactness of the operators $\mathcal{H}_k$ is provided, for instance, by the fulfillment of the conditions

$$|H_k(t_0,t)|<\frac{A}{[s(t_0,t)]^\lambda},\quad k=\overline{0,m},$$

where A, $\lambda\in[0,1)$, are constants and $s(t_0,t)$ is the length of the smallest of two arcs connecting the points t_0 and t on Γ (see [70]).

Following [98, p. 233] this problem will sometimes be called Problem V.

Reduction of the problem V to singular integral equation

Since under the above assumptions the conditions of Theorem 5 are fulfilled, the sought solution Φ is representable by equality (23) for $m=0$ and by equality (24) for $m\ge1$.

Assuming first that $m\ge1$, we calculate the derivatives of the function Φ given by equality (24) and substitute them into (34). Thus we obtain (see [98, pp. 234–235]) that the function φ satisfies the condition

$$N\varphi=A(t_0)\varphi(t_0)+\int\limits_\Gamma N(t_0,t)\varphi(t)\,ds=f(t_0)-d\sigma(t_0),\tag{35}$$

where

$$A(t_0)=\operatorname{Re}\left[(-1)^m(m-1)!\pi i t_0^{1-m}t_0'a_m(t_0)\right],\tag{36}$$

$$\sigma(t_0)=\operatorname{Re}\left[ia_0(t_0)+i\int\limits_\Gamma H_0(t_0,t)\,ds\right],\tag{37}$$

$$
\begin{aligned}
N(t_0,t) &= \sum_{l=0}^{m} \operatorname{Re}\left[a_l(t_0)N_l(t_0,t) + \int_\Gamma H_l(t_0,t_1)N_l(t_1,t)ds_1\right] \\
&\quad + \operatorname{Re}\left[(-1)^m(m-1)!\pi i h_m(t_0,t)t^{1-m}\overline{t'}\right], \qquad (38)\\
N_0(t_0,t) &= \left(1-\frac{t_0}{t}\right)^{m-1}\ln\left(1-\frac{t_0}{t}\right)+1,\\
N_l(t_0,t) &= (-1)^l\,\frac{(m-1)\cdots(m-l)}{t^l}\left(1-\frac{t_0}{t}\right)^{m-l-1}\\
&\quad\times\left(\ln\left(1-\frac{t_0}{t}\right)+\frac{1}{m-1}+\cdots+\frac{1}{m-l}\right), \quad l=\overline{1,m-1},\\
N_m(t_0,t) &= \frac{(-1)^m(m-1)!}{t^{m-1}(t-t_0)}.
\end{aligned}
$$

It is evident that

$$N\varphi = N^0\varphi + T\varphi,$$

where

$$
\begin{aligned}
N^0\varphi &= A(t_0)\varphi(t_0) + \frac{B(t_0)}{\pi i}\int_\Gamma \frac{\varphi(t)\,dt}{t-t_0},\\
B(t_0) &= \frac{1}{2}(-1)^m(m-1)!\pi i\left[t_0^{1-m}\overline{t_0'}a_m(t_0)+\overline{t_0}^{1-m}t_0'\overline{a_m(t_0)}\right], \qquad (39)\\
T\varphi &= \int_\Gamma \sum_{l=0}^{m-1}\operatorname{Re}\left[a_l(t_0)N_l(t_0,t)+\int_\Gamma H_l(t_0,t_1)N_l(t_1,t)\,ds_1\right]\varphi(t)\,ds\\
&\quad+\int_\Gamma \operatorname{Re}\left[(-1)^m(m-1)!\pi i H_m(t_0,t)t^{1-m}\overline{t'}\right]\varphi(t)\,ds.
\end{aligned}
$$

Moreover, (36) implies that the function $A(t_0)$ can be written in the form

$$A(t_0) = \frac{1}{2}(-1)^m(m-1)!\pi i\left[t_0^{1-m}\overline{t_0'}a_m(t_0) - \overline{t_0^{1-m}}t_0'\overline{a_m(t_0)}\right]. \qquad (40)$$

By virtue of our assumptions about Γ, the coefficients $a_0(t_0),\dots,a_{m-1}(t_0)$ and operators $\mathcal{H}_k$ and using the above-mentioned result from [70] it is not difficult to establish that the operator T is compact in $L^{p(\cdot)}(\Gamma;\omega)$.

When $m=0$, the boundary condition (34) takes the form

$$\operatorname{Re}\left[a_0(t_0)\Phi(t_0)+\int_\Gamma H_0(t_0,t)\Phi(t)\,ds\right] = f(t_0).$$

Applying representation (23) we come to equation (35), where $A(t)$, $\sigma(t)$, $N(t_0,t)$ being calculated by the formulas (36)-(38), where we should take $m=0$, $(-1)^m(m-1)!=1$ and $N_0(t_0,t)=t(t-t_0)^{-1}$ (see [98, pp. 235–236]).

Let us calculate the index of the operator N or, which is the same, of the operator N^0. This is equivalent to the calculation of the index of the problem

$$\Psi^+(t) = \frac{A(t) - B(t)}{A(t) + B(t)} \Psi^-(t), \quad \Psi(\infty) = const \tag{41}$$

in the class $\widetilde{K}_0^{p(\cdot)}(\Gamma; \omega)$ (see Remark 3).

From (39) and (40) we obtain

$$G(t) = \frac{A(t) - B(t)}{A(t) + B(t)} = -\frac{\overline{t^{1-m}t'}\overline{a_m(t)}}{t^{1-m}\overline{t'}a_m(t)}.$$

Denote by B_k, $k = \overline{1, \lambda}$, the discontinuity points of this function and let $b_k = w(B_k)$. Among the points B_k there are all points A_k, $k = \overline{1, i}$.

Put

$$-\frac{t^{m-1}t'}{\overline{t^{m-1}}\,\overline{t'}} = e^{-\vartheta(s)}, \quad \frac{\overline{a_m(t)}}{a_m(t)} = e^{i\theta(s)}$$

and write equality (41) in the form

$$\Psi^+(t) = \exp\{i(\vartheta(s) + \theta(s))\}\,\Psi^-(t).$$

The function $G(t) = G(t(s)) = \exp\{i(\vartheta(s) + \theta(s))\}$ is piecewise continuous. Let

$$G(B_k-)/G(B_k+) = \exp(2\pi i u_k).$$

For problem (41) the index $\varkappa$ (see Remark 3), which depends on Γ, p, ω and the numbers u_k, is defined in the class $\widetilde{K}_0^{p(\cdot)}(\Gamma; \omega)$.

The operator

$$(N'g)(t) = A(t)g(t) + \int_\Gamma N(t_0, t)g(s_0)\,ds_0$$

considered in the space $L^{p'(\cdot)}(\Gamma; \omega^{-1})$ is the conjugate operator to N.

Let for a pair of operators N and N' the Noether theorems hold (see Chapter 4, Theorem 23). From this theorem it follows particulary the following statement.

Let $\omega = \exp\left\{\frac{i}{2} S_\Gamma \psi\right\}$, where ψ is a real piecewise-constant function,

$$G_\psi(t) = G(t)\exp(i\psi(t)) \quad \left(G(t) = \frac{\overline{t^{1-m}t'}\overline{a_m(t)}}{t^{1-m}\overline{t'}a_m(t)} \right),$$

and $B_{k,\psi}$, $k = \overline{1, \lambda}$, are all discontinuity points of this function and $G(B_{k,\psi}-) \times [G(B_{k,\psi}+)]^{-1} = \exp(2\pi i \lambda_k)$, $\lambda_k = \lambda_{k,1} + i\lambda_{k,2}$, $k = \overline{1, n_\psi}$.

If $\{\lambda_{k,1}\} \neq \frac{1}{p'(B_{k,\psi})}$, $k = \overline{1, n_\psi}$, then the operator N is Noetherian in $L^{p(\cdot)}(\Gamma; \omega)$ and $\operatorname{ind} N = \operatorname{ind} G_\psi(t) = \varkappa$.

Theorem 6. *For Problem V to be solvable in the class $K^{p(\cdot)}_{D^+,m}(\Gamma;\omega)$ it is necessary and sufficient that for some real d the function $\widetilde{f}(t_0)=f(t_0)-d\sigma(t_0)$ should satisfy the conditions*

$$\int_\Gamma \widetilde{f}(t_0)g_k(s_0)\,ds_0=0,\quad k=\overline{1,n'},$$

where $g_1,\dots,g_{n'}$ are linearly independent solutions from the class $L^{p'(\cdot)}(\Gamma;\omega^{-1})$ of the equation $N'g=0$, where N' is the adjoint operator to the operator N.

In order that Problem V has a solution, for any right-hand part of f it is necessary and sufficient that $n'=0$ or $n'=1$ and in the latter case the solution g of the equation $N'g=0$ must satisfy the condition

$$(g,\sigma)=\int_\Gamma g(t_0)\sigma(t_0)\,ds_0\neq 0.$$

In both cases the homogeneous problem has $\varkappa+1$ linearly independent solutions (where $\varkappa\geq -1$).

Moreover if $(g_k,\sigma)=0$ for any $k=\overline{1,n'}$, then the homogeneous problem has $\varkappa+n'$ linearly independent solutions, and if among the numbers (g_k,σ) there is at least one nonzero number, then it has $\varkappa+n'+1$ solutions.

If $\sigma(t)=0$, then problem (34) is solvable for any right-hand part of $f(t_0)$ if and only if $n'=0$; in that case the homogeneous problem has $\varkappa+1$ linearly independent solutions.

We omit the proof of this theorem because it is analogous to that of the respective I. Vekua's theorem ([129], see also [98, pp. 238–240]).

The notion of the kernel and its application

If we assume

$$\Psi(z)=\int_\Gamma g(t)\Omega^*(t,z)\,ds,\quad z\in D^-,\tag{42}$$

where

$$\Omega^*(t_0,z)=\sum_{k=0}^{m}\Big\{a_k(t_0)N_k(t_0,z)+\int_\Gamma H_k(t_0,t)N_k(t,z)\,ds\Big\},$$

then the equation $N'g=0$ can be written in the form

$$\operatorname{Re}\Psi^-(t_0)=0.\tag{43}$$

It can be easily verified that $\Psi(z)$ belongs to $\widetilde{K}^{p'(\cdot)}_{D^-}(\Gamma;\omega^{-1})$. Thus the function Ψ defined by equality (42) is a solution of the class $\widetilde{K}^{p'(\cdot)}_{D^-}(\Gamma;\omega^{-1})$ of problem (43). In order to apply to the case in which $p(t)$ is replaced by $p'(t)$ and $\omega(t)$ by $[\omega(t)]^{-1}$, we have to calculate the index $\varkappa = \varkappa_0 + \varkappa_1$ of problem (43). Since for the considered Riemann–Hilbert problem $a(t) + ib(t) = 1$, we have $\varkappa_1 = 0$. As to $\varkappa_0$, it is equal to the order at the point $z = \infty$ of the rational function $Q(w) = \prod_{k=1}^{j}(w - w_k)^{\gamma_k}$. Here the numbers γ_k are defined from relations (6) in which the function $\ell(t)$ is replaced by the function $\ell'(t)$ and the numbers α_k by the numbers $(-\alpha_k)$. The number $\varkappa_0$ is calculated by the equality

$$\varkappa_0 = \mathcal{N}\{A_k : p'(A_k) < \nu_k\} + \mathcal{N}\left\{t_k = A_k : \frac{p'(A_k)}{1 - \alpha_k p'(A_k)} < \nu_k < \frac{2p'(A_k)}{1 - \alpha_k p'(A_k)}\right\}, \quad (44)$$

where $\mathcal{N}(E)$ stands for the number of elements of the set E (see Section 5.6).

Problem (43) has $\varkappa_0 + 1$ linearly independent solutions. We denote them by $\Psi_0(z), \Psi_1(z), \dots, \Psi_{\varkappa_0}(z)$.

Let

$$\widetilde{\Omega}(t_0, z) = \Omega^*(t_0, z) - \Psi_0(z) - \cdots - \Psi_{\varkappa_0}(z) \quad (45)$$

and

$$\Omega(t_0, z) = \widetilde{\Omega}(t_0, z) - i \operatorname{Im} \widetilde{\Omega}(t_0, \infty). \quad (46)$$

Then from the definitions of the functions Ω^*, $\Psi_0, \dots, \Psi_{\varkappa_0}$ it follows that for an arbitrary solution g of the equation $N'g = 0$ there holds the equality

$$\int_\Gamma g(t)\Omega(t, z)\, ds = 0 \quad \forall z \in D^-. \quad (47)$$

By analogy with [98, p. 242] we call the function $\Omega(t_0, z)$ a kernel.

Now, following I. Vekua ([128], see also [98, pp. 242–243]), condition (47) can be replaced by an equivalent condition

$$\int_\Gamma g(t)\omega_j(t)\, ds = 0,$$

where $\{\omega_j(t)\}_0^\infty$ can be understood as any of the following systems of functions:

(B_1)
$$\omega_j(t) = \Omega(t, z_j), \quad z_j \in D^-,$$

where $z_0, z_1, \dots$ is an arbitrary sequence having a limit point in D^-;

(B_2)
$$\omega_j(t) = \left[\frac{d^j \Omega(t, z)}{dz^j}\right]_{z=z_0}, \quad z_0 \in D^-;$$

(B_3)
$$\omega_j(t) = X_j(t), \quad j = 0, 1, 2, \dots,$$

where $X_0(t) = \operatorname{Re}\left[a_0(t_0) + \int_\Gamma H_0(t_0, t) ds\right]$, $X_k(t_0) = L\psi$, $\psi = t^k$, $k = 1, 2, \dots$,

$$L\psi = \sum_{j=0}^{m} \left\{a_j(t_0)\psi^{(j)}(t_0) + \int_\Gamma H_j(t_0, t)\psi^{(j)}(t)\, ds\right\}.$$

If k' is a maximal number of linearly independent real functions $g_j(t)$ orthogonal to the kernel $\Omega(t,z)$ or to all elements of the sequence $\{\omega_j(t)\}$, then k' is called the defect of the kernel or of the sequence $\{\omega_j(t)\}$. In the considered case, k' coincides with n' which is the number of linearly independent solutions in $L^{p'(\cdot)}(\Gamma;\omega^{-1})$ of the equation $N'g=0$. Hence the obtained results can be formulated as follows:

Theorem 7. *For Problem V to have a solution for any right-hand part of $f(t)$ it is necessary and sufficient that the defect of the sequence $\{\omega_j(t)\}$ (or of the kernel $\Omega(t,z)$) be equal to 0 or 1. In the latter case the function $g(t)$, which is orthogonal to all elements of the sequence $\{\omega_j(t)\}$, must not be orthogonal to the function $\sigma(t)$.*

7.4 Some particular cases

Poincaré Problem. We will consider this problem formulated as follows:

Find, in the domain D^+, a harmonic function u from the set

$$e^{p(\cdot)}_{D^+,1}(\Gamma;\omega)=\Big\{u:\ u=\operatorname{Re}\Phi,\quad \Phi\in K^{p(\cdot)}_{D^+,1}(\Gamma;\omega)\Big\},$$

for which a.e. on Γ we have

$$\alpha(s)\,\frac{\partial u}{\partial n}+\beta(s)\,\frac{\partial u}{\partial s}+\gamma(s)u=f(s). \tag{48}$$

Here $\alpha(s)$, $\beta(s)$, $\gamma(s)$, $f(s)$ are the real functions given on Γ, $s=s(t)$ is an arc abscissa of point $t\in\Gamma$, $\frac{\partial u}{\partial n}$ is normal derivative. It is assumed that $p\in\widetilde{\mathcal{P}}(\Gamma)$, $\Gamma\in C^1(A_1,\dots,A_i;\nu_1,\dots,\nu_i)$, $0<\nu_k<2$, $k=\overline{1,i}$; α and β belong to Hölder's class, while $\gamma(s)$ and $f(s)$ belong to $L^{p(\cdot)}(\Gamma;\omega)$.

Let

$$a(t)=-\alpha(s)\sin\vartheta(s)+\beta(s)\cos\vartheta(s),\quad b(t)=\alpha(s)\cos\vartheta(s)+\beta(s)\sin\vartheta(s),$$

where $\vartheta(s)$ is the angle formed between the tangent to Γ at the point $t(s)$ and the axis of abscissa. Then condition (48) takes the form

$$a(t)\,\frac{\partial u}{\partial x}+b(t)\,\frac{\partial u}{\partial y}+\gamma(s(t))u=f(t),\quad t\in\Gamma,$$

which can be rewritten as

$$\operatorname{Re}\big[(a(t_0)+ib(t_0))\Phi'(t_0)+\gamma(t_0)\Phi(t_0)\big]=f(t_0),\quad t_0\in\Gamma.$$

Using representation (24), for $m=1$ we obtain the equality

$$\begin{aligned}N\varphi&=\operatorname{Re}\Big[-\pi i\overline{t_0'}(a(t_0)+ib(t_0))\Big]\varphi(t_0)\\&+\int_\Gamma\operatorname{Re}\left[\gamma(t_0)\ln e\left(1-\frac{t_0}{t}\right)-\frac{a(t_0)+ib(t_0)}{t-t_0}\right]\varphi(t)\,ds=f(t_0).\end{aligned}$$

Let us assume that $(a^2+b^2)>0$ (or, which is the same, $\alpha^2+\beta^2>0$). This time we have

$$G(t) = -\frac{t'}{\overline{t'}}\frac{\alpha(-\sin\vartheta - i\cos\vartheta)+\beta(\cos\vartheta - i\sin\vartheta)}{\alpha(-\sin\vartheta + i\cos\vartheta)+\beta(\cos\vartheta + i\sin\vartheta)}$$
$$= -\frac{t'}{\overline{t'}}\frac{-\alpha i\overline{t'}+\beta\overline{t'}}{\alpha i t'+\beta t'} = -\frac{t'}{\overline{t'}}\frac{\overline{t'}(\beta-\alpha i)}{t'(\beta+\alpha i)} = \frac{\alpha+\beta i}{\alpha-\beta i}.$$

Assume that

$$n = \frac{1}{2\pi}\left[\arg(\alpha(t)+i\beta(t))\right]_\Gamma. \tag{49}$$

In that case the index of the operator N is calculated by the equality $\varkappa=\varkappa_0+\varkappa_1$, where $\varkappa_1=2n$ and

$$\varkappa_0 = \mathcal{N}\left\{A_k : A_k \notin \bigcup\{t_j\},\ \nu_k > p(A_k)\right\}$$
$$+\mathcal{N}\left\{t_k = A_k : \frac{p(A_k)}{1+\alpha_k p(A_k)} < \nu_k < \frac{2p(A_k)}{1+\alpha_k p(A_k)}\right\}. \tag{50}$$

The kernel $\Omega(t_0,z)$ is given by equalities (45)–(46), where

$$\Omega^*(t_j,z) = \frac{a(t_0)+ib(t_0)}{t_0-z} + \gamma(t_0)\ln e\left(1-\frac{t_0}{z}\right).$$

Besides, in our case $\sigma(t_0)=0$ and therefore by virtue of the above results we come to an analogue of I. Vekua's theorem ([128], see also [98, p. 245]).

Theorem 8. *For the Poincaré problem to have a solution in the class* $e^{p(\cdot)}_{D^+,1}(\Gamma;\omega)$ *for any right-hand part of* $f(t)$ *it is necessary and sufficient that the equation*

$$N'g = \mathrm{Re}\left\{-\pi i\overline{t'_0}\left[a(t_0)+ib(t_0)\right]\right\}g(t_0)$$
$$+\int_\Gamma \mathrm{Re}\left\{\gamma(t_0)\ln e\left(1-\frac{t_0}{t}\right)+\frac{a(t)+ib(t)}{t-t_0}\right\}g(t)\,ds = 0$$

would not have nonzero solutions in the class $L^{p'(\cdot)}(\Gamma;\frac{1}{\omega})$.

When this condition is fulfilled, the problem has $\varkappa+1$ *linearly independent solutions, where* $\varkappa=2n+\varkappa_0$.

Neumann problem in the class $e^{p(\cdot)}_{D^+,1}(\Gamma;\omega)$**.** The Neumann problem

$$\frac{\partial u}{\partial n} = f \tag{51}$$

for the Laplace equation is a particular case of the Poincaré problem. It reduces to the problem

$$\mathrm{Re}\left[it'_0\Phi'(t_0)\right] = f(t_0)$$

considered in the class $K^{p(\cdot)}_{D^+,1}(\Gamma;\omega)$.

The boundary value problem (43) reduces to the problem

$$\Psi^+(t) = \Psi^-(t), \quad \Psi(\infty) = const,$$

whose conjugate problem is

$$F^+(t) = F^-(t), \quad F \in \widetilde{K}^{p'(\cdot)}_0(\Gamma;\frac{1}{\omega}).$$

If $p'(A_k) \neq \nu_k$ for all k, the latter problem has $\varkappa(p';\frac{1}{\omega})$ solutions, where

$$\varkappa(p';\frac{1}{\omega}) = \mathcal{N}\Big\{A_k : A_k \notin \bigcup\{t_j\},\ p'(A_k) < \nu_k\Big\} \\ +\mathcal{N}\Big\{t_k = A_k : \frac{p'(A_k)}{1-\alpha_k p'(A_k)} < \nu_k < \frac{2p'(A_k)}{1-\alpha_k p'(A_k)}\Big\}$$

(see Section 5.6). Thus $\varkappa(p';\frac{1}{\omega})+1$ conditions must be fulfilled in order that problem (51) be solvable.

If $\omega = 1$, this number of conditions is equal to $\varkappa(p';1) = \mathcal{N}\{A_k : p'(A_k) < \nu_k\}$.

In that case for the Neumann problem we have $\varkappa(p';1)+1$ solvability conditions. Assuming that (5) holds and the solvability conditions are fulfilled, a general solution of problem (51) contains $\varkappa(p;1) = \mathcal{N}\{A_k : p(A_k) < \nu_k\}$ real constants.

7.5 Notes and comments on Chapter 7

For the boundary value problem when boundary condition contains not only the sought function, but also its derivatives up to certain order, it is useful to have formulas giving an integral representation of this analytic function. One form of such a representation, quite convinient for applications, was proposed by I. Vekua [127], [129]. N. Muskhelishvili expounded them in his book [98, pp. 224–232], where it is called as I. Vekua's integral representation. It was considered the case when an analytic function in bounded domain with a Lyapunov boundary is such that its high order derivatives and the boundary function belong to the Hölder class. I. Vekua used his representation to solve quite general BVP, namely, the Riemann-Hilbert-Poincaré problem.

Subsequently, B. Khvedelidze [42] gave a generalization of representation mentioned above for analytic functions when high order derivative is representable by a Cauchy type integral with a density from weighted Lebesgue space, where a weight function is a power type. For the solution of the problem in the case of domains with some non-smooth boundaries we refer the reader to [3].

The exposition of Chapter 7 is based on the paper [63].

Bibliography

[1] R. Abdulaev, On one theorem of S Warschawskiĭ. *Georgian Math. J.* **10**(2003), No. 1, 1–15.

[2] T. Alvarez and C. Pérez, Estimates with A8 weights for various singular integral operators. *Boll. Un. Mat. Ital A* (7)8, No. 1, 123–133 (1994).

[3] K. Aptsiauri, On the generalized Riemann-Hilbert-Poincaré problem in domains with piecewise smooth boundaries. *Proc. A. Razmadze Math. Inst.* **114**(1997), 3–26.

[4] V. Atkinson, Normal solvability of linear equations in normed spaces. (Russian) *Mat. Sb.* **28**(1951), No. 1, 3–14.

[5] A. Böttcher and Y. I. Karlovich, *Carleson curves, Muckenhoupt weights and Toeplitz operators.* Bikrhäuser Verlag, Basel-Boston-Berlin, 1997.

[6] A. P. Calderon, Cauchy integrals on Lipschitz curves and related operators. *Proc. Nat. Acad. Sci. USA* **74**(1977), No. 4, 1324–1327.

[7] T. Carleman, Sur la résolution des certaines équations intégrales. *Ark. Mat. Astr. Fys.* **16**(1922), No. 26, 1–19.

[8] M. A. Cotlar, A unified theory of Hilbert transforms and ergodic theorems. *Rev. Mat. Guyana* **1**(1955), No. 2, 105–167.

[9] D. Cruz-Uribe, L. Diening and P. Hästö. The maximal operator in weighted variable exponent Lebeasgue spaces. *Fract. Calc. and Appl. Anal.* **14**(2001), No. 3, 361–374.

[10] D. Cruz-Uribe, A. Fiorenza, and C.J. Neugebauer. The maximal function on variable L^p spaces. *Ann. Acad. Sci. Fenn. Math.*, **28**(2003), No. 1, 223–228.

[11] E. A. Danilov, Factorization of a positive function on a contour of unbounded turning. (Russian). *Dokl. Akad. Nauk SSSR*, 283(5):1051–1053, 1985.

[12] I. I. Danilyuk, *Irregular Boundary Value Problems in the Plane.* (Russian) Nauka, Moscow, 1975.

[13] M. Dauge, *Elliptic Boundary Value Problems in Corner Domains.* Lecture Notes in Math. 1341, Springer-Verlag, Berlin-Heidelberg-New-York, 1997.

[14] G. David, Opérateurs intégraux singuliers sur certaines courbes du plan complexe. *Ann. Sci. École Norm. Sup.* **17**(1984), 157–189.

[15] L. Diening. Maximal functions on generalized Lebesgue spaces $L^{p(x)}$. *Math. Inequal. and Appl*, **7**(2004), $N_0$2, 245-253.

[16] L. Diening, P. Harjulehto, P. Hästö, and M. Růžička, *Lebesgue and Sobolev Spaces with Variable Exponent.* Lecture Notes in Math., vol. 2017, Springer-Verlag, 2011.

[17] L. Diening and M. Ružička. Calderon-Zygmund operators on generalized Lebesgue spaces $L^{p(x)}$ and problems related to fluid dynamics. *J. Reine Angew. Math*, 563:197–220, 2003.

[18] R. Duduchava and B. Silberman. *Boundary Value Problems in Domains with Peaks.* Mem. Differential Equations Math. Phys, **21**(2000), 1–122.

[19] P. Duren, Theory of H^p spaces. *Academic Press, New York,* 1970.

[20] E. M. Dynkin, Methods of the theory of singular integrals. (Hilbert transforms and Calderon–Zygmund theory.) (Russian) *Itogi Nauki i Tekhniki, Sovrem. Probl. Mat. Fund. Naprav.* 15, *Moscow,* 1987, 197–292.

[21] E. M. Dynkin and B. P. Osilinker, Weighted estimates for singular integrals and their applications. (Russian) *Itogi Nauki i Tekhniki, Matem. Anal.* 21, *Moscow,* 1983, 42–129.

[22] D. E. Edmunds, V. Kokilashvili and A. Meskhi, Two-weight estimates in $L^{p(\cdot)}$ spaces with applications to Fourier series. *Houston J. Math.* **35**(2009), $n_0$2, 665–689.

[23] A. Fiorenza, B. Gupta and P. Jain, The maximal theorem in weighted grand Lebesgue spaces, *Studia Math.* **188**(2008), No.2, 123–133.

[24] F. D. Gakhov, On Riemann boundary value problem. (Russian) *Mat. Sb.* **2(44)**(1937), 673–683.

[25] F. D. Gakhov, *Boundary Value Problems.* Pergamon Press, London, 1966 (Extended Russian edition: Nauka, Moscow, 1977).

[26] I. Ts. Gokhberg and N. Ya. Krupnik, *Introduction to the Theory of One-dimensional Singular Integral Operators.* (Russian) Shtiintsa, Kishinyov, 1973.

[27] G. M. Goluzin, *Geometric Theory of Functions of a Complex Variable.* (Russian) Nauka, Moscow, 1966.

[28] E. G. Gordadze, On Riemann–Privalov problem in case of nonsmooth boundary curve. (Russian) *Trudy Tbiliss. Mat. Inst. Razmadze Akad. Nauk Gruzin. SSR* **33**(1967), 25–31.

[29] L. Greco, T. Iwaniec, and C. Sbordone, *Inverting the p-harmonic operator.* Manuscripta Math. **92** (1997), No. 2, 249–258.

[30] C. Haseman, *Anwendung der Theorie der Integralleichungen und Randwertaufgaben.* Gottingen, 1907.

[31] V. P. Havin, Boundary properties of Cauchy type integrals and harmonically conjugated functions in domains with rectifiable boundary. (Russian) *Mat. Sb.* **68**(1965), No. 4, 499–517.

[32] H. Helson and G. Szegö, A problem in prediction theory. *Ann. Math. Pura Appl.* **51**(1960), 107–138.

[33] D. Hilbert, *Grundzüge einen Allgemeiner Theorie der Linearen Integralglichungen.* Leipzig-Berlin, 1912.

[34] R. Hunt, B. Muckenhoupt, and R. Wheeden, Weighted norm inequalities for the conjugate function and Hilbert transform. *Trans. Amer. Math. Soc.* **176**(1973), No. 2, 227–251.

[35] T. Iwaniec and C. Sbordone, On the integrability of the Jacobian under minimal hypotheses. *Arch. Rational Mech. Anal.* **119** (1992), no. 2, 129–143.

[36] A. Yu. Karlovich. Fredholmness of singular integral operators with piecewise continuous coefficients on weighted Banach function spaces. *J. Integral Equations Appl.* **15** (2003), No. 3, 263–320.

[37] A.Yu. Karlovich, Algebras of singular integral operators on Nakano spaces with Khvedelidze weights over Carleson curves with logarithmic whirl points. *In: "Pseudodifferential Equations and Some Problems of Mathematical Physics" (in Russian). The special issue of "Izvestia VUZ, Severo-Kavkazskii Region, Estestvennye Nauki" dedicated to I. B. Simonenko's 70th birthday, Rostov-on-Don, Russia,* 2005, 135-142.

[38] G. Khuskivadze and V. Paatashvili, On conformal mapping of a circle onto the domain with piecewise smooth boundary. *Proc. A. Razmadze Math. Inst. Georgian Acad. Sci.* **116**(1998), 123–132.

[39] G. Khuskivadze and V. Paatashvili, On the Dirichlet problem for harmonic functions from Smirnov classes in doubly-connected domains. *Proc. A. Razmadze Math. Inst.* **144**(2007), 41–60.

[40] G. Khuskivadze, V. Kokilashvili, and V. Paatashvili, *Boundary Value Problems for Analytic and Harmonic Functions in Domains with Nonsmooth Boundaries. Applications to Conformal Mappings.* Mem. Different. Equat. Math. Phys., **14**, 1998.

[41] G. Khuskivadze, V. Kokilashvili, and V. Paatashvili, The Dirichlet problem for variable exponent Smirnov class harmonic functions in doubly-connected domains. *Mem. Different. Equat. Math. Phys.* **52** (2011), 131–156.

[42] B. V. Khvedelidze, Linear discontinuous boundary value problems of function theory, singular integral equations, and some of their applications. (Russian) *Trudy Tbiliss. Mat. Inst. Razmadze Akad. Nauk Gruzin. SSR* **23**(1956), 3–158.

[43] B. V. Khvedelidze, The method of Cauchy type integrals for discontinuous boundary value problems of the theory of holomorphic functions of one complex variable. *Itogi Nauki i Tekhniki, Sovrem. Probl. Mat.* **7**(1975), 5–162, *J. Sov. Math.* **7**(1977), 309–414.

[44] V. M. Kokilashvili, On boundedness of singular integral operators in the space L^p with weight. (Russian) *In: Proc. of Symposium on Continuum Mechanics and Related Problems of Analysis* (*Tbilisi*, 23–29.IX.1971), v. I, *Metsniereba, Tbilisi,* 1973, 125–141.

[45] V. Kokilashvili, A survey of recent results of Georgian mathematicians on boundary value problems for holomorphic functions. *Mem. Differential Equations Math. Phys.* **23**(2001), 85–138.

[46] V. Kokilashvili, On a progress in the theory of integral operators in weighted Banach Function Spaces. *In: "Function Spaces, Differential Operators and Nonlinear Analysis", Proc. Conf. in honour of A. Kufner, Svrtka, May 27 - June 1, 2004; Math. Inst. Acad. Sci. of Czech Republic, Praha,* 2005, 152–175.

[47] V. Kokilashvili, On the Calderon singular integrals in weighted variable Lebesgue spaces. *Bull. of the Georgian National Academy of Sci.* 2(2008), *N*o4, 3–5.

[48] V. Kokilashvili, Boundedness criterion for the Cauchy singular integral operator and maximal functions in weighted grand Lebesgue spaces. *Bull. Georgian Nat. Academy of Sci.* **3**(2009), No. 3, 5–7.

[49] V. Kokilashvili, The Riemann boundary value problem analytic functions in the frame of grand spaces. *Bull. Georgian Nat. Academy of Sci.* **4**(2010), No.1, 5–8.

[50] V. Kokilashvili, Boundedness criteria for singular integrals in weighted grand Lebesgue spaces. *Journal of Math. Anal. (Springer, New-York)*, **170**(2010), No.1, 20–33.

[51] V. Kokilashvili and A. Meskhi, A note on the boundedness of the Hilbert transform in weighted grand Lebesgue spaces. *Georgian Math.J.* **16**(2009), No.3, 547–551.

[52] V. Kokilashvili, Z. Meshveliani and V. Paatashvili, *Boundary value problems for analytic and harmonic functions of Smirnov classes in domains with non-smooth boundaries.* Factorization, singular operators and related problems (Funchal, 2002), 175-196, Kluwer Acad. Publ., Dordrecht, 2003.

[53] V. Kokilashvili and V. Paatashvili. On boundary value problem of linear conjugation with measurable coefficients. (Russian). *Trudy Tbiliss. Mat. Inst. Razmadze*, 55:59–92, 1977.

[54] V. Kokilashvili and V. Paatashvili, On the Riemann–Hilbert problem in the domain with a nonsmooth boundary. *Georgian Math. J.* **4**(1997), No. 3, 279–302.

[55] V. Kokilashvili and V. Paatashvili, On Hardy classes of analytic functions with a variable exponent. *Proc. A. Razmadze Math. Inst.* **142** (2006), 134–137.

[56] V. Kokilashvili and V. Paatashvili, The Dirichlet problem for harmonic functions in the Smirnov class with variable exponent. *Georgian Math. J.* **14**(2007), No. 2, 289–299.

[57] V. Kokilashvili and V. Paatashvili, On variable Hardy and Smirnov classes of analytic functions. *Georgian International Journal of Sciences. Nova Science Publishers, Inc.* **1** (2008), No. 2, 67–81.

[58] V. Kokilashvili and V. Paatashvili, On the convergence of sequences of functions in Hardy classes with a variable exponent. *Proc. A. Razmadze Math. Inst.* **146**(2008), 124–126.

[59] V. Kokilashvili and V. Paatashvili, The Riemann–Hilbert problem in weighted classes of Cauchy type integrals with density from $L^{p(\cdot)}(\Gamma)$. *Complex Anal. Oper. Theory* **2**(2008), no. 4, 569–591.

[60] V. Kokilashvili and V. Paatashvili, The Riemann–Hilbert problem in a domain with piecewise-smooth boundaries in weighted classes of Cauchy type integrals with a density from variable exponent Lebesgue spaces. *Georgian Math J.* **16**(2009), no. 4, 289–299.

[61] V. Kokilashvili and V. Paatashvili, The Dirichlet problem for harmonic functions from variable exponent Smirnov classes in domains with piecewise smooth boundary. *J. Math. Sci. Springer, New-York*, **172**(2011), no. 3, 1–21.

[62] V. Kokilashvili and V. Paatashvili, Weighted Hardy and Smirnov classes and the Dirichlet problem for a ring within the framework of variable exponent analysis. *Complex Var. Elliptic Equ.*, DOI:10.1080/17476933.2011.557153, 2011.

[63] V. Kokilashvili and V. Paatashvili. Generalization of I. Vekua's integral representations of holomorphic functions and their application to the Riemann-Hilbert-Poincar problem. *J. Funct. Spaces Appl.*, (accepted).

[64] V. Kokilashvili and V. Paatashvili. The Riemann and Dirichlet problems with data from the grand Lebesgue spaces. (submitted).

[65] V. Kokilashvili, V. Paatashvili and S. Samko, Boundary value problems for analytic functions in the class of Cauchy type integrals with density in $L^{p(\cdot)}(\Gamma)$. *Boundary Value Problems* **1**(2005), 43–71.

[66] V. Kokilashvili, V. Paatashvili and S. Samko, Boundedness in Lebesgue spaces with variable exponent of the Cauchy singular integral operator on Carleson curves. *In: "Modern Operator Theory And Applications: Advances and Applications". Operator Theory: Advances and Applications*, **170**(2006), 167–186.

[67] V. Kokilashvili, N. Samko and S. Samko. Singular operators in $L^{p(\cdot)}(\omega, \rho)$ with oscillation weights. *Math. Nachr.* **280**(2007), No. 9–10, 1–12.

[68] V. Kokilashvili and S. Samko, Singular integral equations in the Lebesgue spaces with variable exponent. *Proc. A. Razmadze Math. Inst.* **131**(2003), 61–78.

[69] V. Kokilashvili and S. Samko, Singular integrals in weighted Lebesgue spaces with variable exponent. *Georgian Math. J.* **10**(2003), No. 1, 146–156.

[70] V. Kokilashvili and S. Samko, Maximal and fractional operators in weighted $L^{p(x)}$ spaces. *Rev. Mat. Iberoamericana*, **20** (2004), 493–515.

[71] V. Kokilashvili and S. Samko, Weighted boundedness in Lebesgue spaces with variable exponents of classical operators on Carleson curves. *Proc. A. Razmadze Math. Inst.* **138**(2005), 106–110.

[72] V. Kokilashvili and S. Samko, Boundedness in Lebesgue spaces with variable exponent of maximal, singular and potential operators. *Izv. Visshikh Uchebn. Zaved. Severo-Kavk. Region*, (2005), 152–157.

[73] V. Kokilashvili and S. Samko, Boundedness of maximal operators and potential operators on Carleson curves in Lebesgue spaces with variable exponent. *Acta Math. Sinica*, **23**(2007), No 6, 965–972.

[74] V. Kokilashvili and S. Samko, Operators of harmonic analysis in weighted spaces with non-standard growth. *J. Math. Anal. Appl.* **352**(2009), 15–34.

[75] V. Kokilashvili and S. Samko, Singular operators and Fourier multipliers in weighted Lebesgue spaces with variable exponent. *(Russian) Vestnik St. Petersburg Univ. Mathematics*, **41**(2008), N0. 2, 134–144.

[76] V. Kokilashvili and S. Samko, Boundedness of weighted singular integrals in grand Lebesgue spaces. *Georgian Math. J.*, **18**(2011), 259-269.

[77] A.N. Kolmogorov. Zur normierbarkeit eines allgemeinen topologischen linearen Räumes. *Studia Math.*, 5(1934), 29–33.

[78] V. A. Kondrat'ev and O. A. Oleĭnik, Boundary value problems for PDE in nonsmooth domains. *Uspekhi Mat. Nauk* **38**(1983), No. 2, 3–76.

[79] O. Kováčik and J. Rákosník, On spaces $L^{p(x)}$ and $W^{k,p(x)}$. *Czechoslovak Math. J.* **41**(116) (1991), 592–618.

[80] V. G. Kravchenko and G. S. Litvinchuk, *Introduction to the Theory of Singular Integral Operator with Shift.* Kluwer Academic Publishers, Dordrecht, Boston, London, 1994.

[81] D. A. Kveselava, The solutions of a boundary problem of the theory of functions. (Russian) *DAN SSSR* **53** (1946), No. 8, 683–686.

[82] D. A. Kveselava, Some boundary problems of the theory of functions. (Russian) *Akad. Nauk Gruzin. SSR, Trudy Tbiliss. Mat. Inst. Razmadze* **16** (1948), 39–80.

[83] G. S. Litvinchuk, *Boundary Value Problems and Singulat Integral Equations with Shift.* (Russian) Nauka, Moscow, 1977.

[84] G. S. Litvinchuk and I. M. Spitkovski, *Factorization of Measurable Matrix Functions.* Birkhäuser Verlag, Basel, 1987.

[85] W. A. J. Luxemburg. *Banach Function Spaces.* Thesis. Technische Hogescool to Delft, 1955.

[86] G. F. Manjavidze, *Boundary Value Problems of Conjugation with a Displacement for Analytic and Generalized Function.* (Russian) Izdat. Tbiliss. Univ, 1990.

[87] G. F. Manjavidze and B. V. Khvedelidze, On the Riemann–Privalov problem with continuous coefficients. (Russian) *Dokl. Akad. Nauk SSSR* **123**(1958), No. 5, 791–794.

[88] V. Maz'ya and A. Soloviev, L_p-theory of boundary integral equations on a contour with outward peak. *Integral Equations Operator Theory* **32** (1998), no. 1, 75–100.

[89] V. Maz'ya and A. Soloviev, A direct method for boundary integral equations on a contour with a peak. *Georgian Math. J.* **10** (2003), no. 3, 573–593.

[90] V. Maz'ya, A. Soloviev., Boundary Integral Equations on Contours with Peaks, *Birkhuser*, 2010.

[91] Z. Meshveliani, V. Paatashvili, On Smirnov classes of harmonic functions, and the Dirichlet problem. *Proc. A. Razmadze Math. Inst.* **123** (2000), 61–91.

[92] M. Morse, *Topological Methods of the Theory of Complex Variable Function.* (Russian) Inostr. Lit., Moscow, 1951.

[93] J. Musielak. *Orlicz Spaces and Modular Spaces,* volume 1034 of Lecture Notes in Mathematics. Springer-Verlag, Berlin, 1983.

[94] J. Musielak and W. Orlicz. On modular spaces. *Studia Math.*, 18:49–65, 1959.

[95] N. I. Muskhelishvili, *Applications Des intégrales Analogues à celles de Cauchy à Quelques Problèmes de la Physique Mathématique.* Tbilisi University Press, Tbilisi, 1922.

[96] N. I. Muskhelishvili, *Application of Cauchy Type Integrals to a Class of Singular Integral Equations.* (Russian) Trudy Tbiliss. Mat. Inst. **10**(1941), 1–48.

[97] N. I. Muskhelishvili, *Singular Integral Equations.* (Translation from Russian) P. Noordhoff, Gröningen, 1953.

[98] N. I. Muskhelishvili, *Singular Integral Equations.* Boundary Value Problems and Some of their Applications to Math. Phys., Izd. 3. (Russian) Nauka, Moscow, 1968.

[99] H. Nakano. *Topology of Linear Topological Spaces.* Maruzen Co. Ltd., Tokyo, 1951.

[100] W. Orlicz. Über Konjugirte Exponentenfolgen. *Studia Math.*, **3**(1931), 200–211.

[101] V. A. Paatashvili, On singular Cauchy integrals. (Russian) *Soobshch. Akad. Nauk Gruzin. SSR* **53**(1969), No. 3, 529–532.

[102] V. A. Paatashvili, A certain property of singular Cauchy integral. (Russian) *Soobshch. Akad. Nauk Gruzin. SSR* **61**(1971), No. 1, 15–16.

[103] V. Paatashvili, Some properties of singular integrals and cauchy type integrals. (Russian). *Trudy Tbiliss. Mat. Inst. Razmadze*, **47**(1975), 34–52.

[104] V. Paatashvili, Haseman's problem in the class of functions representable by Cauchy type integral with density in $L^{p(\cdot)}(\Gamma;\rho)$. *Mem. Different. Equat. Math. Phys.* **50**(2010), 139–162.

[105] V. Paatashvili, The Haseman's problem with piecewise continuous coefficients in classes of functions representable by Cauchy type integral with density in $L^{p(\cdot)}(\Gamma,\rho)$. *Proc. A. Razmadze Math. Inst.*. **154**(2010), 111–126.

[106] J. Plemelj, Ein Ergänzungssatz zur Cauchyschen Integraldarstellungen analytischer Funktionen, Randwerte belreffend. *Monatsh. Math. Phys.* **19**(1908), 205–210.

[107] Ch. Pommerenke, *Boundary Behavior of Conformal Maps.* Grundlehren der Mathematischen Wissenschaften 299, Springer-Verlag, Berlin, 1992.

[108] I.I. Privalov, *Boundary Properties of Analytic Functions.* (Russian). Second edition. Gosudarstv. Izdat. Tehn.-Teor. Lit., Moscow-Leningrad, 1950.

[109] I.I. Privalov, *Randeigenschaften Analytischer Funktionen.* Zweite, unter Redaktion von A. I. Markuschewitsch überarbeitete und ergänzte Auflage. Hochschulbücher für Mathematik, Bd. 25. VEB Deutscher Verlag der Wissenschaften, Berlin, 1956.

[110] V. Rabinovich, B.W. Schulze and N. Tarkhanov, Calculus of boundary value Problems in domains with non-Lipschitz singular points, *Math. Nachr.* **215**(2000), No. 1, 115–160.

[111] B. Riemann. *Grundlagen Für Eine Allgemeine Theorie der Funktionen Einer Veränderlichen Komplexen Grösse.* Worke, Leipzig, 1867.

[112] B. Riemann, *Essays.* (Translated in Russian from German) Moscow, 1948.

[113] M. Riesz, Sur les fonctions conjuguées. *Math. Z.* **27**(1927), No. 2, 218–244.

[114] M. Ružička. *Electroreological Fluids: Modeling and Mathematical Theory.* Springer, Lecture Notes in Math., vol. 1748, 2000.

[115] S. Samko. On a progress in the theory of Lebesgue spaces with variable exponent: maximal and singular operators, *Integr. Transf. and Spec. Funct.*, **16**(2005), No. 5-6, 461–482.

[116] S. Samko. On some classical operators of variable order in variable exponent analysis, *Operator Theory: Advances and Applications*, Vol. 193, 281-301, *Birkhäuser*, Basel, 2009.

[117] I. I. Sharapudinov. The topology of the space $L^{p(t)}([0,1])$. (Russian) *Math. Zametki*, **26**(1979), $N_0$4, 613–632.

[118] I. B. Simonenko, The Riemann and Riemann-Hazeman boundary value problems with continuous coefficients. (Russian) *In: Issled. po Sovrem. Probl. Teor. Funkts. Kompl. Perem., Fizmatgiz, Moscow,* 1961, pp. 380–389.

[119] I. B. Simonenko, The Riemann boundary value problem for L^p pairs of functions with measurable coefficients and its application to the investigation of singular integrals in the spaces L^p with weight. (Russian) *Izv. Akad. Nauk SSSR, Ser. Mat.* **28**(1964),277–306.

[120] I. B. Simonenko, Some general questions of the theory of the Riemann boundary value problem. (Russian) *Izv. Akad. Nauk. SSSR, Ser. Mat.* **2**(1968), No. 5, 1091–1099.

[121] I. B. Simonenko, Stability of weight properties of functions, with respect to a singular Cauchy integral. (Russian) *Mat. Zametki* **33**(1983), No. 3, 409–416.

[122] V. I. Smirnov, Sur les valeurs limites des fonctions regulieres a l'interier d'un circle. *Zh. Leningradskogo Fiz.-Mat. Obshch.* **2**(1929), No. 2, 22–37.

[123] V. I. Smirnov, Sur les formules de Cauchy et de Green et quelques problemes qui s'y rattachent. *Izv. Akad. Nauk SSSR, Ser. Fiz. Mat.* **7** (1932), No. 3, 337–372.

[124] E. M. Stein, *Singular Integrals and Differentiability Properties of Functions.* Princeton University Press, Princeton, New Jersey, 1970.

[125] E. M. Stein, *Harmonic Analysis: Real Variables Methods, Orthogonality and Oscillatory Integrals.* Princeton University Press, Princeton, New Jersey, 1993.

[126] G. Ts. Tumarkin and S. Ja. Havinson, Classes of analytic functions in multiply-connected domains. *In: Investigations in the Current Problems of the Theory of Functions of Complex Variable.* (Russian), Gos. Izd. Fiz-Mat Literat., Moscow (1960), 45–76.

[127] I. Vekua, On a new integral representation of analytic functions and its applications, (Russian) *Bull. Acad. Sci. Georgian SSR*, vol. 2, pp. 477–484, 1941.

[128] I. N. Vekua, On the theory of singular integral equations. (Russian) *Soobshch. Akad. Nauk Gruzin. SSR* **9**(1942), 869–876.

[129] I. N. Vekua, On the Riemann boundary value problem. (Russian) *Trudy Tbiliss. Mat. Inst. Razmadze Akad. Nauk Gruzin. SSR* **11**(1942), 109–139.

[130] I. N. Vekua, Generalized Analytic Functions. (Russian) *Moscow,* 1988.

[131] N. P. Vekua, *Systems of Singular Integral Equations and Certain Boundary Problems.* (Russian) Nauka, Moscow, 1970.

[132] J. L. Walsh. *Interpolation and Approximation by Rational Functions in the Complex Domain.* Third edition. American Mathematical Society Colloquium Publications, Vol. XX. American Mathematical Society, Providence, R.I., 1960.

[133] S. E. Warschawskiĭ, Über das Randverhatlen der Abletung der Abbildungsfunktion der konformer Abbildung. *Math. Zeitschr.* **35**(1932), No. 3-4, 321–456.

[134] V. V. Zhikov, *Averaging of functionals of the calculus of variations and elasticity theory.* (Russian) Izv. Acad. Nauk SSSR Ser. Mat **50** (1986), No. 4, 675–710.

[135] V. V. Zhikov, S. E. Pastukhova, *On the improved integrability of the gradient of solutions of elliptic equations with a variable nonlinearity exponent.* (Russian) Mat. Sb. **199** (2008), No. 12, 19–52; translation in Sb. Math. **199** (2008), No. 11-12, 1751–1782.

[136] A. Zygmund, *Trigonometric Series,* I, II. Cambridge Univ. Press, London and New-York, 1968.

Index